Random Graphs

WILEY-INTERSCIENCE
SERIES IN DISCRETE MATHEMATICS AND OPTIMIZATION

A complete list of titles in this series appears at the end of this volume.

Random Graphs

SVANTE JANSON
Uppsala University
Sweden

TOMASZ ŁUCZAK
Adam Mickiewicz University
Poland

ANDRZEJ RUCINSKI
Emory University
Atlanta, Georgia

A Wiley-Interscience Publication
JOHN WILEY & SONS, INC.
New York • Chichester • Weinheim • Brisbane • Singapore • Toronto

A NOTE TO THE READER

This book has been electronically reproduced from digital information stored at John Wiley & Sons, Inc. We are pleased that the use of this new technology will enable us to keep works of enduring scholarly value in print as long as there is reasonable demand for them. The content of this book is identical to previous printings.

This text is printed on acid-free paper. ♾

For ordering and customer service, call 1-800-CALL-WILEY.

Library of Congress Cataloging-in-Publication Data is available.

ISBN 0-471-17541-2

Printed in the United States of America

10 9 8 7 6 5 4 3 2 1

Preface

The theory of random graphs originated in a series of papers published in the period 1959–1968 by two outstanding Hungarian mathematicians, Paul Erdős and Alfred Rényi. Over the forty years that have passed since then, the theory has developed into an independent and fast-growing branch of discrete mathematics, located at the intersection of graph theory, combinatorics and probability theory, with applications to theoretical computer science, reliability of transportation and communication networks, natural and social sciences and to discrete mathematics itself. Aside from applications, random graphs continue to serve as nontrivial, but simple enough models for other, more complex random structures, paving the road for more advanced theories.

In the early days, the literature on the subject was scattered around several probabilistic, combinatorial and general mathematics journals. In the late seventies, Béla Bollobás became the leading scientist in the field and contributed dozens of papers, which gradually made up a framework for his excellent, deep and extensive monograph *Random Graphs*, printed in 1985. The appearance of that book stimulated the research even further, shaping up a new theory.

Two other ingredients that added to this trend were the ongoing series of international conferences on random graphs and probabilistic methods in combinatorics held biennially in Poznań, Poland, since 1983, and the journal, *Random Structures and Algorithms*, launched by Wiley in 1990. Both have established a forum for the exchange of ideas and cooperation in the theory of random graphs and related fields.

It is not accidental then that tremendous progress has been made since 1985. Over the last decade several new, beautiful results have been proved and numerous fine techniques and methods have been introduced. Our goal is to present many of these new developments, including results on threshold functions (Ch. 1), small subgraphs (Ch. 3), generalized matchings (Ch. 4), phase transition (Ch. 5), limit distributions (Ch. 6), chromatic number (Ch. 7), partition and extremal properties (Ch. 8), Hamiltonian cycles in random regular graphs (Ch. 9), and zero-one laws (Ch. 10). We emphasize new techniques and tools such as the martingale, Talagrand and correlation inequalities (Ch. 2), the orthogonal decomposition (Ch. 6), the Regularity Lemma of Szemerédi (Ch. 8), the Contiguity Theorem (Ch. 9), and the analysis of variance (Ch. 9).

In a sense, our book can be viewed as an update on Bollobás's 1985 book. However, the topics selected for the book reflect the interest of its authors and do not pretend to exhaust the entire field. In fact, in order not to duplicate Bollobás's work, we do not include subjects which are covered there, on which only a little progress has been made. In particular, we have no sections on degree sequences, long paths and cycles, automorphisms, and the diameter. Moreover, we restrict ourselves to the main core of the theory and focus on the basic models of random graphs, making no attempt to present such rapidly developing areas as random walks on graphs, randomized algorithms or complexity of Boolean functions. Likewise, we exclude random cubes, directed graphs and percolation.

It has been our goal to make the book accessible to graduate students in mathematics and computer science. This has led to simplifications of some statements and proofs, which, we hope, result in better clarity of exposition. The book may be used as a textbook for a graduate course or an honors course for undergraduate senior mathematics and computer science majors. Although we do not provide problems and exercises separately, we often leave to the reader to complete parts of proofs or to provide proofs of results analogous to those proven. These instances, marked by the parenthetic phrase "(Exercise!)", can easily be picked up by the instructor and turned into homework assignments. The prerequisites are limited to basic courses in graph theory or combinatorics, elementary probability and calculus. We believe that the book will also be used by scientists working in the broad area of discrete mathematics and theoretical computer science. It is both an introduction for newcomers and a source of the most recent developments for those working in the field for many years.

We would like to thank several friends and colleagues, without whom this book would be *a.a.s.* worse than it is. Among those whose insightful remarks and suggestions led to improvements of earlier drafts are: Andrzej Czygrinow, Dwight Duffus, Ehud Friedgut, Johan Jonasson, Michał Karoński, Yoshiharu Kohayakawa, Michael Krivelevich, Justyna Kurkowiak, Jiří Matoušek, Brendan Nagle, Yuejian Peng, Joanna Polcyn, Vojtěch Rödl, Jozef Skokan, Joel Spencer, Edyta Szymańska, Michelle Wagner, and Julie White.

Special thanks are due to Penny Haxell and Izolda Gorgol. Penny spent several days correcting our English. Without her tedious work the text would probably need subtitles to be understood by an American reader. Izolda generously exercised her editing skills providing us with electronic files of all figures.

Jessica Downey, the Wiley editor, earned our deep appreciation for her continuous enthusiasm and support of the project.

Finally, the three authors would like to thank each other for patience, mutual encouragement and persistence in negotiations, the compromising effect of which is now in your hands.

SVANTE JANSON
TOMASZ ŁUCZAK
ANDRZEJ RUCIŃSKI

Uppsala, Poznań, and Atlanta

Contents

1

Preliminaries

1.1 MODELS OF RANDOM GRAPHS

The notion of a random graph originated in a paper of Erdős (1947), which is considered by some as the first conscious application of the probabilistic method. It was used there to prove the existence of a graph with a specific Ramsey property.

The model introduced by Erdős is very natural and can be described as choosing a graph at random, with equal probabilities, from the set of all $2^{\binom{n}{2}}$ graphs whose vertex set is $[n] = \{1, 2, \ldots, n\}$. In other words, it can be described as the probability space $(\Omega, \mathcal{F}, \mathbb{P})$, where Ω is the set of all graphs with vertex set $[n]$, $\mathcal{F}$ is the family of all subsets of Ω, and for every $\omega \in \Omega$

$$\mathbb{P}(\omega) = 2^{-\binom{n}{2}}.$$

This probability space can also be viewed as the product of $\binom{n}{2}$ binary spaces. In simple words, it is a result of $\binom{n}{2}$ independent tosses of a fair coin, where "turning up heads" means "drawing an edge".

Generally speaking, a random graph is a graph constructed by a random procedure. In accordance with standard definitions in probability theory, this is formalized by representing the "random procedure" by a probability space $(\Omega, \mathcal{F}, \mathbb{P})$ and the "construction" by a function from the probability space into a suitable family of graphs. The *distribution* of a random graph is the induced probability distribution on the family of graphs; for many purposes this is the only relevant feature of the construction and we usually do not distinguish between different random graphs with the same distribution. Indeed, it is often convenient to define a random graph by specifying its distribution; that is, we specify a family of graphs and a probability distribution on it. Note, however, that it is not sufficient to formally define a random graph as a probability distribution only, as is sometimes done in the literature; an important case in which this would not do is when several random graphs are considered at once, for example, in the two-round exposure described at the end of this section.

The word "model" is used rather loosely in the theory of random graphs. It may refer to a specific class of random graphs, defined as above, or perhaps to a specific distribution. Usually, however, there is also a parameter involved which measures the size of the graphs and typically it tends to infinity; there may also be other parameters. Needless to say, the whole theory of random graphs is thus asymptotic in its nature.

Two basic models

Nowadays, among several models of random graphs, there are two basic ones, the binomial model and the uniform model, both originating in the simple model introduced by Erdős (1947). In this book we will mainly restrict ourselves to studying these two models.

Given a real number p, $0 \le p \le 1$, the *binomial random graph*, denoted by $\mathbb{G}(n,p)$, is defined by taking as Ω the set of all graphs on vertex set $[n]$ and setting

$$\mathbb{P}(G) = p^{e_G}(1-p)^{\binom{n}{2}-e_G} ,$$

where $e_G = |E(G)|$ stands for the number of edges of G. It can be viewed as a result of $\binom{n}{2}$ independent coin flippings, one for each pair of vertices, with the probability of success (i.e., drawing an edge) equal to p. For $p = 1/2$ this is the model of 1947. However, most of the random graph literature is devoted to cases in which $p = p(n) \to 0$ as $n \to \infty$.

The binomial model is a special case of a *reliability network*. In this more general model, Ω is the family of all spanning subgraphs of a given graph F and $\mathbb{P}(G) = p^{e_G}(1-p)^{e_F - e_G}$. By a *spanning subgraph* we mean a graph G such that $V(G) = V(F)$ and $E(G) \subseteq E(F)$. Thus, in a reliability network, the edges of a given graph (network) are independently destroyed, each with failure probability $1-p$. One can generalize this model even further, by allowing different probabilities of failure at different edges. (Binomial models are sometimes called *Bernoulli.*)

Taking $F = K_n$, the complete graph on n vertices, we obtain the model $\mathbb{G}(n,p)$. Taking $F = K_{m,n}$, the complete bipartite graph (here either m is a function of n, or they are two independent parameters, typically both tending to infinity), we obtain the bipartite random graph $\mathbb{G}(m,n,p)$. Other popular models, not discussed here, are those in which the initial graph F is the hypercube or the $n \times n$ square lattice. The reliability network based on the infinite square lattice belongs to percolation theory (Grimmett 1992a) which too, as all infinite models, is beyond the scope of this book.

The main advantage of the binomial model $\mathbb{G}(n,p)$ is the independence of presence of edges, but the drawback is that the number of edges is not fixed; it varies according to a binomial distribution with expectation $\binom{n}{2}p$. If one conditions on the event that $|E(\mathbb{G}(n,p))| = M$, then a uniform space is obtained. This space can be defined directly.

Given an integer M, $0 \le M \le \binom{n}{2}$, the *uniform random graph*, denoted by $\mathbb{G}(n, M)$, is defined by taking as Ω the family of all graphs on the vertex set $[n]$ with exactly M edges, and as $\mathbb{P}$ the uniform probability on Ω,

$$\mathbb{P}(G) = \binom{\binom{n}{2}}{M}^{-1}, \qquad G \in \Omega.$$

This model, closely related to enumerative combinatorics, was apparently considered already in 1939 in an unpublished work of Erdős and Whitney on the connectedness of almost all graphs with n vertices and about $M = \frac{1}{2}n \log n$ edges. This was the model used throughout by Erdős and Rényi in their series of papers between 1959 and 1968, which gave rise to the theory of random graphs. (For an account of the contents of these eight fundamental papers, see Karoński and Ruciński (1997).)

The two basic models are in many cases asymptotically equivalent, provided $\binom{n}{2}p$ is close to M (see Section 1.4).

The uniform random graph $\mathbb{G}(n, M)$ belongs to a broad family of uniform random graphs defined by taking the uniform distribution over a family of graphs $\mathcal{F}$. The pioneering model from Erdős (1947) belongs here too, with $\mathcal{F}$ being the family of all graphs on a given set of vertices. Other popular models of this type are random trees (not studied in this book), where $\mathcal{F}$ is the family of all n^{n-2} trees on n labeled vertices, and random r-regular graphs (see Chapter 9), where $\mathcal{F}$ is the family of all graphs on n vertices of equal degree r, provided nr is even. We will use $\mathbb{G}(n, r)$ to denote a uniform random r-regular graph. It may look dangerous to use the notation $\mathbb{G}(n,p)$, $\mathbb{G}(n, M)$ and $\mathbb{G}(n,r)$ for three different things: What is $\mathbb{G}(n, 1)$? In practice, however, the correct meaning is always clear from the context. (As for the three models: $\mathbb{G}(n,p)$ with $p = 1$, $\mathbb{G}(n, M)$ with $M = 1$, and $\mathbb{G}(n,r)$ with $r = 1$, each one is rather dull.)

Both the binomial and the uniform model have their counterparts for directed graphs. Besides these, there are interesting, natural random directed graphs which do not have analogues in the undirected case. Let us mention the k-out model, in which every vertex independently chooses k out-neighbors (including or excluding itself); the case of random mappings (i.e., $k = 1$) is well studied (Kolchin 1986, Aldous and Pitman 1994). Random tournaments, in which every edge of a complete graph assumes randomly one of the two possible orientations, have a broad literature too (Moon 1968, Gruszka, Łuczak and Ruciński 1996, Andersson 1998).

There are still other random graphs which do not fall into either category (binomial or uniform). For instance, in some reliability networks the vertices but not the edges are destroyed. Furthermore, some random graphs result from more complex probabilistic experiments, and here the sky is the limit. Restricted random graph processes constitute an interesting class of such experiments, but we should better define the unrestricted case first.

Random graph processes

In general, a *random graph process* is a stochastic process that describes a random graph evolving in time. In other words, it is a family $\{\mathbb{G}(t)\}_t$ of random graphs (defined on a common probability space) where the parameter t is interpreted as time; the time can be either discrete or continuous. The processes studied here will have a fixed vertex set (typically $[n]$), and they will start without any edges and grow monotonically by adding edges according to some rule but never deleting any.

A simple and important random graph process $\{\mathbb{G}(n, M)\}_M$ (sometimes called *the random graph process*) was introduced by Erdős and Rényi (1959) and has been well studied since then. It begins with no edges at time 0 and adds new edges, one at a time; each new edge is selected at random, uniformly among all edges not already present. Hence this random graph process is a Markov process, with time running through the set $\{0, 1, \dots, \binom{n}{2}\}$. The M-th stage of this process can be identified with the uniform random graph $\mathbb{G}(n, M)$. The process, however, allows one to study the random graph $\mathbb{G}(n, M)$ as it evolves with M growing from 0 to $\binom{n}{2}$. For example, a typical result, meaningful only for random graph processes, says that, with probability approaching 1 as $n \to \infty$, the very edge which links the last isolated vertex with another vertex makes the graph connected (Bollobás and Thomason (1985); see also Bollobás (1985)).

A related *continuous time random graph process* can be defined by assigning a random variable T_e to each edge e of the complete graph K_n, such that the $\binom{n}{2}$ variables T_e are independent with a common continuous distribution, and then defining the edge set of $\{\mathbb{G}(t)\}_t$ to consist of all e with $T_e \le t$. Clearly, the resulting random graph $\{\mathbb{G}(t)\}_{t_0}$ at a fixed time t_0 can be identified with the binomial random graph $\mathbb{G}(n, p)$, where $p = \mathbb{P}(T_e \le t_0)$. Furthermore, since almost surely no two values of the random variables T_e coincide, we may define $T_{(i)}$ as the random time at which the i-th edge is added. Then, by symmetry, $\mathbb{G}(T_{(i)})$ is the uniform random graph $\mathbb{G}(n, i)$, and the sequence $\{\mathbb{G}(T_{(i)})\}$ for $i = 1, \dots, \binom{n}{2}$, equals the ordinary random graph process $\{\mathbb{G}(n, M)\}_M$ defined above. Hence, this continuous time random graph process is a joint generalization of the binomial random graph, the uniform random graph and the standard discrete time random graph process.

Clearly, different choices of the distribution of T_e affect the model only trivially, by a change in the time variable. The continuous time evolving model was introduced by Stepanov (1970) with T_e exponentially distributed; we prefer the uniform distribution over the interval $[0,1]$, in which case $p = \mathbb{P}(T_e \le t) = t$, $0 \le t \le 1$. Thus, we may unambiguously use the notation $\{\mathbb{G}(n,t)\}_t$.

Recently, a number of *restricted random graph processes* have been studied. In general, such a process can be defined as a random graph process in which edges are chosen one by one uniformly from a dynamically modified set of available pairs of vertices until this set becomes empty. More formally, consider a Markov chain of random edge sets $E_0 = \emptyset, E_1, \dots, E_s$, where $E_i = \{e_1, \dots, e_i\}$ and e_i is chosen uniformly from a set A_i which depends only on the set E_{i-1}.

In one of these restricted models, studied by Ruciński and Wormald (1992), the maximum degree is bounded from above by a given integer d. Thus, the set A_i contains only those pairs whose addition to the set E_{i-1} does not create a vertex of degree $d+1$. The graph at the end of the process may not be d-regular, though it is shown to be so with probability approaching 1. See also Wormald (1999a), where, moreover, further related processes are defined and studied.

Another restricted process is studied by Erdős, Suen and Winkler (1995), in which it is not allowed to create a triangle. In this model it is even an open problem to determine the length of a typical process, measured by the number of edges in the final graph. It is only known that with high probability the process takes more than $c_1 n^{3/2}$ but fewer than $c_2 n^{3/2} \log n$ steps, where c_1 and c_2 are positive constants. Recently, this result was generalized to a wide class of forbidden subgraphs by Osthus and Taraz (2000+).

By forbidding cycles, one obtains a process which creates a non-uniform random tree (Aldous 1990), while forbidding components with more than one cycle leads to a random graph which still is to be studied.

Random subsets

The two basic models of random graphs fall into the framework of random subsets of a set. Monotonicity, equivalence and threshold behavior of the probabilities of properties of random graphs can often be proved at no extra cost in this general setting. Other principal examples of random subsets of a set include random sets of integers and random hypergraphs. In the remaining sections of this chapter (as well as in parts of Chapter 2) we will mainly study this more general random set framework. For an arbitrary set X and an integer k, let $[X]^k$ stand for the family of all k-element subsets of X. If $X = [n]$, we will simplify this notation to $[n]^k$.

Let Γ be a finite set, $|\Gamma| = N$, let $0 \le p \le 1$ and $0 \le M \le N$. Then the random subset Γ_p of Γ is obtained by flipping a coin, with probability p of success, for each element of Γ to determine whether the element is to be included in Γ_p; the distribution of Γ_p is the probability distribution on $\Omega = 2^\Gamma$ given by $\mathbb{P}(F) = p^{|F|}(1-p)^{|\Gamma|-|F|}$ for $F \subseteq \Gamma$. Similarly, let Γ_M be a randomly chosen element of $[\Gamma]^M$; that is, Γ_M has the uniform distribution $\mathbb{P}(F) = \binom{N}{M}^{-1}$ for $F \in [\Gamma]^M$.

Taking $\Gamma = [n]^2$ we obtain the two basic models of random graphs defined above, $\mathbb{G}(n,p)$ and $\mathbb{G}(n,M)$.

The binomial model Γ_p can be generalized to $\Gamma_{p_1,\dots,p_N}$, where the element i is included with probability p_i, independently for all $i = 1,\dots,N$.

Two-round exposure

The two-round exposure is a successful proof technique applicable to the binomial model. It relies on viewing Γ_p as a union of two independent random subsets Γ_{p_1} and Γ_{p_2}, where p_1 and p_2 are such that $p = p_1 + p_2 - p_1p_2$. (It is easy to see that this union indeed is distributed as Γ_p – Exercise!) In the special case of random graphs we first generate a random graph $\mathbb{G}(n,p_1)$ and then, independently, another random graph $\mathbb{G}(n,p_2)$ on the same vertex set. By replacing double edges by single ones, we obtain $\mathbb{G}(n,p)$.

An argument typically used in applications of the two-round exposure can be expressed in the following general form. Let $\mathbb{P}_1$ be the probability distribution associated with Γ_{p_1}, and let $\mathbb{P}_F$ be the conditional probability in Γ_p under the condition $\Gamma_{p_1} = F$. Then for any two families $\mathcal{A}$ and $\mathcal{B}$ of subsets of Γ

$$\mathbb{P}(\mathcal{A}) \geq \sum_{F\in\mathcal{B}} \mathbb{P}_F(\mathcal{A})\,\mathbb{P}_1(F) \geq \mathbb{P}_{F_0}(\mathcal{A})\,\mathbb{P}_1(\mathcal{B}), \tag{1.1}$$

where F_0 minimizes the probability $\mathbb{P}_F(\mathcal{A})$ over all $F \in \mathcal{B}$. Thus, knowing that $\mathbb{P}_1(\mathcal{B}) \to 1$, in order to prove that also $\mathbb{P}(\mathcal{A}) \to 1$, it is enough to show that $\mathbb{P}_F(\mathcal{A}) \to 1$, for every $F \in \mathcal{B}$. In practice, computing the last probability means fixing an instance of $\Gamma_{p_1} \in \mathcal{B}$ and throwing in new elements independently with probability p_2 (the second round of exposure).

1.2 NOTES ON NOTATION AND MORE

Graph theory

All graphs are simple and undirected, unless otherwise stated. We use standard notation for graphs. For example, $V(G)$ is the vertex set of a graph G, $E(G)$ is the edge set, $v_G = |V(G)|$ is the number of vertices and $e_G = |E(G)|$ is the number of edges; for typographical reasons we sometimes write the latter two as $v(G)$ and $e(G)$. In this book the *size* of G always means $v(G)$ (and not $e(G)$ as sometimes used by other authors). However, we also will call $v(G)$ the *order* of G.

Moreover, let $d(G) = e_G/v_G$ be the *density* and $m(G) = \max_{H\subseteq G} d(H)$ the *maximum density* of G. (Note that $d(G)$ equals half the average degree of G, and that some authors define $d(G)$ as the average degree, which is twice our value.) Another measure of the density of a graph G, ranging between 0 and 1, is defined as $\rho(G) = e(G)/\binom{v(G)}{2}$. (It is sometimes called the *relative density* of G.)

Furthermore, $\delta(G)$ is the *minimum degree*, $\Delta(G)$ is the *maximum degree*, $\chi(G)$ is the *chromatic number*, $D(G) = \max_{H \subseteq G} \delta(H)$ is the *degeneracy number*, $\alpha(G)$ is the *stability number* (the size of the largest stable, or independent, set of vertices), and $\operatorname{aut}(G)$ is the number of automorphisms of G.

We let $N(v) = N_G(v)$ denote the *neighborhood* of a vertex v in G, that is, the set $\{w \in V(G) : vw \in E(G)\}$. Its size is called the *degree* of v and is denoted by $\deg(v) = \deg_G(v)$. Similarly, if $S \subseteq V(G)$, its neighborhood $N_G(S) = \bigcup_{v \in S} N_G(v) \setminus S$ is the set of all vertices outside S adjacent to at least one vertex in S. Moreover, we let $\overline{N}_G(v) = N_G(v) \cup \{v\}$ and $\overline{N}_G(S) = N_G(S) \cup S$ denote the corresponding *closed neighborhoods*, which include v and S, respectively.

Any graph without edges will be called *empty*, while the graph with no vertices (and thus no edges) will be called the *null graph* and denoted by $\emptyset$.

Some special graphs are: the complete graph K_n on n vertices, the complete bipartite graph $K_{m,n}$ on $m+n$ vertices, the cycle C_k with k vertices, and the path P_k with k edges and thus $k+1$ vertices. A star is any graph $K_{1,n}$, $n \geq 0$. We let jG denote the union of j vertex-disjoint copies of G. A matching is a forest consisting of isolated edges only (i.e., a graph of the form jK_2, $j \geq 0$).

If G is a graph and $V \subseteq V(G)$, then $G[V]$ denotes the restriction of G to V, defined as the graph with vertex set V and edge set $E(G) \cap [V]^2$; similarly, if $E \subseteq [V(G)]^2$, $G[E]$ denotes the graph with vertex set $V(G)$ and edge set $E(G) \cap E$. A subgraph of G of the type $G[V]$ is called *induced* or *spanned* by V, while a subgraph of the type $G[E]$ is called *spanning*. The number of edges in the subgraph $G[V]$ is sometimes denoted by $e_G(V) = e(V)$, while for two disjoint subsets $A, B \subset V(G)$, the quantity $e_G(A,B)$ counts the number of edges of G with one endpoint in A and the other in B.

By a *copy of a given graph* G inside another graph F we mean any, not necessarily induced, subgraph of F which is isomorphic to G. If the subgraph happens to be induced, we call it an *induced copy* of G.

Although we define our random graphs as labelled, we are mainly interested in properties that are independent of the labelling, that is, properties that depend on the isomorphism type only. Such properties are called *graph properties*. (In contrast, "vertex 1 is isolated" is not a graph property; such properties will occasionally be studied too.)

Probability

We use $\operatorname{Bi}(n,p)$, $\operatorname{Be}(p) = \operatorname{Bi}(1,p)$, $\operatorname{Po}(\lambda)$ and $\operatorname{N}(\mu, \sigma^2)$ to denote the binomial, Bernoulli, Poisson and normal distributions, respectively. We further write $X \in \mathcal{L}$, meaning that X is a random variable with distribution $\mathcal{L}$ (e.g., $X \in$

N(0, 1)). The distribution of a random variable X is occasionally denoted by $\mathcal{L}(X)$.

We denote by $\mathbf{1}[\mathcal{E}]$ the indicator function of the event $\mathcal{E}$, which equals 1 if $\mathcal{E}$ occurs and 0 otherwise. We will often consider random variables that are the indicator functions of some events; such random variables will be called *indicator* or *zero-one random variables*. They clearly have Bernoulli distributions with $p = \mathbb{P}(\mathcal{E})$, where $\mathcal{E}$ is the corresponding event.

The expected value and the variance of a random variable X (if they exist) will be denoted by $\mathbb{E}\,X$ and $\operatorname{Var} X$, respectively. Thus, the well-known *Chebyshev's inequality*, which will be frequently used throughout the book, can be stated in the following, standard form. If $\operatorname{Var} X$ exists, then

$$\mathbb{P}(|X - \mathbb{E}\,X| \geq t) \leq \frac{\operatorname{Var} X}{t^2}, \qquad t > 0. \tag{1.2}$$

Similarly, *Markov's inequality* states that, if $X \geq 0$ a.s., then

$$\mathbb{P}(X \geq t) \leq \frac{\mathbb{E}\,X}{t}, \qquad t > 0. \tag{1.3}$$

We denote the covariance of two random variables X and Y by $\operatorname{Cov}(X, Y)$. Recall that the variance of a (finite) sum of random variables is given by $\operatorname{Var}\big(\sum_i X_i\big) = \sum_i \sum_j \operatorname{Cov}(X_i, X_j)$.

The conditional expectation of X given an event $\mathcal{E}$ is denoted by $\mathbb{E}(X \mid \mathcal{E})$. We similarly write $\mathbb{E}(X \mid Y_1, \dots, Y_k)$ for the conditional expectation of X given some random variables $Y_1, \dots, Y_k$; note that this conditional expectation is a function of $(Y_1, \dots, Y_k)$ and thus itself a random variable. When using martingales (Section 2.4), we will more generally denote by $\mathbb{E}(X \mid \mathcal{G})$ the conditional expectation of X given a sub-σ-algebra $\mathcal{G}$ of $\mathcal{F}$.

Quite frequently our proofs will rely on the elementary *law of total probability* which states that for any partition of the probability space $\Omega = \mathcal{E}_1 \cup \mathcal{E}_2 \dots$ and any random variable X defined on Ω,

$$\mathbb{E}\,X = \sum_i \mathbb{E}(X \mid \mathcal{E}_i)\,\mathbb{P}(\mathcal{E}_i)\,.$$

In particular, if $X = \mathbf{1}[\mathcal{E}]$, then $\mathbb{P}(\mathcal{E}) = \sum_i \mathbb{P}(\mathcal{E} \mid \mathcal{E}_i)\,\mathbb{P}(\mathcal{E}_i)$.

If $X_1, X_2, \dots$ are random variables and a is a constant, we say that X_n *converges in probability* to a as $n \to \infty$, and write $X_n \xrightarrow{p} a$, if $\mathbb{P}(|X_n - a| > \varepsilon) \to 0$ for every $\varepsilon > 0$; see, for example, Gut (1995, Chapter VI).

One similarly defines $X_n \xrightarrow{p} Y$, where Y is another random variable, but then Y and every X_n have to be defined on the same probability space; this can be reduced to the preceding case, since $X_n \xrightarrow{p} Y$ if and only if $X_n - Y \xrightarrow{p} 0$.

Let $X_1, X_2, \dots$ and Z be random variables. We say that X_n *converges in distribution* to Z as $n \to \infty$, and write $X_n \xrightarrow{d} Z$, if $\mathbb{P}(X_n \leq x) \to \mathbb{P}(Z \leq x)$ for every real x that is a continuity point of $\mathbb{P}(Z \leq x)$ (Billingsley 1968, Gut 1995).

If $X_1, X_2, \ldots$ and Z are integer-valued then, equivalently, $X_n \xrightarrow{d} Z$ if and only if $\mathbb{P}(X_n = k) \to \mathbb{P}(Z = k)$ for every integer k.

Note that convergence in distribution is really a property of the distributions of the random variables and does not require the variables to be defined on the same probability space. Nevertheless, it is customary (and convenient) to talk about convergence of random variables. We also use hybrid notation such as $X_n \xrightarrow{d} \mathrm{N}(0,1)$, which means $X_n \xrightarrow{d} Z$ for some (and thus every) random variable $Z \in \mathrm{N}(0,1)$.

An important special case is one in which Z is a (non-random) real constant. It is easily shown that convergence in distribution to a constant is the same as convergence in probability, that is, $X_n \xrightarrow{d} a$ if and only if $X_n \xrightarrow{p} a$ for $a \in \mathbb{R}$. A useful fact is that if $X_n \xrightarrow{d} Z$ and $Y_n \xrightarrow{p} a$, where a is a constant, then $X_n + Y_n \xrightarrow{d} Z + a$ and $Y_n X_n \xrightarrow{d} aZ$ (*Cramér's theorem*), see, for example, Gut (1995, Theorem VI.7.5).

The definition of convergence in distribution extends to random vectors with values in $\mathbb{R}^k$ for every fixed k; this is also expressed as *joint convergence in distribution* of the components of the vectors. A powerful method for extending results on the real random variables to the vector-valued ones is known as the *Cramér–Wold device* (Billingsley 1968, Theorem 7.7). It states that $(X_{n1}, \ldots, X_{nk}) \xrightarrow{d} (Z_1, \ldots, Z_k)$ if and only if $\sum_i t_i X_{ni} \xrightarrow{d} \sum_i t_i Z_i$ for every sequence of real numbers $t_1, \ldots, t_k$. For more details, as well as for the convergence of random variables with values in even more general spaces, see Billingsley (1968).

Remark 1.1. Convergence in distribution does *not*, in general, imply convergence of the sequence of means or variances. However, in many specific applications we find that these sequences do, in fact, converge to the mean and variance of the limit distribution.

Asymptotics

We will often use the following standard notation for the asymptotic behavior of the relative order of magnitude of two sequences of numbers a_n and b_n, depending on a parameter $n \to \infty$. The same notation is also used in other situations, for example, for functions of a variable ε that tends to 0. We will often omit the phrase "as $n \to \infty$" when there is no risk of confusion. For simplicity we assume $b_n > 0$ for all sufficiently large n.

- $a_n = O(b_n)$ as $n \to \infty$ if there exist constants C and n_0 such that $|a_n| \le C b_n$ for $n \ge n_0$, i.e., if the sequence a_n / b_n is bounded, except possibly for some small values of n for which the ratio may be undefined.

- $a_n = \Omega(b_n)$ as $n \to \infty$ if there exist constants $c > 0$ and n_0 such that $a_n \ge c b_n$ for $n \ge n_0$. If $a_n \ge 0$, this is equivalent to $b_n = O(a_n)$.

- $a_n = \Theta(b_n)$ as $n \to \infty$ if there exist constants $C, c > 0$ and n_0 such that $cb_n \le a_n \le Cb_n$ for $n \ge n_0$, i.e., if $a_n = O(b_n)$ and $a_n = \Omega(b_n)$. This is sometimes expressed by saying that a_n and b_n are of the *same order of magnitude*.

- $a_n \asymp b_n$ if $a_n = \Theta(b_n)$.

- $a_n \sim b_n$ if $a_n/b_n \to 1$.

- $a_n = o(b_n)$ as $n \to \infty$ if $a_n/b_n \to 0$, i.e., if for every $\varepsilon > 0$ there exists n_ε such that $|a_n| < \varepsilon b_n$ for $n \ge n_\varepsilon$.

- $a_n \ll b_n$ or $b_n \gg a_n$ if $a_n \ge 0$ and $a_n = o(b_n)$.

Since most results in this book are asymptotic, we will be frequently assuming in the proofs that n is sufficiently large, sometimes without explicitly saying so.

Probability asymptotics

We say that an event $\mathcal{E}_n$, describing a property of a random structure depending on a parameter n, holds *asymptotically almost surely* (abbreviated *a.a.s.*), if $\mathbb{P}(\mathcal{E}_n) \to 1$ as $n \to \infty$.

Remark 1.2. In many publications on random structures the phrase "almost surely" or a.s. is used. However, we wish to reserve that phrase for what it normally means in probability theory, i.e. that the probability of an event equals exactly 1. It seems that the first paper where the phrase *a.a.s.* and not *a.s.* was used is Shamir and Upfal (1981). (Some authors use the phrase "almost every" or a.e. which we reject for the same reason as "almost surely". Others write "with high probability", or **whp**.)

When discussing asymptotics of random variables, we avoid expressions like "$X_n = O(1)$ *a.a.s.*" or "$X_n = o(1)$ *a.a.s.*", which may be ambiguous, since they combine two asymptotic notions. As a substitute we give probabilistic versions of some of the symbols above, denoting them with a subscript p or C. Let X_n be random variables and a_n positive real numbers. We then define:

- $X_n = O_p(a_n)$ as $n \to \infty$ if for every $\delta > 0$ there exist constants C_δ and n_0 such that $\mathbb{P}(|X_n| \le C_\delta a_n) > 1 - \delta$ for every $n \ge n_0$.

- $X_n = O_C(a_n)$ as $n \to \infty$ if there exists a constant C such that *a.a.s.* $|X_n| \le Ca_n$.

- $X_n = \Theta_p(a_n)$ as $n \to \infty$ if for every $\delta > 0$ there exist constants $c_\delta > 0$, $C_\delta > 0$ and n_0 such that $\mathbb{P}(c_\delta a_n \le X_n \le C_\delta a_n) > 1 - \delta$ for every $n \ge n_0$.

- $X_n = \Theta_C(a_n)$ as $n \to \infty$ if there exist positive constants c and C such that *a.a.s.* $ca_n \le X_n \le Ca_n$.

- $X_n = o_p(a_n)$ as $n \to \infty$ if for every $\varepsilon > 0$, *a.a.s.* $|X_n| < \varepsilon a_n$.

Note that $X_n = O_C(a_n)$ implies $X_n = O_p(a_n)$, but not conversely; indeed, $X_n = O_C(a_n)$ if and only if the constant C_δ in the definition of O_p can be chosen independently of δ. For example, any sequence X_n of identically distributed random variables is $O_p(1)$, but such a sequence is $O_C(1)$ only if the common distribution has support in a finite interval.

Similarly, $X_n = \Theta_C(a_n)$ implies $X_n = \Theta_p(a_n)$, but not conversely. On the other hand, $X_n = o_p(a_n)$ implies $X_n = O_C(a_n)$.

Remark 1.3. It is easy to verify (Exercise!) that $X_n = O_p(a_n)$ if and only if for every function $\omega(n) \to \infty$, $|X_n| \le \omega(n)a_n$ *a.a.s.* Similarly, $X_n = o_p(a_n)$ if and only if for some function $\omega(n) \to \infty$, $|X_n| \le a_n/\omega(n)$ *a.a.s.*

Such notation with an unspecified sequence $\omega(n)$ is common in publications on random structures, but we believe that the equivalent notation O_p and o_p is clearer.

It is an immediate consequence of the definitions (Exercise!) that $X_n = o_p(a_n)$ if and only if $X_n/a_n \xrightarrow{p} 0$. Conversely, $X_n \xrightarrow{p} a$ if and only if $X_n = a + o_p(1)$ (and $X_n \xrightarrow{p} Y$ if and only if $X_n = Y + o_p(1)$).

Remark 1.4. The symbol O_p can also be expressed by equivalent standard probabilistic concepts. In fact, a sequence X_n is *bounded in probability*, or *tight*, if $X_n = O_p(1)$. Hence, $X_n = O_p(a_n)$ if and only if the sequence X_n/a_n is bounded in probability (or tight).

Dependency graphs

Let $\{X_i\}_{i\in\mathcal{I}}$ be a family of random variables (defined on a common probability space). A *dependency graph* for $\{X_i\}$ is any graph L with vertex set $V(L) = \mathcal{I}$ such that if A and B are two disjoint subsets of $\mathcal{I}$ with $e_L(A, B) = 0$, then the families $\{X_i\}_{i\in A}$ and $\{X_i\}_{i\in B}$ are mutually independent.

Dependency graphs will be used several times in this book. They are particularly useful when they are sparse, meaning that there is a lot of independence in the family $\{X_i\}$.

Example 1.5. In a standard situation, there is an underlying family of independent random variables $\{Y_\alpha\}_{\alpha\in\mathcal{A}}$, and each X_i is a function of the variables $\{Y_\alpha\}_{\alpha\in A_i}$ for some subset $A_i \subseteq \mathcal{A}$. Let $\mathcal{S} = \{A_i : i \in \mathcal{I}\}$. Then the graph $L = L(\mathcal{S})$ with vertex set $\mathcal{I}$ and edge set $\{ij : A_i \cap A_j \ne \emptyset\}$ is a dependency graph for the family $\{X_i\}_{i\in\mathcal{I}}$ (Exercise!).

Example 1.6. As a special case of the preceding example, let $\{H_i\}_{i\in\mathcal{I}}$ be given subgraphs of the complete graph K_n and let X_i be the indicator that H_i appears as a subgraph in $\mathbb{G}(n,p)$, that is, $X_i = \mathbf{1}[H_i \subseteq \mathbb{G}(n,p)]$, $i \in \mathcal{I}$. Then $L(\mathcal{S})$, with $\mathcal{S} = \{E(H_i) : i \in \mathcal{I}\}$, is a natural dependency graph with edge set $\{ij : E(H_i) \cap E(H_j) \ne \emptyset\}$ (Exercise!).

Remark 1.7. In particular, if L is a dependency graph for $\{X_i\}$, then two variables X_i and X_j are independent unless there is an edge in L between i and j. Note, however, that this is only a necessary condition, and does not imply that L is a dependency graph (Exercise!).

Remark 1.8. Another context, outside the scope of this book, in which dependency graphs are used is the Lovász Local Lemma (Erdős and Lovaśz (1975); see also Alon and Spencer (1992)). There it actually suffices to use a slightly weaker definition, considering only singletons B in the definition above.

Remark 1.9. In our applications, there exists a natural dependency graph, but it should be observed that, in general, there is no canonical choice and the dependency graph is not unique, even if it is required to be minimal (Exercise!).

The subsubsequence principle

It is often convenient to use the well-known *subsubsequence principle*, which states that if for every subsequence of a sequence there is a subsubsequence converging to a limit a, then the entire sequence must converge to the same limit. This holds for sequences of real numbers, vectors, random variables (both for convergence in probability and for convergence in distribution) and, in general, for sequences in any topological space.

For example, this means that if we want to prove a limit theorem for $\mathbb{G}(n,p)$, we may without loss of generality assume that an expression such as $n^a p^b$ converges to some $c \leq \infty$ (provided, of course, that the result we want to prove does not depend on the limit c).

We will be using this principle throughout the book (see, e.g., the proof of Proposition 1.15), sometimes without explicitly mentioning it.

And finally . . .

The base of all logarithms is e, unless specified otherwise.

1.3 MONOTONICITY

A family of subsets $\mathcal{Q} \subseteq 2^\Gamma$ is called *increasing* if $A \subseteq B$ and $A \in \mathcal{Q}$ imply that $B \in \mathcal{Q}$. A family of subsets is *decreasing* if its complement in 2^Γ is increasing, or, equivalently, if the family of the complements in Γ is increasing. A family which is either increasing or decreasing is called *monotone*. A family $\mathcal{Q}$ is *convex* if $A \subseteq B \subseteq C$ and $A, C \in \mathcal{Q}$ imply $B \in \mathcal{Q}$. We identify properties of subsets of Γ with the corresponding families of all subsets having the property; we thus use the same notation and terminology for properties.

In the special case in which $\Gamma = [n]^2$, any family $\mathcal{Q} \subseteq 2^{\Gamma}$ is a family of graphs and, if it is closed under isomorphism, it can be identified with a graph property. Some examples of increasing graph properties are "being connected", "containing a triangle" and "having a perfect matching". Automatically, the negations of all of them are decreasing. Natural decreasing graph properties include "having at least k isolated vertices", "having at most k edges" and "being planar". The property of "having exactly k isolated vertices" is an example of a convex but not monotone property, whereas "the largest component is a tree" is not even convex (Exercise!).

It is reasonable to expect that the probability of a random set falling into an increasing family of sets gets larger when the (expected) size of the random set does. This is indeed the case. Lemma 1.10 below appeared first in Bollobás (1979).

Lemma 1.10. *Let $\mathcal{Q}$ be an increasing property of subsets of Γ, $0 \le p_1 \le p_2 \le 1$ and $0 \le M_1 \le M_2 \le N$. Then*

$$\mathbb{P}(\Gamma_{p_1} \in \mathcal{Q}) \le \mathbb{P}(\Gamma_{p_2} \in \mathcal{Q})$$

and

$$\mathbb{P}(\Gamma_{M_1} \in \mathcal{Q}) \le \mathbb{P}(\Gamma_{M_2} \in \mathcal{Q}).$$

Proof. To prove the first inequality we employ a simple version of the two-round exposure technique (see Section 1.1). Let $p_0 = (p_2 - p_1)/(1 - p_1)$. Then Γ_{p_2} can be viewed as a union of two independent random subsets, Γ_{p_1} and Γ_{p_0}. As then $\Gamma_{p_1} \subseteq \Gamma_{p_2}$ and $\mathcal{Q}$ is increasing, the event "$\Gamma_{p_1} \in \mathcal{Q}$" implies the event "$\Gamma_{p_2} \in \mathcal{Q}$", completing the proof of first inequality.

For the second inequality, we construct a *random subset process* $\{\Gamma_M\}_M$, similar to the random graph process defined in Section 1.1, by selecting the elements of Γ one by one in random order. Clearly, Γ_M can be taken as the M-th subset in the process. Then $\Gamma_{M_1} \subseteq \Gamma_{M_2}$, and, as in the first part, the event "$\Gamma_{M_1} \in \mathcal{Q}$" implies the event "$\Gamma_{M_2} \in \mathcal{Q}$", which completes the proof. ■

Trivially, each monotone property is convex. In a special case this can be, in a sense, reversed: if $\mathcal{Q}$ is convex, and for some M we have $[\Gamma]^M \subseteq \mathcal{Q}$ then, for $M' < M$, $\mathcal{Q}$ behaves like an increasing property, and in particular $\mathbb{P}(\Gamma_{M'} \in \mathcal{Q}) \le \mathbb{P}(\Gamma_{M''} \in \mathcal{Q})$ for all $M' \le M'' \le M$ (Exercise!). Similarly, for $M'' > M$, $\mathcal{Q}$ can be treated as decreasing. A probabilistic version of this simple observation is stated in the next lemma.

Lemma 1.11. *Let $\mathcal{Q}$ be a convex property of subsets of Γ, and let M_1, M, M_2 be three integer functions of N satisfying $0 \le M_1 \le M \le M_2 \le N$. Then*

$$\mathbb{P}(\Gamma_M \in \mathcal{Q}) \ge \mathbb{P}(\Gamma_{M_1} \in \mathcal{Q}) + \mathbb{P}(\Gamma_{M_2} \in \mathcal{Q}) - 1.$$

Hence, if $\mathbb{P}(\Gamma_{M_2} \in \mathcal{Q}) \to 1$ as $N \to \infty$, then $\mathbb{P}(\Gamma_{M_1} \in \mathcal{Q}) \le \mathbb{P}(\Gamma_M \in \mathcal{Q}) + o(1)$. In particular, if $\mathbb{P}(\Gamma_{M_i} \in \mathcal{Q}) \to 1$ as $N \to \infty$, $i = 1, 2$, then $\mathbb{P}(\Gamma_M \in \mathcal{Q}) \to 1$.

Proof. The following simple proof was observed by Johan Jonasson (personal communication). It is easily seen that a convex property $\mathcal{Q}$ is the intersection of an increasing property $\mathcal{Q}'$ and a decreasing property $\mathcal{Q}''$. (Exercise! – Note that the converse is obvious.) Thus

$$\begin{aligned}\mathbb{P}(\Gamma_M \in \mathcal{Q}) &\geq \mathbb{P}(\Gamma_M \in \mathcal{Q}') + \mathbb{P}(\Gamma_M \in \mathcal{Q}'') - 1 \\ &\geq \mathbb{P}(\Gamma_{M_1} \in \mathcal{Q}') + \mathbb{P}(\Gamma_{M_2} \in \mathcal{Q}'') - 1 \\ &\geq \mathbb{P}(\Gamma_{M_1} \in \mathcal{Q}) + \mathbb{P}(\Gamma_{M_2} \in \mathcal{Q}) - 1. \qquad \blacksquare\end{aligned}$$

1.4 ASYMPTOTIC EQUIVALENCE

In this section we examine the *asymptotic equivalence* of the two models Γ_p and Γ_M; recall that this includes the random graphs $\mathbb{G}(n,p)$ and $\mathbb{G}(n,M)$ as a special case. Our goal is to establish conditions under which convergence of $\mathbb{P}(\Gamma_p \in \mathcal{Q})$ implies convergence of $\mathbb{P}(\Gamma_M \in \mathcal{Q})$ to the same limit and vice versa. One expects such equivalence when M is near Np. Since Γ_p is a mixture of Γ_M's for different M, the above implication is more straightforward in the direction from the uniform to the binomial model and then does not require any restriction on $\mathcal{Q}$. The only tools we use are the elementary law of total probability and Chebyshev's inequality. Most results in this section are based on Łuczak (1990a); in the case in which the limit is one they already appeared in Bollobás (1979, 1985).

Let $\Gamma(n)$ be a sequence of sets of size $N(n) = |\Gamma(n)| \to \infty$. (In the example of main concern to us, *viz.* random graphs, $\Gamma(n) = [n]^2$ and thus $N(n) = \binom{n}{2}$.) We further consider a property $\mathcal{Q}$ of subsets of these sets; formally the property corresponds to a sequence $\mathcal{Q}(n) \subseteq 2^{\Gamma(n)}$ of families of subsets of $\Gamma(n)$, $n = 1, 2, \dots$. Finally, $p(n)$ is a given sequence of real numbers with $0 \leq p(n) \leq 1$, and $M(n)$ is a sequence of integers with $0 \leq M(n) \leq N(n)$. We usually omit the argument n and write Γ, N, $\mathcal{Q}$, p and M; moreover, we let $q = 1 - p$.

Proposition 1.12. *Let $\mathcal{Q}$ be an arbitrary property of subsets of $\Gamma = \Gamma(n)$ as above, $p = p(n) \in [0,1]$ and $0 \leq a \leq 1$. If for every sequence $M = M(n)$ such that $M = Np + O(\sqrt{Npq})$ it holds that $\mathbb{P}(\Gamma_M \in \mathcal{Q}) \to a$ as $n \to \infty$, then also $\mathbb{P}(\Gamma_p \in \mathcal{Q}) \to a$ as $n \to \infty$.*

Proof. Let C be a large constant and define (for each n)

$$\mathcal{M}(C) = \{M : |M - Np| \leq C\sqrt{Npq}\}.$$

Furthermore, let M_* be the element M of $\mathcal{M}(C)$ that minimizes $\mathbb{P}(\Gamma_M \in \mathcal{Q})$. By the law of total probability,

$$\begin{aligned}\mathbb{P}(\Gamma_p \in \mathcal{Q}) &= \sum_{M=0}^{N} \mathbb{P}(\Gamma_p \in \mathcal{Q} \mid |\Gamma_p| = M)\, \mathbb{P}(|\Gamma_p| = M) \\ &= \sum_{M=0}^{N} \mathbb{P}(\Gamma_M \in \mathcal{Q})\, \mathbb{P}(|\Gamma_p| = M) \\ &\geq \sum_{M \in \mathcal{M}(C)} \mathbb{P}(\Gamma_{M_*} \in \mathcal{Q})\, \mathbb{P}(|\Gamma_p| = M) \\ &= \mathbb{P}(\Gamma_{M_*} \in \mathcal{Q})\, \mathbb{P}\big(|\Gamma_p| \in \mathcal{M}(C)\big).\end{aligned}$$

By assumption, $\mathbb{P}(\Gamma_{M_*} \in \mathcal{Q}) \to a$, and using Chebyshev's inequality (1.2), we also have $\mathbb{P}\big(|\Gamma_p| \notin \mathcal{M}(C)\big) \leq \operatorname{Var}|\Gamma_p| / \big(C\sqrt{Npq}\big)^2 = 1/C^2$. Consequently,

$$\liminf_{n\to\infty} \mathbb{P}(\Gamma_p \in \mathcal{Q}) \geq a \liminf_{n\to\infty} \mathbb{P}\big(|\Gamma_p| \in \mathcal{M}(C)\big) \geq a(1 - C^{-2}).$$

Similarly, if M^* maximizes $\mathbb{P}(\Gamma_M \in \mathcal{Q})$ among $M \in \mathcal{M}(C)$,

$$\mathbb{P}(\Gamma_p \in \mathcal{Q}) \leq \mathbb{P}(\Gamma_{M^*} \in \mathcal{Q}) + \mathbb{P}\big(|\Gamma_p| \notin \mathcal{M}(C)\big) \leq \mathbb{P}(\Gamma_{M^*} \in \mathcal{Q}) + C^{-2},$$

and

$$\limsup_{n\to\infty} \mathbb{P}(\Gamma_p \in \mathcal{Q}) \leq a + C^{-2}.$$

The result follows by letting $C \to \infty$. ∎

In the other direction no asymptotic equivalence can be true in such generality. The property of containing exactly M edges serves as a simplest counterexample (Exercise!). However, the additional assumption of monotonicity of $\mathcal{Q}$ suffices.

Proposition 1.13. *Let $\mathcal{Q}$ be a monotone property of subsets of $\Gamma = \Gamma(n)$ as above, $0 \leq M \leq N$ and $0 \leq a \leq 1$. If for every sequence $p = p(n) \in [0,1]$ such that $p = M/N + O\big(\sqrt{M(N-M)/N^3}\big)$ it holds that $\mathbb{P}(\Gamma_p \in \mathcal{Q}) \to a$, then $\mathbb{P}(\Gamma_M \in \mathcal{Q}) \to a$ as $n \to \infty$.*

Proof. We consider only the case in which $\mathcal{Q}$ is increasing (the decreasing case is similar). Let C be a large constant, $p_0 = M/N$, $q_0 = 1 - p_0$, and define $p_+ = \min(p_0 + C\sqrt{p_0 q_0/N}, 1)$ and $p_- = \max(p_0 - C\sqrt{p_0 q_0/N}, 0)$. Arguing as in the proof of Proposition 1.12 and using Lemma 1.10, we have

$$\begin{aligned}\mathbb{P}(\Gamma_{p_+} \in \mathcal{Q}) &\geq \sum_{M' \geq M} \mathbb{P}(\Gamma_{M'} \in \mathcal{Q})\, \mathbb{P}(|\Gamma_{p_+}| = M') \\ &\geq \mathbb{P}(\Gamma_M \in \mathcal{Q})\, \mathbb{P}(|\Gamma_{p_+}| \geq M) \\ &\geq \mathbb{P}(\Gamma_M \in \mathcal{Q}) - \mathbb{P}(|\Gamma_{p_+}| < M) \qquad (1.4)\end{aligned}$$

and similarly

$$\mathbb{P}(\Gamma_{p_-} \in \mathcal{Q}) \leq \mathbb{P}(\Gamma_M \in \mathcal{Q}) + \mathbb{P}(|\Gamma_{p_-}| > M). \tag{1.5}$$

The cases $M = 0$ and $M = N$ are trivial (Exercise!), so we may further assume $1 \leq M \leq N - 1$, and thus $Np_0q_0 = M(N - M)/N \geq 1/2$. Since $|\Gamma_{p_-}|$ has the binomial distribution with mean Np_- and variance

$$Np_-(1 - p_-) \leq M(1 - p_0 + C\sqrt{p_0q_0/N}) \leq Np_0q_0 + C\sqrt{Np_0q_0},$$

Chebyshev's inequality (1.2) yields, with $\delta(C) = C^{-2} + \sqrt{2}C^{-1}$,

$$\mathbb{P}(|\Gamma_{p_-}| > M) \leq \frac{Np_-(1 - p_-)}{(Np_0 - Np_-)^2} \leq \frac{Np_0q_0 + C\sqrt{Np_0q_0}}{C^2Np_0q_0} \leq \delta(C),$$

and similarly $\mathbb{P}(|\Gamma_{p_+}| < M) \leq \delta(C)$. Since

$$\lim_{n\to\infty} \mathbb{P}(\Gamma_{p_+} \in \mathcal{Q}) = \lim_{n\to\infty} \mathbb{P}(\Gamma_{p_-} \in \mathcal{Q}) = a$$

by assumption, the inequalities (1.4) and (1.5) yield

$$a - \delta(C) \leq \liminf_{n\to\infty} \mathbb{P}(\Gamma_M \in \mathcal{Q}) \leq \limsup_{n\to\infty} \mathbb{P}(\Gamma_M \in \mathcal{Q}) \leq a + \delta(C),$$

and the result follows by letting $C \to \infty$, which implies $\delta(C) \to 0$. ∎

Remark 1.14. In the above proof one can relax the monotonicity of $\mathcal{Q}$ and instead require only that in the range $M' = M + O(\sqrt{M(N - M)/N})$

$$\mathbb{P}(\Gamma_{M'} \in \mathcal{Q}) \leq \mathbb{P}(\Gamma_M \in \mathcal{Q}) + o(1)$$

for $M' \leq M$, and

$$\mathbb{P}(\Gamma_{M'} \in \mathcal{Q}) \geq \mathbb{P}(\Gamma_M \in \mathcal{Q}) + o(1)$$

for $M' \geq M$. By Lemma 1.11, these conditions are satisfied whenever $\mathcal{Q}$ is convex and for some M' with $M' - M \gg \sqrt{M(N - M)/N}$, it holds that $\lim_{n\to\infty} \mathbb{P}(\Gamma_{M'} \in \mathcal{Q}) = 1$ (Exercise!).

The next result simplifies Proposition 1.13 for $a = 1$ by showing that for convex properties $\mathcal{Q}$, we have *a.a.s.* $\Gamma_M \in \mathcal{Q}$ provided *a.a.s.* $\Gamma_{M/N} \in \mathcal{Q}$.

Proposition 1.15. *Let $\mathcal{Q}$ be a convex property of subsets of Γ and let $0 \leq M \leq N$. If $\mathbb{P}(\Gamma_{M/N} \in \mathcal{Q}) \to 1$ as $n \to \infty$, then $\mathbb{P}(\Gamma_M \in \mathcal{Q}) \to 1$.*

Proof. We assume for simplicity that $M(N - M)/N \to \infty$, leaving the cases in which M or $N - M$ is bounded to the reader (Exercise!). (Note that the subsubsequence principle implies that it suffices to consider these three cases – Exercise!)

Let M_1 and M_2 maximize $\mathbb{P}(\Gamma_{M'} \in \mathcal{Q})$ among $M' \leq M$ and $M' \geq M$, respectively. Arguing as in the proof of Proposition 1.12, we then have

$$\mathbb{P}(\Gamma_{M/N} \in \mathcal{Q}) \leq \mathbb{P}(\Gamma_{M_1} \in \mathcal{Q})\,\mathbb{P}(|\Gamma_{M/N}| \leq M) + \mathbb{P}(|\Gamma_{M/N}| > M)$$

and thus, since $\mathbb{P}(|\Gamma_{M/N}| \leq M) \to 1/2$ by the central limit theorem,

$$1 = \lim_{n\to\infty} \mathbb{P}(\Gamma_{M/N} \in \mathcal{Q}) \leq \tfrac{1}{2} \liminf_{n\to\infty} \mathbb{P}(\Gamma_{M_1} \in \mathcal{Q}) + \tfrac{1}{2},$$

which implies that $\lim_{n\to\infty} \mathbb{P}(\Gamma_{M_1} \in \mathcal{Q}) = 1$. Similarly, $\lim_{n\to\infty} \mathbb{P}(\Gamma_{M_2} \in \mathcal{Q}) = 1$. Since $M_1 \leq M \leq M_2$, Lemma 1.11 yields $\mathbb{P}(\Gamma_M \in \mathcal{Q}) \to 1$. ∎

If the convergence of $\mathbb{P}(\Gamma_p \in \mathcal{Q})$ to 1 is reasonably fast, then the passage from the binomial to the uniform model can be made without any restriction on $\mathcal{Q}$. Indeed, by the law of total probability, for any M,

$$\begin{aligned}\mathbb{P}(\Gamma_p \notin \mathcal{Q}) &= \sum_{k=0}^{N} \mathbb{P}(\Gamma_k \notin \mathcal{Q}) \binom{N}{k} p^k (1-p)^{N-k} \\ &\geq \mathbb{P}(\Gamma_M \notin \mathcal{Q}) \binom{N}{M} p^M (1-p)^{N-M},\end{aligned}$$

from which it easily follows (Exercise!) that, taking $p = M/N$,

$$\mathbb{P}(\Gamma_M \notin \mathcal{Q}) \leq 3\sqrt{M}\,\mathbb{P}(\Gamma_{M/N} \notin \mathcal{Q}). \tag{1.6}$$

This inequality (Bollobás 1985, p. 35) is a slight sharpening of a result by Pittel (1982), and is therefore known as Pittel's inequality.

The following simple corollary of Propositions 1.15 and 1.12 (Exercise!) is stated here for future reference.

Corollary 1.16. *Let $\mathcal{Q}$ be an increasing property of subsets of Γ, and let $M = M(n) \to \infty$. Assume further that $\delta > 0$ is fixed and $0 \leq (1\pm\delta)M/N \leq 1$.*

(i) *If $\mathbb{P}(\Gamma_{M/N} \in \mathcal{Q}) \to 1$, then $\mathbb{P}(\Gamma_M \in \mathcal{Q}) \to 1$.*

(ii) *If $\mathbb{P}(\Gamma_{M/N} \in \mathcal{Q}) \to 0$, then $\mathbb{P}(\Gamma_M \in \mathcal{Q}) \to 0$.*

(iii) *If $\mathbb{P}(\Gamma_M \in \mathcal{Q}) \to 1$, then $\mathbb{P}(\Gamma_{(1+\delta)M/N} \in \mathcal{Q}) \to 1$.*

(iv) *If $\mathbb{P}(\Gamma_M \in \mathcal{Q}) \to 0$, then $\mathbb{P}(\Gamma_{(1-\delta)M/N} \in \mathcal{Q}) \to 0$.* ∎

Remark 1.17. The results of this section indicate that in a vast majority of cases the properties of random graphs $\mathbb{G}(n,p)$ and $\mathbb{G}(n,M)$, where $M \sim \binom{n}{2}p$, are very similar to each other. Even if the equivalence statements do not apply directly, typically repeating a proof step by step leads to an analogous result for the other model. Thus, in this book we very often state and prove theorems only in one of the two basic models. However, one should bear in mind that there are exceptions to this "equivalence rule of thumb" (compare, e.g., Theorem 3.9 with Theorem 3.11, or Theorem 6.52 with Theorem 6.58).

1.5 THRESHOLDS

The most intriguing discovery made by Erdős and Rényi in the course of investigating random graphs is the phenomenon of thresholds. For many graph properties the limiting probability that a random graph possesses them jumps from 0 to 1 (or vice versa) very rapidly, that is, with a rather small increase in the (expected) number of edges. This behavior is not just a feature of random graphs; as shown by Bollobás and Thomason (1987), it holds for monotone properties of arbitrary random subsets (Theorem 1.24 below).

We consider in this section, as in the previous one, a property $\mathcal{Q}$ of random subsets of a sequence $\Gamma(n)$ of sets, with $N(n) = |\Gamma(n)|$. Throughout we assume that we exclusively deal with properties that are neither always true nor always false.

For an increasing property $\mathcal{Q}$, a sequence $\widehat{p} = \widehat{p}(n)$ is called a *threshold* if

$$\mathbb{P}(\Gamma_p \in \mathcal{Q}) \to \begin{cases} 0 & \text{if} \quad p \ll \widehat{p}, \\ 1 & \text{if} \quad p \gg \widehat{p}. \end{cases} \tag{1.7}$$

Thresholds $\widehat{M} = \widehat{M}(n)$ for the uniform model are defined analogously by

$$\mathbb{P}(\Gamma_M \in \mathcal{Q}) \to \begin{cases} 0 & \text{if} \quad M \ll \widehat{M}, \\ 1 & \text{if} \quad M \gg \widehat{M}. \end{cases} \tag{1.8}$$

There is really no need to insist that $\widehat{M}$ is an integer, but we can always replace $\widehat{M}$ by $\lceil \widehat{M} \rceil$. In order to avoid trivial complications we assume $\widehat{M} \geq 1$, or at least $\inf \widehat{M}(n) > 0$.

Thresholds for decreasing families are defined as the thresholds of their complements.

Throughout the book we will often refer to the first line of (1.7) or (1.8) as the *0-statement* and to the second line as the *1-statement* of the respective threshold result.

Remark 1.18. Corollary 1.16 implies that $\widehat{p}$ is a threshold for a monotone property if and only if $\widehat{M} = \widehat{p}|\Gamma|$ is (Exercise!). Hence it does not matter which of the two basic models for random subsets we use.

Remark 1.19. Strictly speaking, a threshold is not uniquely determined since if $\widehat{p}$ is a threshold and $\widehat{p}' \asymp \widehat{p}$, then $\widehat{p}'$ is a threshold too (and similarly for $\widehat{M}$). Nevertheless, it is customary to talk about *the* threshold; this is convenient but it should be remembered that the threshold really is defined only within constant factors.

Example 1.20. If $\Gamma = [n]$ then Γ_p and Γ_M are random subsets of integers. Let $\mathcal{Q}$ be the property of containing a 3-term arithmetic progression. We will show in Example 3.2 that $\widehat{p} = n^{-2/3}$ is the threshold for $\mathcal{Q}$ in Γ_p, and so $\widehat{M} = n^{1/3}$ is the threshold for $\mathcal{Q}$ in Γ_M.

Example 1.21. In the case in which $\Gamma = [n]^2$ we deal with random graphs. We will soon learn (Theorem 3.4) that the threshold for containing a triangle is $\widehat{p} = 1/n$ in $\mathbb{G}(n,p)$, and thus $\widehat{M} = n$ in $\mathbb{G}(n,M)$.

Remark 1.22. Suppose that we construct a random subset sequentially by adding random elements one by one; in other words, we consider the random subset process $\{\Gamma_M\}_0^N$ as in the proof of Lemma 1.10. Define a random variable $\widetilde{M}$ as the number of elements selected when the random set first satisfies a given increasing property $\mathcal{Q}$; $\widetilde{M}$ is often called the *hitting time* of $\mathcal{Q}$. Then $\widetilde{M} \leq M$ if and only if $\Gamma_M \in \mathcal{Q}$, and thus

$$\mathbb{P}(\widetilde{M} \leq M) = \mathbb{P}(\Gamma_M \in \mathcal{Q}). \tag{1.9}$$

Hence $\widehat{M}$ is a threshold if and only if $\widetilde{M} = \Theta_p(\widehat{M})$ (Exercise!).

In order to investigate thresholds further, we introduce some more notation. For a given increasing property $\mathcal{Q}$ and $0 < a < 1$, we define $p(a)$ as the number in $(0,1)$ for which

$$\mathbb{P}(\Gamma_{p(a)} \in \mathcal{Q}) = a.$$

(The existence and uniqueness of this number follow because $p \mapsto \mathbb{P}(\Gamma_p \in \mathcal{Q})$ is a continuous, strictly increasing function; *cf.* Lemma 1.10. – Exercise!) We similarly define

$$M(a) = \min\{M : \mathbb{P}(\Gamma_M \in \mathcal{Q}) \geq a\}$$

(in this case, of course, we should not expect to have $\mathbb{P}(\Gamma_{M(a)} \in \mathcal{Q}) = a$); it follows that

$$\mathbb{P}(\Gamma_{M(a)-1} \in \mathcal{Q}) < a \leq \mathbb{P}(\Gamma_{M(a)} \in \mathcal{Q}). \tag{1.10}$$

Since $\mathcal{Q}$ and Γ depend on a parameter n, we also write $p(a;n)$ and $M(a;n)$.

Proposition 1.23. *Suppose that $\mathcal{Q}$ is an increasing property of subsets of $\Gamma = \Gamma(n)$. Then $\widehat{p}(n)$ is a threshold if and only if $p(a;n) \asymp \widehat{p}(n)$ as $n \to \infty$, for every $a \in (0,1)$. Similarly, $\widehat{M}(n)$ is a threshold if and only if $M(a;n) \asymp \widehat{M}(n)$ for every $a \in (0,1)$.*

Proof. Suppose first that $\widehat{M}$ is a threshold. If $0 < a < 1$ but $M(a) \not\asymp \widehat{M}$, then there exists a subsequence $s = (n_1, n_2, \ldots)$, along which either $M(a)/\widehat{M} \to 0$ or $M(a)/\widehat{M} \to \infty$. In the first case, by (1.8), $\mathbb{P}(\Gamma_{M(a)} \in \mathcal{Q}) \to 0$ along s, which contradicts (1.10). In the second case, along s, $M(a) - 1 \gg \widehat{M}$ and thus (1.8) yields $\mathbb{P}(\Gamma_{M(a)-1} \in \mathcal{Q}) \to 1$, which again contradicts (1.10). Consequently, $M(a) \asymp \widehat{M}$ holds for every $a \in (0,1)$.

Conversely, suppose that $\widehat{M}$ is not a threshold. Then there exists a sequence $M = M(n)$ such that either $M/\widehat{M} \to 0$ and $\liminf \mathbb{P}(\Gamma_M \in \mathcal{Q}) > 0$, or $M/\widehat{M} \to \infty$ and $\limsup \mathbb{P}(\Gamma_M \in \mathcal{Q}) < 1$.

In the first case, there exist $a > 0$ and a subsequence along which $\mathbb{P}(\Gamma_M \in \mathcal{Q}) \geq a$, and thus $M(a) \leq M \ll \widehat{M}$; in the second case, similarly there exist $a < 1$ and a subsequence along which $M(a) > M \gg \widehat{M}$. In both cases $M(a) \not\asymp \widehat{M}$.

The proof for p is almost identical so we omit it here. ∎

Theorem 1.24. *Every monotone property has a threshold.*

Proof. Without loss of generality assume that $\mathcal{Q}$ is increasing. Let $0 < \varepsilon < 1$, and let m be an integer such that $(1-\varepsilon)^m \leq \varepsilon$. Consider m independent copies $\Gamma^{(1)}, \ldots, \Gamma^{(m)}$ of $\Gamma_{p(\varepsilon)}$. Their union is $\Gamma_{p'}$, with $p' = 1 - (1-p(\varepsilon))^m \leq mp(\varepsilon)$, and hence by Lemma 1.10

$$\mathbb{P}(\Gamma^{(1)} \cup \cdots \cup \Gamma^{(m)} \in \mathcal{Q}) \leq \mathbb{P}(\Gamma_{mp(\varepsilon)} \in \mathcal{Q}).$$

On the other hand, since $\mathcal{Q}$ is increasing, if any $\Gamma^{(i)} \in \mathcal{Q}$, then $\Gamma^{(1)} \cup \cdots \cup \Gamma^{(m)} \in \mathcal{Q}$, and thus

$$\begin{aligned}\mathbb{P}(\Gamma^{(1)} \cup \cdots \cup \Gamma^{(m)} \notin \mathcal{Q}) \leq \mathbb{P}(\Gamma^{(i)} \notin \mathcal{Q} \text{ for every } i) &= \big(1 - \mathbb{P}(\Gamma_{p(\varepsilon)} \in \mathcal{Q})\big)^m \\ &= (1-\varepsilon)^m \leq \varepsilon.\end{aligned}$$

Consequently,

$$\mathbb{P}(\Gamma_{mp(\varepsilon)} \in \mathcal{Q}) \geq \mathbb{P}(\Gamma^{(1)} \cup \cdots \cup \Gamma^{(m)} \in \mathcal{Q}) \geq 1 - \varepsilon$$

and thus $p(1-\varepsilon) \leq mp(\varepsilon)$. Hence, if $0 < \varepsilon < 1/2$,

$$p(\varepsilon) \leq p(1/2) \leq p(1-\varepsilon) \leq mp(\varepsilon),$$

with m depending on ε but not on the parameter n; this implies that $p(\varepsilon) \asymp p(1/2) \asymp p(1-\varepsilon)$, and Proposition 1.23 shows that, for example, $p(1/2)$ is a threshold.

The existence of a threshold $\widehat{M}$ for $\mathbb{G}(n, M)$ can be proved similarly; it follows also by Remark 1.18. ∎

For non-monotone properties one adopts a "local" version of the definition of a threshold, with (1.7) being satisfied only in the vicinity of $\widehat{p}$. Observe that a property may have no threshold at all or it may have countably many thresholds (see Spencer (1991) for more on this). Convex properties have at most two thresholds, one of the 0–1 form and one in reverse.

1.6 SHARP THRESHOLDS

We end this chapter with a discussion of some recent general results on the widths of thresholds. We continue with the assumptions of the preceding section, and let $\delta(\varepsilon) = p(1-\varepsilon) - p(\varepsilon)$, $0 < \varepsilon < 1/2$. We should think here of ε

as fixed and very small; then $\delta(\varepsilon)$ is a measure of the width of the threshold. We may similarly define $\delta_M(\varepsilon) = M(1-\varepsilon) - M(\varepsilon)$ for the uniform model; note that by (1.9), δ_M also measures the concentration of the random variable $\widetilde{M}$ (Exercise!).

Theorem 1.24 shows that for every fixed ε, $\delta(\varepsilon) = O(\widehat{p})$. More precisely, the proof of Theorem 1.24 implies that, for any increasing property $\mathcal{Q}$ and $0 < \varepsilon < 1/2$,

$$1 \leq p(1-\varepsilon)/p(\varepsilon) \leq \lceil \varepsilon^{-1} \log \varepsilon^{-1} \rceil,$$

and hence $p(\varepsilon)/p(1/2)$ is bounded from above and below by universal constants for every fixed $\varepsilon \in (0,1)$. (For the uniform model Γ_M we similarly have $1 \leq M(1-\varepsilon)/M(\varepsilon) \leq \lceil \varepsilon^{-1} \log \varepsilon^{-1} \rceil$.)

However, certain monotone properties enjoy sharper thresholds than those guaranteed by Theorem 1.24. Sometimes

$$\mathbb{P}(\Gamma_p \in \mathcal{Q}) \to \begin{cases} 0 & \text{if} \quad p \leq (1-\eta)\widehat{p} \\ 1 & \text{if} \quad p \geq (1+\eta)\widehat{p} \end{cases}$$

for every $\eta > 0$; in this case $\widehat{p}$ is called a ***sharp threshold***. Note that while thresholds in general are defined up to the asymptotic relation $\asymp$ (see Remark 1.19), we have defined sharp thresholds up to $\sim$. The existence of a sharp threshold is equivalent to $p(\varepsilon; n)/p(1/2; n) \to 1$ as $n \to \infty$ for every ε with $0 < \varepsilon < 1$, and further to $\delta(\varepsilon) = o(\widehat{p})$ for every fixed ε (Exercise!).

In contrast, if there exists $\varepsilon > 0$ such that $\delta(\varepsilon) = \Theta(\widehat{p})$, then the threshold is called *coarse*.

Similarly, we define sharp and coarse thresholds for Γ_M; it is easily seen by Corollary 1.16 that if $\widehat{p}$ and $\widehat{M} = \widehat{p}|\Gamma|$ are corresponding thresholds for Γ_p and Γ_M, and moreover $\widehat{M} \to \infty$ (to rule out some trivial counterexamples), then $\widehat{p}$ is a sharp threshold if and only if $\widehat{M}$ is. Moreover, using the notation of Remark 1.22, $\widehat{M}$ is a sharp threshold if and only if $\widetilde{M}/M(1/2) \xrightarrow{p} 1$ (Exercise!).

Let us now restrict attention to random graphs $\mathbb{G}(n,p)$ and graph properties.

Example 1.25. A classic example of a sharp threshold is the threshold $\widehat{p} = \log n/n$ for disappearence of isolated vertices; in this case $\delta(\varepsilon) = \Theta(1/n)$, see Corollary 3.31. This coincides with the thresholds for connectivity and for the existence of a perfect matching (for n even); see Chapter 4.

Example 1.26. The property of containing a given graph as a subgraph, studied in detail in Chapter 3, has a coarse threshold; see, e.g., Theorem 3.9.

Remark 1.27. There are quite natural properties with (coarse) thresholds which are sharp on one side but not on the other; for example "$\mathbb{G}(n,p)$ contains a cycle". Another example can be seen in Theorem 8.1. We abstain from giving a formal definition of such "semi-sharp" thresholds.

Friedgut and Kalai (1996) showed that $\delta(\varepsilon) = O(1/\log n)$ for every monotone graph property; this was improved by Bourgain and Kalai (1997) to $O(1/\log^{2-\delta} n)$ for every $\delta > 0$, and it is conjectured that the absolute width is actually $O(1/\log^2 n)$, which is achieved by simple examples. (In view of $\delta(\varepsilon) = O(\widehat{p})$, this result is of interest mainly for thresholds that are constant or tend to 0 very slowly.)

Friedgut and Kalai (1996) gave also the following version, which implies the $O(1/\log n)$ estimate for arbitrary $\widehat{p}$ and improves it for $\widehat{p} \to 0$.

Theorem 1.28. *For every ε with $0 < \varepsilon < 1$, there exists a constant C_ε such that for every monotone graph property*

$$p(1-\varepsilon; n) \leq p(\varepsilon; n) + C_\varepsilon \frac{p(\varepsilon; n)\log(2/p(\varepsilon; n))}{\log n}. \tag{1.11}$$

■

In particular, it follows that a threshold $\widehat{p}$ such that $\log 1/\widehat{p} = o(\log n)$ is always sharp. However, if $\widehat{p}$ decreases as some power of n, as for most properties treated in this book, Theorem 1.28 yields only an $O(\widehat{p})$ estimate, just as the simpler and more general Theorem 1.24. (The assertion in Friedgut and Kalai (1996) that C_ε in (1.11) can be taken as $C \log 1/\varepsilon$ for some universal constant C is not correct; there are counterexamples with rapidly decreasing $\varepsilon = \varepsilon(n)$ and $p(\varepsilon; n)$.)

A recent result by Friedgut (1999) shows that Examples 1.25 and 1.26 are typical, in the sense that, roughly speaking, graph properties that depend on containing a large subgraph have sharp thresholds. (This is not literally true, as is seen by the example "a random graph contains a triangle and has at least $\log n$ edges"; this property is essentially the same as "contains a triangle", and has the same coarse threshold, since the probability of obtaining a triangle in a random graph with fewer than $\log n$ edges is very small.) More precisely, Friedgut's result says that a monotone graph property with a coarse threshold may be approximated by the property of containing at least one of a certain (finite) family of small graphs as a subgraph. A precise formulation (slightly different from Friedgut's) is as follows.

Theorem 1.29. *Suppose that $\varepsilon, \eta > 0$ and $c > 1$. Then there exists $k = k(\varepsilon, \eta, c)$ such that for every monotone graph property $\mathcal{Q}$ and every n for which $p(1-\varepsilon; n)/p(\varepsilon; n) \geq c$, there exists some p with $p(\varepsilon; n) \leq p \leq p(1-\varepsilon; n)$ and a family $G_1, \ldots, G_m$ of graphs with at most k vertices, such that if $\mathcal{Q}'$ is the property "contains a subgraph isomorphic to some G_j", then*

$$\mathbb{P}(\mathbb{G}(n,p) \in \mathcal{Q} \Delta \mathcal{Q}') < \eta.$$

■

Remark 1.30. Note that the theorem is stated for a fixed n rather than as an asymptotic result; this is because, in general, the approximating property $\mathcal{Q}'$ may depend on n, unless we restrict attention to a subsequence. Indeed, nothing prevents us from defining graph properties that depend on, say, the parity of n in some trivial explicit way.

Moreover, the theorem claims only that Q' is a good approximation for some $p \in [p(\varepsilon; n), p(1-\varepsilon; n)]$; it is easy to construct (artificial) examples where different approximations are required for different p, and good approximations are absent for some choices of p.

For a "natural" property, these complications are not to be expected, and it is reasonable to hope that the same Q' works for all n and p.

A related result (indeed, a corollary of Theorem 1.29) by Friedgut (1999) shows that a coarse threshold for a monotone graph property may only be of the type $n^{-\alpha}$ for some rational α, except that again it may be necessary to consider subsequences.

Theorem 1.31. *Suppose that a monotone graph property has a coarse threshold* $\widehat{p}(n)$. *Then there exists a partition of* $\mathbb{N} = \{1, 2, \dots\}$ *into a finite number of sequences* $\mathbb{N}_1, \dots, \mathbb{N}_m$ *and rational numbers* $\alpha_1, \dots, \alpha_m > 0$ *such that* $\widehat{p}(n) \asymp n^{-\alpha_j}$ *for* $n \in \mathbb{N}_j$. ∎

Theorems 1.29 and 1.31 and related results (Friedgut 1999, Bourgain 1999) can be used to show that certain properties have sharp thresholds, by showing that otherwise the conclusion of these results would yield a contradiction; see, for example, Achlioptas and Friedgut (1999) (the property of having at least a given chromatic number) and Friedgut and Krivelevich (2000) (Ramsey properties).

We emphasize that while Theorem 1.24 holds for arbitrary monotone properties of general random subsets, the more refined results discussed here require some symmetry assumptions; we have for simplicity stated them for random graphs and graph properties, that is, for properties that are invariant under permutations of the vertices. We consider the random graphs as random sets of edges, so the graph properties are properties of subsets of $[n]^2$ that are invariant under the permutations induced by the permutations of $[n]$. The same or similar results have been shown for other cases of random subsets with certain symmetry assumptions, including random hypergraphs; see Friedgut and Kalai (1996), Bourgain and Kalai (1997), Friedgut (1999). The results depend on the type of symmetry assumed (Bourgain and Kalai 1997). Related results without symmetry assumptions are given by Talagrand (1994) and Bourgain (1999).

2

Exponentially Small Probabilities

A common feature in many probabilistic arguments is the need to show that a random variable with large probability is not too far from its mean. One simple, but very useful, result of this type is Chebyshev's inequality (1.2), which holds for any random variable with finite variance. In this chapter we give several stronger inequalities valid under more restrictive assumptions, which for suitable random variables X and (positive) real numbers t yield estimates of the probability $\mathbb{P}(X \geq \mathbb{E}\,X + t)$ that decrease exponentially as $t \to \infty$.

In most cases we use the method, going back at least to Bernstein (1924), of applying Markov's inequality (1.3) to $\mathbb{E}\,e^{uX}$. Thus, for every $u \geq 0$,

$$\mathbb{P}(X \geq \mathbb{E}\,X + t) = \mathbb{P}\big(e^{uX} \geq e^{u(\mathbb{E}\,X+t)}\big) \leq e^{-u(\mathbb{E}\,X+t)}\,\mathbb{E}\,e^{uX}, \tag{2.1}$$

and similarly, for every $u \leq 0$,

$$\mathbb{P}(X \leq \mathbb{E}\,X - t) \leq e^{-u(\mathbb{E}\,X-t)}\,\mathbb{E}\,e^{uX}. \tag{2.2}$$

Then the moment generating function (or Laplace transform) $\mathbb{E}\,e^{uX}$ is estimated in some way, and an optimal or near-optimal u is chosen.

An estimate of $\mathbb{P}(|X - \mathbb{E}\,X| \geq t)$ may obviously be obtained by adding estimates of $\mathbb{P}(X \geq \mathbb{E}\,X + t)$ and $\mathbb{P}(X \leq \mathbb{E}\,X - t)$. We will often give only one-sided estimates below, leaving the corresponding two-sided estimates to the reader.

An important special case is estimating $\mathbb{P}(X = 0)$, which can be done by taking $t = \mathbb{E}\,X$ in (2.2) (assuming $\mathbb{E}\,X \geq 0$). We give several such results explicitly.

In the first section we consider random variables X that can be written as sums of independent terms. In the following sections we give extensions in various directions, covering cases with dependent summands.

2.1 INDEPENDENT SUMMANDS

An important case is that in which the random variable X can be expressed as a sum $\sum_1^n X_i$ of independent random variables. Then (2.1) can be written

$$\mathbb{P}(X \geq \mathbb{E}\, X + t) \leq e^{-u(\mathbb{E}\, X+t)}\, \mathbb{E}\, e^{uX} = e^{-u(\mathbb{E}\, X+t)} \prod_{i=1}^{n} \mathbb{E}\, e^{uX_i}, \tag{2.3}$$

and it remains to estimate the individual factors $\mathbb{E}\, e^{uX_i}$. Here we will be mainly interested in the case in which each X_i is a random indicator variable; thus $X_i \in \mathrm{Be}(p_i)$ where $p_i = \mathbb{P}(X_i = 1) = \mathbb{E}\, X_i$. Let $\lambda = \mathbb{E}\, X = \sum_1^n p_i$.

The binomial case

Consider first the case of a binomially distributed random variable $X \in \mathrm{Bi}(n, p)$; this is of the type above with all $p_i = p$. (Thus $\lambda = np$.) Then (2.3) yields

$$\mathbb{P}(X \geq \lambda + t) \leq e^{-u(\lambda+t)}(1 - p + pe^u)^n, \qquad u \geq 0.$$

The right-hand side attains its minimum at $e^u = (\lambda + t)(1 - p)/(n - \lambda - t)p$, assuming $\lambda + t < n$. This yields

$$\mathbb{P}(X \geq \mathbb{E}\, X + t) \leq \left(\frac{\lambda}{\lambda + t}\right)^{\lambda+t} \left(\frac{n - \lambda}{n - \lambda - t}\right)^{n-\lambda-t}, \qquad 0 \leq t \leq n - \lambda; \tag{2.4}$$

for $t > n - \lambda$ the probability is 0. This bound is implicit in Chernoff (1952) and is often called the Chernoff bound. (It appears explicitly in Okamoto (1958).)

For applications, it is usually convenient to replace the right-hand side of (2.4) by a larger but simpler bound. Two such consequences of (2.4) are presented in the next theorem, together with their lower tail counterparts. Any of these bounds or their numerous consequences contained in Corollaries 2.2 – 2.4 will be referred to as to Chernoff's inequality.

Theorem 2.1. *If $X \in \mathrm{Bi}(n, p)$ and $\lambda = np$, then, with $\varphi(x) = (1 + x)\log(1 + x) - x$, $x \geq -1$, (and $\varphi(x) = \infty$ for $x < -1$)*

$$\mathbb{P}(X \geq \mathbb{E}\, X + t) \leq \exp\Big(-\lambda\varphi\Big(\frac{t}{\lambda}\Big)\Big) \leq \exp\Big(-\frac{t^2}{2(\lambda + t/3)}\Big), \qquad t \geq 0; \tag{2.5}$$

$$\mathbb{P}(X \leq \mathbb{E}\, X - t) \leq \exp\Big(-\lambda\varphi\Big(\frac{-t}{\lambda}\Big)\Big) \leq \exp\Big(-\frac{t^2}{2\lambda}\Big), \qquad t \geq 0. \tag{2.6}$$

Proof. We can rewrite (2.4) as

$$\mathbb{P}(X \geq \mathbb{E}\,X + t) \leq \exp\Big(-\lambda\varphi\Big(\frac{t}{\lambda}\Big) - (n-\lambda)\varphi\Big(\frac{-t}{n-\lambda}\Big)\Big), \qquad 0 \leq t \leq n - \lambda.$$

Replacing X by $n - X$, or by a similar argument using (2.2), we obtain also

$$\mathbb{P}(X \leq \mathbb{E}\,X - t) \leq \exp\Big(-\lambda\varphi\Big(\frac{-t}{\lambda}\Big) - (n-\lambda)\varphi\Big(\frac{t}{n-\lambda}\Big)\Big), \qquad 0 \leq t \leq \lambda.$$

Since $\varphi(x) \geq 0$ for every x, we immediately obtain the first inequalities in (2.5) and (2.6). (These inequalities are trivial for $t > n-\lambda$ and $t > \lambda$, respectively.)

Since $\varphi(0) = 0$ and $\varphi'(x) = \log(1+x) \leq x$, we have $\varphi(x) \geq x^2/2$ for $-1 \leq x \leq 0$; hence the second inequality in (2.6) follows.

Similarly, $\varphi(0) = \varphi'(0) = 0$ and

$$\varphi''(x) = \frac{1}{1+x} \geq \frac{1}{(1+x/3)^3} = \Big(\frac{x^2}{2(1+x/3)}\Big)'',$$

whence $\varphi(x) \geq x^2/(2(1+x/3))$. Thus the second inequality in (2.5) follows. ∎

Note that the exponents in the estimates in (2.5) are $\Theta(t^2)$ for small t, say $t \leq \lambda$, but for larger t only $\Theta(t\log t)$ (the first estimate) and $\Theta(t)$ (the second estimate).

For small ratio t/λ, the exponent in (2.5) is almost $t^2/2\lambda$. The following corollary is sometimes convenient (*cf.* Alon and Spencer (1992, Theorem A.11)).

Corollary 2.2. *If $X \in \mathrm{Bi}(n,p)$ and $\lambda = np$, then*

$$\mathbb{P}(X \geq \mathbb{E}\,X + t) \leq \exp\Big(-\frac{t^2}{2\lambda} + \frac{t^3}{6\lambda^2}\Big). \tag{2.7}$$

Proof. The bound follows from (2.5), since $(\lambda + t/3)^{-1} \geq (\lambda - t/3)/\lambda^2$. ∎

Another immediate corollary is the following two-sided estimate.

Corollary 2.3. *If $X \in \mathrm{Bi}(n,p)$ and $\varepsilon > 0$, then*

$$\mathbb{P}(|X - \mathbb{E}\,X| \geq \varepsilon\,\mathbb{E}\,X) \leq 2\exp(-\varphi(\varepsilon)\,\mathbb{E}\,X), \tag{2.8}$$

where $\varphi(\varepsilon) = (1+\varepsilon)\log(1+\varepsilon) - \varepsilon$. In particular, if $\varepsilon \leq 3/2$, then

$$\mathbb{P}(|X - \mathbb{E}\,X| \geq \varepsilon\,\mathbb{E}\,X) \leq 2\exp\Big(-\frac{\varepsilon^2}{3}\,\mathbb{E}\,X\Big). \tag{2.9}$$

Proof. The first estimate is immediate by Theorem 2.1, since $\varphi(-\varepsilon) > \varphi(\varepsilon)$ (Exercise!). The second one follows because $\varphi(\varepsilon) \geq \varepsilon^2/[2(1+\varepsilon/3)] \geq \varepsilon^2/3$. ∎

For larger deviations, we state another handy version of (2.5). Note that λ does not appear explicitly in the estimate.

Corollary 2.4. *If $X \in \mathrm{Bi}(n,p)$, $\lambda = np$ and $c > 1$, then*

$$\mathbb{P}(X \geq x) \leq \exp(-c'x), \qquad x \geq c\lambda, \tag{2.10}$$

where $c' = \log c - 1 + 1/c > 0$. In particular,

$$\mathbb{P}(X \geq x) \leq \exp(-x), \qquad x \geq 7\lambda. \tag{2.11}$$

Proof. Apply (2.5) with $t = x - \lambda$ and note that $\lambda\varphi(x/\lambda - 1) = x\psi(x/\lambda)$, where $\psi(y) = \log y - 1 + 1/y$ is increasing for $y > 1$. Finally, note that $\psi(7) > 1$. ∎

Remark 2.5. Another simple bound is, still assuming $X \in \mathrm{Bi}(n,p)$,

$$\mathbb{P}(X \geq \mathbb{E}\,X + t) \leq \exp\Bigl(-\frac{2t^2}{n}\Bigr), \qquad t \geq 0; \tag{2.12}$$

the same bound holds for $\mathbb{P}(X \leq \mathbb{E}\,X - t)$ by symmetry. For $p = 1/2$ these bounds are better than (2.5) and (2.6), but they are worse for small p (note that the denominator of the exponent is n, not λ). Inequality (2.12) can be derived from (2.4) (Exercise!). It is also a special case of Azuma's inequality in the version given in Remark 2.28 below (Exercise!).

Remark 2.6. As a limiting case, obtained by taking $p = \lambda/n$ for any fixed $\lambda > 0$ and letting $n \to \infty$, (2.5)–(2.12) hold for a Poisson distributed random variable $X \in \mathrm{Po}(\lambda)$ too.

Remark 2.7. The estimate (2.5) would not hold without the term $t/3$ in the denominator; this can be seen by considering a limiting Poisson distribution as in Remark 2.6, in which case $\mathbb{P}(X \geq \mathbb{E}\,X + t) = \exp\bigl(-\Theta(t \log t)\bigr)$ as $t \to \infty$ (Exercise!).

The general case

Now we return to the general case in which $X_i \in \mathrm{Be}(p_i)$ with (possibly) different p_i. Let $Y \in \mathrm{Bi}(n, \bar{p})$ with $\bar{p} = \lambda/n = \sum p_i/n$. It is easily seen, taking the logarithm and using Jensen's inequality, that for every real u,

$$\mathbb{E}\,e^{uX} = \prod_i \bigl(1 + p_i(e^u - 1)\bigr) \leq \bigl(1 + \bar{p}(e^u - 1)\bigr)^n = \mathbb{E}\,e^{uY}.$$

Consequently, every bound for $\mathbb{P}(Y - \mathbb{E}\,Y \geq t)$ derived from (2.1) applies to X too; since $\lambda = \mathbb{E}\,X = \mathbb{E}\,Y$, the following theorem holds.

Theorem 2.8. *If $X_i \in \text{Be}(p_i)$, $i = 1, \dots, n$, are independent and $X = \sum_1^n X_i$, then* (2.5)–(2.12) *hold, with* $\lambda = \mathbb{E}\, X$. ■

Remark 2.9. By the proof above, also (2.4) holds under the conditions of Theorem 2.8, and yields sharper bounds than (2.5) and (2.6). Further similar bounds under the same conditions and, even more generally, for any independent random variables X_i such that $0 \le X_i \le 1$, are given, for example, by Bennett (1962), Hoeffding (1963) and Alon and Spencer (1992, Appendix A). We mention the following which use the variance of X rather than the mean.

Let $\sigma^2 = \operatorname{Var} X = \sum_1^n p_i(1-p_i)$, and let $\varphi(x) = (1+x)\log(1+x) - x$ as above. Then (Bennett 1962, Hoeffding 1963),

$$\mathbb{P}(X \ge \mathbb{E}\, X + t) \le \exp\bigl(-\sigma^2 \varphi(t/\sigma^2)\bigr), \qquad t \ge 0; \tag{2.13}$$

$$\mathbb{P}(X \ge \mathbb{E}\, X + t) \le \exp\Bigl(-\frac{t^2}{2(\sigma^2 + t/3)}\Bigr), \qquad t \ge 0; \tag{2.14}$$

and, by symmetry,

$$\mathbb{P}(X \le \mathbb{E}\, X - t) \le \exp\bigl(-\sigma^2 \varphi(t/\sigma^2)\bigr), \qquad t \ge 0; \tag{2.15}$$

$$\mathbb{P}(X \le \mathbb{E}\, X - t) \le \exp\Bigl(-\frac{t^2}{2(\sigma^2 + t/3)}\Bigr), \qquad t \ge 0. \tag{2.16}$$

The bound (2.14) is due to Bernstein, while the sharper (2.13) is due to Bennett (1962). Note that these are sharper than (2.5) because $\sigma^2 \le \lambda$, but for small p_i, the difference is small and often negligible. For fairly small t, these bounds are quite sharp and not far from what might be hoped for, based on the asymptotics given by the central limit theorem (for $t \asymp \sigma$). However, for large t, they can be substantially improved.

The hypergeometric distribution

Let m, n and N be positive integers with $\max(m,n) \le N$. The *hypergeometric distribution* with parameters N, n and m is the distribution of the random variable X defined by taking a set Γ with $|\Gamma| = N$ and a subset $\Gamma' \subseteq \Gamma$ with $|\Gamma'| = m$, and letting $X = |\Gamma_n \cap \Gamma'|$, where Γ_n is a random subset of Γ with n elements as in Chapter 1. (E.g., we can take $\Gamma = [N]$ and $\Gamma' = [m]$.)

In other words, we draw n elements of Γ without replacement, and count the number of them that belong to Γ'. Note that drawing with replacement would yield a binomial random variable; it seems reasonable that drawing without replacement tends to produce smaller random fluctuations, and indeed the bounds obtained above still hold (Hoeffding 1963).

Theorem 2.10. *Let X have the hypergeometric distribution with parameters N, n and m. Then* (2.5)–(2.12) *hold, with* $\lambda = \mathbb{E}\, X = mn/N$.

Proof. Let $Y \in \text{Bi}(n, m/N)$. It is not difficult to show that $\mathbb{E}\, e^{uX} \le \mathbb{E}\, e^{uY}$ (Hoeffding 1963) for any real u, which yields (2.5) and (2.6), and thus (2.7)–(2.12) too, by the argument above (Exercise!).

Alternatively, as a special case of a result by Vatutin and Mikhailov (1982), X has the same distribution as a certain sum of n independent indicator random variables. (The proof is algebraic, and based on showing that the probability generating function has only real roots; there is no (known) probabilistic interpretation of these random indicators, which, in general, have irrational expectations.) Consequently we can apply Theorem 2.8. ∎

Remark 2.11. The second proof shows that, in fact, all bounds (2.4)–(2.16) hold with $\lambda = \mathbb{E}\,X = mn/N$ and $\sigma^2 = \operatorname{Var} X = nm(N-n)(N-m)/N^2(N-1)$. (The first proof yields all the bounds involving λ, but it only gives weaker versions of (2.13)–(2.16) with $\sigma^2 = \operatorname{Var} Y > \operatorname{Var} X$.)

2.2 BINOMIAL RANDOM SUBSETS

The FKG inequality

We begin by quoting a celebrated correlation inequality known as the FKG inequality (Fortuin, Kasteleyn and Ginibre 1971). (For a simple proof, see Grimmett and Stirzaker (1992, Problem 3.11.18(b)); for a powerful combinatorial generalization, see Ahlswede and Daykin (1978) or Bollobás (1986).) Consider a binomial random subset Γ_p as in Chapter 1, or more generally $\Gamma_{p_1,\dots,p_N}$, which is defined by including the element i with probability p_i, independently of all other elements, $i = 1, \dots, N$ (assuming $\Gamma = [N]$ for notational convenience). We say that a function $f : 2^\Gamma \to \mathbb{R}$ is *increasing* if for $A \subseteq B$, $f(A) \le f(B)$, and *decreasing* if $f(A) \ge f(B)$.

Theorem 2.12. *If the random variables X_1 and X_2 are two increasing or two decreasing functions of $\Gamma_{p_1,\dots,p_N}$, then*

$$\mathbb{E}(X_1X_2) \ge \mathbb{E}(X_1)\,\mathbb{E}(X_2).$$

In particular, if $\mathcal{Q}_1$ and $\mathcal{Q}_2$ are two increasing or two decreasing families of subsets of Γ, then

$$\mathbb{P}(\Gamma_{p_1,\dots,p_N} \in \mathcal{Q}_1 \cap \mathcal{Q}_2) \ge \mathbb{P}(\Gamma_{p_1,\dots,p_N} \in \mathcal{Q}_1)\,\mathbb{P}(\Gamma_{p_1,\dots,p_N} \in \mathcal{Q}_2). \qquad ∎$$

As an important application consider a family $\mathcal{S}$ of non-empty subsets of Γ and for each $A \in \mathcal{S}$ let $I_A = \mathbf{1}[A \subseteq \Gamma_{p_1,\dots,p_N}]$. Note that every I_A is increasing. Finally, let $X = \sum_{A\in\mathcal{S}} I_A$, i.e. X is the number of sets $A \in \mathcal{S}$ that are contained in $\Gamma_{p_1,\dots,p_N}$.

Corollary 2.13. *For $X = \sum_{A\in\mathcal{S}} I_A$ of the form just described,*

$$\mathbb{P}(X = 0) \ge \exp\left\{-\frac{\mathbb{E}\,X}{1 - \max_i p_i}\right\}.$$

Proof. By Theorem 2.12 and induction we immediately obtain

$$\mathbb{P}(X=0) \geq \prod_{A\in\mathcal{S}} (1-\mathbb{E}\, I_A).$$

Now, using the inequalities $1-x \geq e^{-x/(1-x)}$ and $\mathbb{E}\, I_A \leq \max p_i$ we conclude that

$$\mathbb{P}(X=0) \geq \exp\left\{-\frac{\mathbb{E}\, X}{1-\max_A \mathbb{E}\, I_A}\right\} \geq \exp\left\{-\frac{\mathbb{E}\, X}{1-\max_i p_i}\right\}. \qquad \blacksquare$$

We will soon give some similar exponential upper bounds on $\mathbb{P}(X=0)$. First, however, we show a more general large deviation result.

Upper bounds for lower tails

We continue to study random variables of the form $X = \sum_{A\in\mathcal{S}} I_A$ as in the preceding subsection. For the lower tail of the distribution of X, the following analogue of Theorem 2.1 holds (Janson 1990b).

Theorem 2.14. *Let $X = \sum_{A\in\mathcal{S}} I_A$ as above, and let $\lambda = \mathbb{E}\, X = \sum_A \mathbb{E}\, I_A$ and $\overline{\Delta} = \sum\sum_{A\cap B\neq\emptyset} \mathbb{E}(I_A I_B)$. Then, with $\varphi(x) = (1+x)\log(1+x) - x$, for $0 \leq t \leq \mathbb{E}\, X$,*

$$\mathbb{P}(X \leq \mathbb{E}\, X - t) \leq \exp\left(-\frac{\varphi(-t/\lambda)\lambda^2}{\overline{\Delta}}\right) \leq \exp\left(-\frac{t^2}{2\overline{\Delta}}\right).$$

Remark 2.15. Note that the definition of $\overline{\Delta}$ includes the diagonal terms with $A = B$. It is often convenient to treat them separately, and we define

$$\Delta = \tfrac{1}{2} \sum\sum_{A\neq B,\, A\cap B\neq\emptyset} \mathbb{E}(I_A I_B).$$

(The factor $\frac{1}{2}$ reflects the fact that Δ is the sum of $\mathbb{E}(I_A I_B)$ over all *unordered* pairs $\{A,B\} \in [\mathcal{S}]^2$ with $A \cap B \neq \emptyset$.) Thus $\overline{\Delta} = \lambda + 2\Delta$.

Remark 2.16. Clearly, $\Delta \geq 0$ and thus $\overline{\Delta} \geq \lambda$, with equality if and only if the sets A are disjoint and thus the random indicators I_A are independent. In the independent case, the bounds in Theorem 2.14 are the same as (2.6); Theorem 2.14 is thus an extension of (the lower tail part of) Theorem 2.1. More importantly, in a weakly dependent case with, say, $\Delta = o(\lambda)$ and thus $\overline{\Delta} \sim \lambda$, we get almost the same bounds as in the independent case.

Remark 2.17. There can be no corresponding general exponential bound for the upper tail probabilities $\mathbb{P}(X \geq \mathbb{E}\, X + t)$, as is seen by the following example (for another counterexample, see Remark 2.50). Let λ be an integer, let $\Gamma = \{0, \ldots, 2\lambda^2\}$ with $p_0 = \lambda^{-4}$, $p_i = 1 - \lambda^{-4}$ for $1 \leq i \leq \lambda^2$ and

$p_i = \lambda^{-1} - \lambda^{-4} + \lambda^{-8}$ for $\lambda^2 + 1 \le i \le 2\lambda^2$, and consider the family $\mathcal{S}$ of the subsets $A_i = \{0, i\}$ for $1 \le i \le \lambda^2$ and $A_i = \{i\}$ for $\lambda^2 + 1 \le i \le 2\lambda^2$. Then $\mathbb{E}X = \lambda$ and $\Delta < 1$. Nevertheless, for any $c < \infty$ and $\varepsilon > 0$, if λ is large enough,

$$\mathbb{P}(X > c\lambda) \ge \lambda^{-4}(1 - \lambda^{-4})^{\lambda^2} \ge \tfrac{1}{2}\lambda^{-4} > \exp(-\varepsilon\lambda).$$

Some partial results for the upper tail are given in Section 2.6.

Proof of Theorem 2.14. Let $\Psi(s) = \mathbb{E}(e^{-sX})$, $s \ge 0$. We will show first that

$$-(\log \Psi(s))' \ge \lambda e^{-s\overline{\Delta}/\lambda}, \qquad s > 0, \tag{2.17}$$

which implies

$$-\log \Psi(s) \ge \int_0^s \lambda e^{-u\overline{\Delta}/\lambda} du = \frac{\lambda^2}{\overline{\Delta}}(1 - e^{-s\overline{\Delta}/\lambda}). \tag{2.18}$$

In order to do this, we represent $-\Psi'(s)$ in the form

$$-\Psi'(s) = \mathbb{E}(Xe^{-sX}) = \sum_A \mathbb{E}(I_A e^{-sX}) \tag{2.19}$$

and for every $A \in \mathcal{S}$ we split $X = Y_A + Z_A$, where $Y_A = \sum_{B\cap A\neq\emptyset} I_B$. Then, by the FKG inequality (applied to $\Gamma_{p_1,\dots,p_N}$ conditioned of $I_A = 1$, which is a random subset of the same type) and by the independence of Z_A and I_A we get, setting $p_A = \mathbb{E}(I_A)$,

$$\begin{aligned}\mathbb{E}(I_A e^{-sX}) &= p_A\,\mathbb{E}(e^{-sY_A}e^{-sZ_A}|I_A = 1) \ge p_A\,\mathbb{E}(e^{-sY_A}|I_A = 1)\,\mathbb{E}(e^{-sZ_A}) \\ &\ge p_A\,\mathbb{E}(e^{-sY_A}|I_A = 1)\Psi(s).\end{aligned} \tag{2.20}$$

Recall that $\lambda = \sum_A p_A$. From (2.19) and (2.20), by applying Jensen's inequality twice, first to the conditional expectation and then to the sum, we obtain

$$\begin{aligned}-(\log \Psi(s))' &= \frac{-\Psi'(s)}{\Psi(s)} \ge \sum_A p_A\,\mathbb{E}(e^{-sY_A}|I_A = 1) \\ &\ge \lambda \sum_A \frac{1}{\lambda} p_A \exp\{-\,\mathbb{E}(sY_A|I_A = 1)\} \\ &\ge \lambda \exp\left\{-\sum_A \frac{1}{\lambda} p_A\,\mathbb{E}(sY_A|I_A = 1)\right\} \\ &= \lambda \exp\left\{-\frac{s}{\lambda}\sum_A \mathbb{E}(Y_A I_A)\right\} = \lambda e^{-s\overline{\Delta}/\lambda}.\end{aligned}$$

We have shown (2.17) and thus (2.18). Now, by Markov's inequality (2.2) together with (2.18),

$$\log \mathbb{P}(X \le \lambda - t) \le \log \mathbb{E}(e^{-sX}) + s(\lambda - t) \le -\frac{\lambda^2}{\overline{\Delta}}(1 - e^{-s\overline{\Delta}/\lambda}) + s(\lambda - t).$$

The right-hand side is minimized by choosing $s = -\log(1 - t/\lambda)\lambda/\overline{\Delta}$, which yields the first bound (for $t = \lambda$, let $s \to \infty$); the second follows because $\varphi(x) \ge x^2/2$ for $x \le 0$ as shown in the proof of Theorem 2.1. ■

The probability of nonexistence

Taking $t = \mathbb{E} X$ in Theorem 2.14, we obtain an estimate for the probability of no set in $\mathcal{S}$ occuring, which we state separately as part of the following theorem (Janson, Łuczak and Ruciński 1990).

Theorem 2.18. *With* $X = \sum_{A \in \mathcal{S}} I_A$, $\lambda = \mathbb{E} X$ *and* Δ *as above,*

(i) $\mathbb{P}(X = 0) \le \exp(-\lambda + \Delta)$;

(ii) $$\mathbb{P}(X = 0) \le \exp\left(-\frac{\lambda^2}{\lambda + 2\Delta}\right) = \exp\left(-\frac{\lambda^2}{\sum\sum_{A \cap B \ne \emptyset} \mathbb{E}(I_A I_B)}\right).$$

Remark 2.19. Both parts are valid for any λ and Δ, but (i) is uninteresting unless $\Delta < \lambda$. In fact, (i) gives the better bound when $\Delta < \lambda/2$, while (ii) is better for larger Δ (Exercise!).

Proof. Taking $t = \lambda$ in Theorem 2.14, or directly letting $s \to \infty$ in (2.18) and observing that $\lim_{s\to\infty} \Psi(s) = \mathbb{P}(X = 0)$, we immediately obtain (ii).

For (i), we obtain from the proof of Theorem 2.14, with $Y_A' = Y_A - I_A$,

$$-\log \mathbb{P}(X = 0) = -\int_0^\infty (\log \Psi(s))'\, ds \ge \int_0^\infty \sum_A p_A \mathbb{E}(e^{-sY_A} \mid I_A = 1)\, ds$$
$$= \sum_A p_A \mathbb{E}\Big(\frac{1}{Y_A} \Bigm| I_A = 1\Big).$$

When $I_A = 1$, we find $1/Y_A = 1/(1 + Y_A') \ge 1 - \frac{1}{2} Y_A'$ (since Y_A' is an integer), and thus

$$-\log \mathbb{P}(X = 0) \ge \sum_A p_A \mathbb{E}(1 - \tfrac{1}{2} Y_A' \mid I_A = 1)$$
$$= \sum_A (p_A - \tfrac{1}{2} \mathbb{E}(I_A Y_A')) = \lambda - \Delta.$$ ■

Remark 2.20. Boppana and Spencer (1989) gave another proof, resembling the proof of the Lovász Local Lemma, of a version of Theorem 2.18(i), namely,

$$\mathbb{P}(X = 0) \le \exp\{\Delta/(1 - \varepsilon)\} \prod_A (1 - \mathbb{E} I_A) \le \exp\{-\lambda + \Delta/(1 - \varepsilon)\}, \quad (2.21)$$

where $\varepsilon = \max p_A$. See also Spencer (1990) for another proof of Theorem 2.18(ii), but with an extra factor $\frac{1}{2}$ in the exponent. Finally, note that a slightly weaker version of Theorem 2.18(i) with the bound $\exp(-\lambda + 2\Delta)$ follows directly from (ii), because $\lambda^2 \geq (\lambda - 2\Delta)(\lambda + 2\Delta)$.

Remark 2.21. Although the bound in Theorem 2.18 and the first bound in (2.21) are quite close when $\varepsilon = \max p_\alpha$ is small, neither of them dominates the other. It is intriguing to note that the conceivable common improvement $\exp\{\Delta\} \prod_A (1 - \mathbb{E} I_A)$ fails to be an upper bound for $\mathbb{P}(X = 0)$; this is seen by the simple example where $X = I_1 + I_2$ with $I_1 = I_2 \in \mathrm{Be}(p)$, for which $\Delta = p$ and $\mathbb{P}(X = 0) = 1 - p > e^p(1-p)^2$, see Janson (1998) for further discussion.

The quantity Δ is a measure of the pairwise dependence between the I_A's (*cf.* Remark 2.16). If $\Delta = o(\lambda)$, then the exponents in Theorem 2.18 are $-\mathbb{E} X\big(1 + o(1)\big)$, matching asymptotically the lower bound Corollary 2.13, provided further $\max p_i \to 0$.

The development of the exponential bounds in this section were stimulated by the application in which X counts copies of a given graph in the random graph $\mathbb{G}(n,p)$. This will be presented in detail in Chapter 3 (*cf.* Theorem 3.9).

For a generalization of Theorem 2.14 see Roos (1996).

2.3 SUEN'S INEQUALITY

A drawback of the inequalities in Section 2.2 is that they apply only to the sum of random indicator variables with a very special structure. For example, they apply, as stated above, to the number of copies of a given graph in $\mathbb{G}(n,p)$, but they do not apply to the number of induced copies.

An inequality much more general than Theorem 2.18(i), and only slightly weaker, was given by Suen (1990). We do not give Suen's original inequality here, but rather the following related results, proved by Suen's method. For further similar results, see Janson (1998) and Spencer (1998).

The Suen inequalities use the concept of a dependency graph, defined in Section 1.2. Although formally valid for the sum of any family of random indicator variables, the inequalities are useful in cases in which there exists a sparse dependency graph. (No assumption is made on the type of the dependencies.)

Theorem 2.22. *Let $I_i \in \mathrm{Be}(p_i)$, $i \in \mathcal{I}$, be a finite family of Bernoulli random variables having a dependency graph L. Let $X = \sum_i I_i$ and $\lambda = \mathbb{E} X = \sum_i p_i$. Moreover, write $i \sim j$ if $ij \in E(L)$, and let $\Delta = \frac{1}{2}\sum\sum_{i \sim j} \mathbb{E}(I_i I_j)$ and $\delta = \max_i \sum_{k \sim i} p_k$. Then*

(i) $\mathbb{P}(X = 0) \leq \exp\{-\lambda + \Delta e^{2\delta}\}$;

(ii) $\mathbb{P}(X = 0) \leq \exp\left\{-\min\left(\frac{\lambda^2}{8\Delta}, \frac{\lambda}{2}, \frac{\lambda}{6\delta}\right)\right\}$. ∎

Theorem 2.23. *Let $I_i \in \mathrm{Be}(p_i)$, $i \in \mathcal{I}$, be a finite family of Bernoulli random variables having a dependency graph L. Let $X = \sum_i I_i$ and $\lambda = \mathbb{E}\,X$, and let also Δ and δ be as in Theorem 2.22. Moreover, let $\overline{\Delta} = \lambda + 2\Delta = \lambda + \sum\sum_{i\sim j} \mathbb{E}(I_i I_j)$. If $0 \le t \le \lambda$, then*

$$\mathbb{P}(X \le \lambda - t) \le \exp\Bigl(-\min\Bigl(\frac{t^2}{4\overline{\Delta}}, \frac{t}{6\delta}\Bigr)\Bigr). \tag{2.22}$$

For simplicity, we give only the proof of Theorem 2.23 here; the proof of Theorem 2.22 is similar (Janson 1998). Note that taking $t = \lambda$ in Theorem 2.23 we obtain an inequality which is only slightly weaker than Theorem 2.22(ii).

Remark 2.24. We have given results similar to Theorems 2.14 and 2.18, but with somewhat worse constants, and extra terms (typically negligible) involving δ. It is not known whether the terms with δ really are needed; in fact, it is conceivable that the estimates in Section 2.2 hold also under the weaker assumptions in this section.

Proof of Theorem 2.23. Define, for real s, the random function

$$F(s) = e^{s\lambda - sX},$$

and, for each subset $A \subseteq \mathcal{I}$, $X_A = \sum_{i\in A} I_i$ and $F_A(s) = \exp\bigl(s(\mathbb{E}\,X_A - X_A)\bigr)$. We differentiate and obtain

$$F'(s) = \lambda F(s) - \sum_i I_i F(s). \tag{2.23}$$

For each index $i \in \mathcal{I}$, let $N_i = \{i\} \cup \{j \in \mathcal{I} : i \sim j\}$ and $U_i = \mathcal{I} \setminus N_i$; then $X = X_{N_i} + X_{U_i}$ and

$$I_i e^{-sX} = p_i e^{-sX} + (I_i - p_i)e^{-sX_{U_i}} - (I_i - p_i)(1 - e^{-sX_{N_i}})e^{-sX_{U_i}}. \tag{2.24}$$

Now assume that $s \ge 0$. Then, for any set $A \subseteq \mathcal{I}$,

$$0 \le 1 - e^{-sX_A} \le sX_A,$$

and thus, considering the cases $I_i = 0$ and $I_i = 1$ separately,

$$(I_i - p_i)\bigl(1 - e^{-sX_A}\bigr) \le I_i s X_A. \tag{2.25}$$

Choosing $A = N_i$, (2.23), (2.24) and (2.25) imply

$$\begin{aligned} F'(s) &\le \lambda F(s) - \sum_i p_i e^{s\lambda - sX} - \sum_i (I_i - p_i)e^{s\lambda - sX_{U_i}} + \sum_i sI_i X_{N_i} e^{s\lambda - sX_{U_i}} \\ &= -e^{s\lambda}\sum_i (I_i - p_i)e^{-sX_{U_i}} + e^{s\lambda}\sum_i sI_i X_{N_i} e^{-sX_{U_i}}. \end{aligned}$$

Since, by the definition of a dependency graph, I_i and X_{U_i} are independent, $\mathbb{E}\big((I_i - p_i)e^{-sX_{U_i}}\big) = 0$. Moreover, $X_{U_i} \geq X_{U_i \cap U_j}$, which is independent of $I_i I_j$, and

$$\lambda - \mathbb{E}\, X_{U_i \cap U_j} = \sum_{k \notin U_i \cap U_j} p_k \leq \sum_{k \sim i} p_k + \sum_{k \sim j} p_k \leq 2\delta.$$

Hence,

$$\begin{aligned}
\mathbb{E}\, F'(s) &\leq e^{s\lambda} \sum_i \sum_{j \in N_i} s\, \mathbb{E}(I_i I_j e^{-sX_{U_i}}) \leq s e^{s\lambda} \sum_i \sum_{j \in N_i} \mathbb{E}(I_i I_j e^{-sX_{U_i \cap U_j}}) \\
&= s e^{s\lambda} \sum_i \sum_{j \in N_i} \mathbb{E}(I_i I_j)\, \mathbb{E}\, e^{-sX_{U_i \cap U_j}} \\
&\leq s e^{2\delta s} \sum_i \sum_{j \in N_i} \mathbb{E}(I_i I_j)\, \mathbb{E}\, F_{U_i \cap U_j}(s), \qquad s \geq 0.
\end{aligned} \tag{2.26}$$

We claim that

$$\mathbb{E}\, F(s) \leq e^{\frac{1}{2} s^2 \overline{\Delta} e^{2\delta s}}, \qquad s \geq 0. \tag{2.27}$$

In fact, using induction over $|\mathcal{I}|$, the number of random indicator variables, we may assume that the corresponding inequality holds for $\mathbb{E}\, F_A(s)$ for every proper subset A of $\mathcal{I}$ (and all $s \geq 0$). Since the corresponding values $\overline{\Delta}_A$ and δ_A for a subset A satisfy $\overline{\Delta}_A \leq \overline{\Delta}$ and $\delta_A \leq \delta$, it then follows from (2.26) that

$$\begin{aligned}
\mathbb{E}\, F'(s) &\leq s e^{2\delta s} \sum_i \sum_{j \in N_i} \mathbb{E}(I_i I_j) e^{\frac{1}{2} s^2 \overline{\Delta} e^{2\delta s}} = s e^{2\delta s} \overline{\Delta} e^{\frac{1}{2} s^2 \overline{\Delta} e^{2\delta s}} \\
&\leq \frac{d}{ds}\left(e^{\frac{1}{2} s^2 \overline{\Delta} e^{2\delta s}}\right), \qquad s \geq 0.
\end{aligned}$$

Hence (2.27) follows by integration, since it obviously holds for $s = 0$.

Markov's inequality (2.2) and (2.27) yield, for any $s \geq 0$,

$$\mathbb{P}(X \leq \lambda - t) \leq \mathbb{P}\big(F(s) \geq e^{st}\big) \leq e^{-st}\, \mathbb{E}\, F(s) \leq e^{-st + \frac{1}{2} s^2 \overline{\Delta} e^{2\delta s}}. \tag{2.28}$$

We choose here $s = \min(t/2\overline{\Delta}, 1/3\delta)$; then $e^{2\delta s} \leq e^{2/3} < 2$, and thus

$$\tfrac{1}{2} s^2 \overline{\Delta} e^{2\delta s} \leq s^2 \overline{\Delta} \leq st/2.$$

Consequently, (2.28) yields

$$\mathbb{P}(X \leq \lambda - t) \leq e^{-st/2},$$

which is (2.22). ∎

2.4 MARTINGALES

Martingales were first applied to random graphs by Shamir and Spencer (1987), followed by the spectacular success of Bollobás's (1988a) solution of the chromatic number problem. (These results are presented in Chapter 7.)

We begin by recalling the definition of a martingale. Note, however, that for applications to random subsets, and to random graphs in particular, one usually uses Corollary 2.27 below, where martingales are not explicitly mentioned.

Given a probability space $(\Omega, \mathcal{F}, \mathbb{P})$ and an increasing sequence of sub-σ-fields $\mathcal{F}_0 = \{\emptyset, \Omega\} \subseteq \mathcal{F}_1 \subseteq \cdots \subseteq \mathcal{F}_n = \mathcal{F}$, a sequence of random variables $X_0, X_1, \dots, X_n$ (with finite expectations) is called a *martingale* if for each $k = 0, \dots, n-1$, $\mathbb{E}(X_{k+1} \mid \mathcal{F}_k) = X_k$. In this case (with a finite sequence), every martingale is obtained from a random variable X by taking $X_k = \mathbb{E}(X \mid \mathcal{F}_k)$, $k = 0, \dots, n$. Then $X_0 = \mathbb{E}\,X$ and $X_n = X$. Also, we always have $\mathbb{E}(X_{k+1}) = \mathbb{E}(X_k)$.

In combinatorial applications, often Ω is a finite space and $\mathcal{F}$ is the family of all subsets; each $\mathcal{F}_k$ then corresponds to a partition $\mathcal{P}_k$ of Ω, with coarser partitions for smaller k. If $\mathbb{P}$ is the uniform probability measure on Ω, then a sequence (X_k) is a martingale if and only if each X_k is a function $\Omega \to \mathbb{R}$ that is constant on the blocks of the partition $\mathcal{P}_k$, with the value on each block being the average value of X_n on that block. If $\mathbb{P}$ is another probability measure, then this still holds if we use suitably weighted averages.

Azuma's inequality

The following result appears in Azuma (1967) and is often called Azuma's inequality, although it also appears in Hoeffding (1963).

Theorem 2.25. *If $(X_k)_0^n$ is a martingale with $X_n = X$ and $X_0 = \mathbb{E}\,X$, and there exist constants $c_k > 0$ such that*

$$|X_k - X_{k-1}| \le c_k$$

for each $k \le n$, then, for every $t > 0$,

$$\mathbb{P}(X \ge \mathbb{E}\,X + t) \le \exp\Big(-\frac{t^2}{2\sum_1^n c_k^2}\Big), \tag{2.29}$$

$$\mathbb{P}(X \le \mathbb{E}\,X - t) \le \exp\Big(-\frac{t^2}{2\sum_1^n c_k^2}\Big). \tag{2.30}$$

Proof. Set $Y_k = X_k - X_{k-1}$ and $S_k = \sum_{i=1}^k Y_i = X_k - X_0$. For any $u > 0$, by Markov's inequality (2.1), we have

$$\mathbb{P}(X - \mathbb{E}\,X \ge t) = \mathbb{P}(S_n \ge t) \le e^{-ut}\,\mathbb{E}(e^{uS_n}). \tag{2.31}$$

Because S_{n-1} is a $\mathcal{F}_{n-1}$-measurable function we also have

$$\mathbb{E}(e^{uS_n}) = \mathbb{E}[\mathbb{E}(e^{uS_n} \mid \mathcal{F}_{n-1})] = \mathbb{E}[e^{uS_{n-1}}\, \mathbb{E}(e^{uY_n} \mid \mathcal{F}_{n-1})].$$

Now we need the following fact: If a random variable Y satisfies $\mathbb{E}\,Y = 0$ and $-a \leq Y \leq a$ for some $a \geq 0$, then, for any u,

$$\mathbb{E}(e^{uY}) \leq e^{u^2a^2/2}. \tag{2.32}$$

To prove (2.32), note that by the convexity of e^{uy},

$$e^{uY} \leq \frac{a+Y}{2a}e^{ua} + \frac{a-Y}{2a}e^{-ua}.$$

Hence,

$$\mathbb{E}(e^{uY}) \leq \tfrac{1}{2}e^{ua} + \tfrac{1}{2}e^{-ua} \leq e^{u^2a^2/2},$$

where the last inequality follows by comparing the Taylor expansions (see Alon and Spencer (1992, Lemma A.6) for another proof).

Coming back to the proof of Theorem 2.25, we conclude that $\mathbb{E}(e^{uY_n} \mid \mathcal{F}_{n-1}) \leq e^{u^2c_n^2/2}$ and thus

$$\mathbb{E}(e^{uS_n}) \leq e^{u^2c_n^2/2}\, \mathbb{E}(e^{uS_{n-1}}).$$

Iterating this inequality n times, we find $\mathbb{E}(e^{uS_n}) \leq e^{u^2 \sum c_k^2/2}$, and thus by (2.31)

$$\mathbb{P}(X - \mathbb{E}\,X \geq t) \leq e^{-ut}e^{u^2\sum c_k^2/2};$$

substituting $u = t/\sum c_k^2$ we find (2.29).

The inequality (2.30) follows by symmetry. ∎

Remark 2.26. After a minor modification, (2.29) extends to supermartingales and (2.30) to submartingales; see Wormald (1999a).

Combinatorial setting

In the applications to random graphs, we will use the following consequence of Theorem 2.25. Note that the notion of a martingale has disappeared from the statement. In most applications, one simply has $c_k = 1$. (We tacitly assume that the function f is measurable; in the case of finite sets, this holds trivially.)

Corollary 2.27. *Let $Z_1, \dots, Z_N$ be independent random variables, with Z_k taking values in a set Λ_k. Assume that a function $f : \Lambda_1 \times \Lambda_2 \times \cdots \times \Lambda_N \to \mathbb{R}$ satisfies the following Lipschitz condition for some numbers c_k:*

(L) *If two vectors* $z, z' \in \prod_1^N \Lambda_i$ *differ only in the kth coordinate, then* $|f(z) - f(z')| \le c_k$.

Then, the random variable $X = f(Z_1, \ldots, Z_N)$ *satisfies, for any* $t \ge 0$,

$$\mathbb{P}(X \ge \mathbb{E}\, X + t) \le \exp\Bigl(-\frac{t^2}{2\sum_1^N c_k^2}\Bigr), \tag{2.33}$$

$$\mathbb{P}(X \le \mathbb{E}\, X - t) \le \exp\Bigl(-\frac{t^2}{2\sum_1^N c_k^2}\Bigr). \tag{2.34}$$

Proof. Let us define $\mathcal{F}_k$ to be the σ-field generated by $Z_1, \ldots, Z_k$ and consider the corresponding martingale defined by $X_k = \mathbb{E}\bigl(f(Z_1, \ldots, Z_N) \mid \mathcal{F}_k\bigr)$, $k = 0, \ldots, N$. The assumption about f implies that X_k and X_{k-1} differ by at most c_k (Exercise!). The corollary now follows from Theorem 2.25. ∎

Remark 2.28. A more careful proof along the same lines shows that (2.33) and (2.34) hold with $\exp\{-2t^2/\sum_1^N c_k^2\}$ on the right-hand side, that is, the exponents in the estimates may be multiplied by a factor 4 (McDiarmid 1989).

Returning to the random set Γ_p, one typically defines the random variables Z_k via the random indicators $I_\gamma = \mathbf{1}[\gamma \in \Gamma_p]$, $\gamma \in \Gamma$. Given a partition $A_1, \ldots, A_N$ of Γ, each Z_k is then taken as the random vector $(I_\gamma : \gamma \in A_k) \in \{0,1\}^{A_k}$, and for a given function $f : 2^\Gamma \to \mathbb{R}$, the Lipschitz condition (L) in Corollary 2.27 is equivalent to saying that for any two subsets $A, B \subseteq \Gamma$, $|f(A) - f(B)| \le c_k$ whenever the symmetric difference of the sets A and B is contained in A_k. (We identify the set of subsets 2^Γ and the set of sequences $\{0,1\}^\Gamma$.)

When $\Gamma = [n]^2$ and so $\Gamma_p = \mathbb{G}(n,p)$, there are two common choices of the partition $[n]^2 = A_1 \cup \cdots \cup A_N$. The *vertex exposure* martingale (used by Shamir and Spencer (1987)) corresponds to the choice $N = n$ and $A_k = [k]^2 \setminus [k-1]^2$. The *edge exposure* martingale (used by Bollobás (1988a)) is one in which $N = \binom{n}{2}$ and $|A_k| = 1$ for each k. Note that vertex exposure requires a stronger condition on the function f than edge exposure, but it also gives a stronger result when applicable. (With $c_k = 1$, edge exposure is applicable provided the random variable X changes by at most 1 if a single edge is added or deleted, while vertex exposure is applicable provided the random variable changes by at most 1 if any number of edges incident to a single vertex are added and/or deleted.)

For further similar results (and applications), see the surveys by Bollobás (1988b) and McDiarmid (1989).

2.5 TALAGRAND'S INEQUALITY

Talagrand (1995) has given several inequalities yielding exponential estimates under various conditions. In particular, one of his results leads to estimates that are similar to those obtained by Azuma's inequality in the preceding section, but often much stronger.

Here we will treat only one of Talagrand's inequalities. Moreover, the general version (Theorem 2.37, below) is rather technical; we thus begin with a special case which is easily applied in a number of combinatorial settings. For further results and many applications, see Talagrand (1995). Other proofs of Theorems 2.37 and Theorem 2.39 (and of further related inequalities) are given by Marton (1996) and Dembo (1997).

Combinatorial setting

In the sequel we assume, as for Corollary 2.27, that $N \geq 1$ is an integer and that $Z_1, \ldots, Z_N$ are some independent random variables, taking values in some sets $\Lambda_1, \ldots, \Lambda_N$, respectively. (In many applications, $\Lambda_1 = \cdots = \Lambda_N$ and the Z_i are identically distributed, but that is not necessary.) We write $z = (z_i)_{i=1}^N$ for an element of the product space $\Lambda = \prod_1^N \Lambda_i$. (To be precise, the sets Λ_i are measurable spaces, that is, sets equipped with σ-fields of subsets, and the function f is tacitly assumed to be measurable; in the case of finite sets, this assumption is trivially true.) The two common choices of Z_k and Λ_k in random graph theory are given by vertex exposure and edge exposure, just as discussed for martingales at the end of the preceding section.

Recall that a median of a (real valued) random variable X is a number m such that $\mathbb{P}(X < m) \leq 1/2$ and $\mathbb{P}(X > m) \leq 1/2$. A median always exists, but it is not always unique.

Theorem 2.29. *Suppose that $Z_1, \ldots, Z_N$ are independent random variables taking their values in some sets $\Lambda_1, \ldots, \Lambda_N$, respectively. Suppose further that $X = f(Z_1, \ldots, Z_N)$, where $f : \Lambda_1 \times \cdots \times \Lambda_N \to \mathbb{R}$ is a function such that, for some constants c_k, $k = 1, \ldots, N$, and some function ψ, the following two conditions hold:*

(L) *If $z, z' \in \Lambda = \prod_1^N \Lambda_i$ differ only in the kth coordinate, then $|f(z) - f(z')| \leq c_k$.*

(C) *If $z \in \Lambda$ and $r \in \mathbb{R}$ with $f(z) \geq r$, then there exists a set $J \subseteq \{1, \ldots, N\}$ with $\sum_{i \in J} c_i^2 \leq \psi(r)$, such that for all $y \in \Lambda$ with $y_i = z_i$ when $i \in J$, we have $f(y) \geq r$.*

Then, for every $r \in \mathbb{R}$ and $t \geq 0$,

$$\mathbb{P}(X \leq r - t)\, \mathbb{P}(X \geq r) \leq e^{-t^2/4\psi(r)}. \qquad (2.35)$$

In particular, if m is a median of X, then for every $t \geq 0$,

$$\mathbb{P}(X \leq m - t) \leq 2e^{-t^2/4\psi(m)} \qquad (2.36)$$

and

$$\mathbb{P}(X \geq m + t) \leq 2e^{-t^2/4\psi(m+t)}. \tag{2.37}$$

Remark 2.30. The Lipschitz condition (L) is the same as the condition in Corollary 2.27.

Remark 2.31. Note that the set J in (C) generally depends on z and r. The vector $(z_i)_{i\in J}$, which forces $f \geq r$, is called a *certificate* (of $f \geq r$).

We postpone the proof of the theorem until the end of this section and first discuss some consequences and applications. Note that the function ψ formally may be chosen arbitrarily such that (C) holds; however, we want to find a small ψ since the bounds the theorem yields are better the smaller ψ is.

Remark 2.32. In most applications, $c_k = 1$ for all k. In this case, the first condition on J in (C) is $|J| \leq \psi(r)$; thus, $\psi(r) = N$ will always do, but smaller bounds on $|J|$ give better estimates.

Comparison with Azuma's inequality

Every function f trivially satisfies (C) with $\psi(r) = \sum_1^N c_i^2$ for all r; just take $J = \{1, \dots, N\}$. Thus Theorem 2.29 yields, for example, the estimate

$$\mathbb{P}(|X - m| \geq t) \leq 4e^{-t^2/4\sum c_i^2}, \qquad t \geq 0, \tag{2.38}$$

for any function f satisfying (L). This is very similar to Corollary 2.27. The conditions are the same and the conclusions differ only in that here we get worse constants and that the median is used instead of the mean. These differences are typically not important; note that Corollary 2.27 implies that if $a > \mathbb{E}\,X + \left(2\log 2 \sum c_k^2\right)^{1/2}$, then $\mathbb{P}(X \geq a) < 1/2$ and thus $m < a$, which, together with a similar lower bound, yields

$$|\,\mathbb{E}\,X - m| \leq \left(2\log 2 \sum c_k^2\right)^{1/2}.$$

(This, with another constant, follows also from (2.38), using arguments as in Example 2.33 below. – Exercise!) The constant may be improved by using Remark 2.28.)

In many applications, (C) holds with a much smaller ψ; this leads to stronger estimates that significantly surpass Azuma's inequality.

Example 2.33. In several interesting cases, X assumes non-negative integer values only, (L) holds with $c_k = 1$, and (C) holds with $\psi(r) = r$ for integers $r \geq 1$. (Equivalently, (C) holds with $\psi(r) = \lceil r \rceil$ for $r \geq 0$.) In this case, (2.35) yields

$$\mathbb{P}(X \leq a)\,\mathbb{P}(X \geq r) \leq e^{-(r-a)^2/4r} \tag{2.39}$$

for every integer $r \geq 1$ and real $a \leq r$. Since $(r-a)^2/r$ is an increasing function of $r \geq a$ and $\mathbb{P}(X \geq r) = \mathbb{P}(X \geq \lceil r \rceil)$, (2.39), in fact, holds for any real a and r with $a \leq r$. (The case $a < 0$ is trivial.)

Consequently, if m is a median of X, then

$$\mathbb{P}(X \leq m - t) \leq 2e^{-t^2/4m}, \qquad t \geq 0, \tag{2.40}$$

and

$$\mathbb{P}(X \geq m + t) \leq 2e^{-t^2/4(m+t)} \leq \begin{cases} 2e^{-t^2/8m}, & 0 \leq t \leq m, \\ 2e^{-t/8}, & t > m. \end{cases} \tag{2.41}$$

Hence,

$$\mathbb{P}(|X - m| \geq t) \leq \begin{cases} 4e^{-t^2/8m}, & 0 \leq t \leq m, \\ 2e^{-t/8}, & t > m. \end{cases} \tag{2.42}$$

In particular, it follows that

$$\begin{aligned} |\mathbb{E}\, X - m| \leq \mathbb{E}\, |X - m| &= \int_0^\infty \mathbb{P}(|X - m| > t)\, dt \\ &\leq \int_0^m 4e^{-t^2/8m}\, dt + \int_m^\infty 2e^{-t/8}\, dt \leq 2\sqrt{8\pi m} + 16 \end{aligned}$$

(the constants can be improved). Hence, using also $\frac{1}{2}m \leq m\,\mathbb{P}(X \geq m) \leq \mathbb{E}\, X$,

$$|\mathbb{E}\, X - m| = O(\sqrt{\mathbb{E}\, X}),$$

which implies estimates similar to (2.40)–(2.42) for $X - \mathbb{E}\, X$; for example, for some universal constant $\gamma > 0$,

$$\mathbb{P}(|X - \mathbb{E}\, X| \geq t) \leq 4e^{-\gamma t^2/(\mathbb{E}\, X + t)}, \qquad t \geq 0. \tag{2.43}$$

We see that if m (or, equivalently, $\mathbb{E}\, X$) is much smaller than N, then Theorem 2.29 yields much stronger estimates than Corollary 2.27.

Example 2.34. A simple instance of the situation in Example 2.33 is a binomial random variable, or, more generally, a sum of independent Bernoulli random variables. In this case, we let the sets Λ_i equal $\{0, 1\}$ and let $X = f(Z_1, \ldots, Z_N) = Z_1 + \cdots + Z_N$. It is easy to see that (L) and (C) hold with $c_i = 1$ and $\psi(r) = r$ when r is a non-negative integer, and thus (2.39)–(2.43) hold. This yields estimates similar to those given in Theorem 2.1 (although with inferior constants).

This example shows that it is not possible to improve the estimate (2.41) to $\mathbb{P}(Z \geq m+t) \leq 2e^{-t^2/4m}$, or something similar with other constants; consider for example the limiting case of a random variable with distribution Po(1) (or the family Bi$(n, 1/n)$) and large t; compare with Remark 2.7.

Example 2.35. A more interesting application is obtained by letting X be the stability number $\alpha(\mathbb{G}(n,p))$ of the random graph $\mathbb{G}(n,p)$, that is, the order of the largest independent set of vertices. It is easily seen that, using vertex exposure, the conditions in Example 2.33 are satisfied (Exercise!); a certificate of $\alpha \geq r$ (for an integer $r \geq 1$) is just any independent set of order r. Consequently (2.39)–(2.43) hold. We will return to this application in Theorem 7.4.

The same applies to the clique number of $\mathbb{G}(n,p)$, that is, the order of the largest complete subgraph (which is just the independence number of the complement of the graph).

Remark 2.36. In general, say that a function $f : \prod_1^N \Lambda_i \to \{0, 1, \dots\}$ is a *configuration function* if for each $J \subseteq [N]$ there exists a set $A_J \subseteq \prod_{j \in J} \Lambda_j$ of "configurations" such that:

(i) If $(x_j)_{j \in J} \in A_J$ and $J' \subset J$ then $(x_j)_{j \in J'} \in A_{J'}$;

(ii) $f(z) = \max\{|J| : (z_j)_{j \in J} \in A_J\}$.

In other words, the configurations are certain sequences $(z_{j_1}, \dots, z_{j_n})$, a subsequence of a configuration is a configuration and f is the size of the largest configuration included in $(z_1, \dots, z_N)$. The independence and clique numbers in Example 2.35 above are obvious examples.

Every configuration function satisfies (L) and (C) with $c_k = 1$ and $\psi(r) = r$ for integers $r \geq 0$, so the conclusions in Example 2.33 hold. (Conversely, it may be shown that every such f with values in $\{0, \dots, N\}$ is a configuration function. See also Talagrand (1995, (7.1.7)) for yet another characterization.)

Moreover, Boucheron, Lugosi and Massart (2000+) have recently shown that for configuration functions, the inequalities (2.5) and (2.6) hold, which yield somewhat sharper (and simpler) estimates than the inequalities above.

General form of Talagrand's inequality

In order to state the general form of Talagrand's inequality, we need more notation.

Assume, as above, that $\Lambda_1, \dots, \Lambda_N$ are sets. Assume further that $\mu_1, \dots, \mu_N$ are probability measures on $\Lambda_1, \dots, \Lambda_N$, respectively, and let $\mathbb{P}$ be the product measure $\mu_1 \times \cdots \times \mu_N$ on $\Lambda = \Lambda_1 \times \cdots \times \Lambda_N$.

We define a kind of distance between a point $x \in \Lambda$ and a subset $A \subseteq \Lambda$ in the following way. We first define two subsets $U_A(x)$ and $V_A(x)$ of $\mathbb{R}^N$:

$$U_A(x) = \left\{(s_i)_1^N \in \{0,1\}^N : \exists y \in A \text{ such that } x_i = y_i \text{ for all } i \text{ with } s_i = 0\right\}$$

and $V_A(x)$ is the convex hull of $U_A(x)$. Thus $U_A(x)$ contains the vectors $(\mathbf{1}[x_i \neq y_i])_1^N$, $y \in A$, but also vectors with more 1's. We then define

$$d(A, x) = \inf\{\|v\|_2 : v \in V_A(x)\},$$

where $\|v\|_2 = (\sum v_i^2)^{1/2}$ is the usual Euclidean norm in $\mathbb{R}^N$; thus $d(A, x)$ is the Euclidean distance from 0 to $V_A(x)$. (If $A = \emptyset$, then $U_A(x) = V_A(x) = \emptyset$ and we set $d(A, x) = \infty$. On the other hand, if A is non-empty, then $d(A, x)$ is finite for every x because at least $(1, \dots, 1) \in U_A(x)$. Moreover, the infimum in the definition of $d(A, x)$ is attained, since $U_A(x)$ is a finite set and thus $V_A(x)$ is compact.)

Note that

$$d(A, x) = 0 \iff 0 \in V_A(x) \iff 0 \in U_A(x) \iff x \in A.$$

With this notation, we may give the general form of Talagrand's inequality.

Theorem 2.37. *For every (measurable) subset A of Λ,*

$$\int_\Lambda e^{\frac{1}{4}d^2(A,x)}\, d\,\mathbb{P}(x) \leq \frac{1}{\mathbb{P}(A)}. \tag{2.44}$$

Remark 2.38. We assume that the set A is measurable, but even so the function $d(A, x)$ is, in general, not measurable, so the integral in (2.44) is not always defined as an ordinary Lebesgue integral. Of course, there is no problem for finite sets, and it is easy to give further sufficient conditions for $d(A, x)$ to be measurable, but a simpler and more general approach is to allow $d(A, x)$ to be non-measurable and interpret the integral in (2.44) as an *iterated upper integral* $\int^* \cdots \int^*$. The theorem is then valid without any further assumptions (by the proof given below and simple properties of the upper integral). Moreover, there is no problem in using this version in applications such as the proof of Theorem 2.39, below. (Recall that the upper integral $\int^* g$ of a non-negative function g is defined to be the infimum of $\int h$ over all measurable functions h with $h \geq g$; this infimum is always attained.)

Before proving Theorem 2.37, we use it to prove Theorem 2.29. We begin with a simple corollary of Theorem 2.37 (Talagrand 1995).

Theorem 2.39. *Suppose that A and B are two (measurable) subsets of Λ such that for some $t \geq 0$ the following separation condition holds:*

(S) *For every $z \in B$, there exists a non-zero vector $\alpha = (\alpha_i)_1^N \in \mathbb{R}^N$ such that for every $y \in A$,*

$$\sum_{i: y_i \neq z_i} \alpha_i \geq t \Big(\sum_1^N \alpha_i^2\Big)^{1/2}.$$

Then

$$\mathbb{P}(A)\,\mathbb{P}(B) \le e^{-t^2/4}.$$

Remark 2.40. Note that Condition (S) is not symmetric in A and B; the vector α may depend on z but not on y.

Proof. Suppose that $z \in B$ and let α be as in Condition (S). We may assume that $\alpha_i \ge 0$ for every i; otherwise we replace α_i by $|\alpha_i|$. Then, denoting the scalar product in $\mathbb{R}^N$ by $\langle \cdot, \cdot \rangle$,

$$\langle \alpha, s \rangle \ge t\|\alpha\|_2$$

for every $s \in U_A(z)$. Since $\alpha \mapsto \langle \alpha, s \rangle$ is a linear functional, this extends to all s in the convex hull $V_A(z)$, and thus, by the Cauchy–Schwarz inequality, for every $s \in V_A(z)$,

$$t\|\alpha\|_2 \le \langle \alpha, s \rangle \le \|s\|_2 \|\alpha\|_2$$

and hence $t \le \|s\|_2$. Consequently $d(A, z) \ge t$, for every $z \in B$, which, together with Theorem 2.37, yields

$$e^{t^2/4}\,\mathbb{P}(B) \le \int_\Lambda e^{\frac{1}{4}d^2(A,z)}\,d\,\mathbb{P}(z) \le \frac{1}{\mathbb{P}(A)}. \qquad \blacksquare$$

Next, we use Theorem 2.39 to prove Theorem 2.29.

Proof of Theorem 2.29. Let $A = \{z \in \Lambda : f(z) \le r - t\}$ and $B = \{z \in \Lambda : f(z) \ge r\}$. For $z \in B$, let J be as in (C) and define

$$\alpha_i = \begin{cases} c_i, & i \in J, \\ 0, & i \notin J; \end{cases}$$

thus, by (C), $\|\alpha\|_2 \le \sqrt{\psi(r)}$. If, furthermore, $y \in A$, define $y' \in \Lambda$ by

$$y'_i = \begin{cases} z_i, & i \in J, \\ y_i, & i \notin J. \end{cases}$$

Then $f(y') \ge r$ by (C), and thus $f(y') - f(y) \ge t$, while (L) implies

$$|f(y) - f(y')| \le \sum_{i \in J : y_i \ne z_i} c_i = \sum_{i : y_i \ne z_i} \alpha_i.$$

Consequently,

$$\sum_{i : y_i \ne z_i} \alpha_i \ge t \ge t\psi(r)^{-1/2}\|\alpha\|_2.$$

If $t > 0$, this also shows that $\alpha \neq 0$, and (2.35) follows by applying Theorem 2.39 with t replaced by $t/\sqrt{\psi(r)}$. The case $t = 0$ is trivial.

Finally (2.36) and (2.37) follow by taking $r = m$ and $r = m+t$ in (2.35). ■

Remark 2.41. The conclusion of Theorem 2.39 can be improved to

$$\sqrt{\log(1/\mathbb{P}(A))} + \sqrt{\log(1/\mathbb{P}(B))} \geq t/\sqrt{2}$$

(Talagrand 1995, Corollary 4.2.5), which, by the argument above, implies, for example, that (2.36) can be improved to the smaller, but more complicated, bound

$$\mathbb{P}(X \leq m - t) \leq \exp\Bigl(-\Bigl(\frac{t}{\sqrt{2\psi(m)}} - \sqrt{\log 2}\Bigr)^2\Bigr), \qquad t \geq \sqrt{2\log 2\,\psi(m)}.$$

It remains to prove Theorem 2.37. We follow Talagrand (1995), and begin with a simple lemma.

Lemma 2.42. *Suppose that* $0 \leq r \leq 1$. *Then*

$$\inf_{0 \leq \tau \leq 1} e^{\tau^2/4} r^{\tau - 1} \leq 2 - r.$$

Proof. Taking $\tau = \min\bigl(2\log(1/r), 1\bigr)$, it suffices to show that if $e^{-1/2} \leq r \leq 1$, then

$$e^{\log^2 r} r^{-2\log r - 1} \leq 2 - r.$$

Substituting $r = e^{-t}$ and taking logarithms, we have to show that

$$h(t) = \log(2 - e^{-t}) + t^2 - t \geq 0$$

for $0 \leq t \leq 1/2$. But this, in fact, holds for all $t \geq 0$ because elementary calculations yield $h(0) = h'(0) = 0$ and $h''(t) = 2 - 2e^{-t}/(2 - e^{-t})^2 \geq 0$ for $t \geq 0$. ■

Proof of Theorem 2.37. We use induction in N, starting with the simple case $N = 1$. (The bold reader may start with the really trivial case $N = 0$ instead.)

If $N = 1$, and A is any non-empty subset of $\Lambda = \Lambda_1$, then, as is easily seen from the definition, $d(A, z) = 0$ when $z \in A$ and $d(A, z) = 1$ when $z \notin A$. Consequently, using $e^{1/4} \leq 2$ and $t(2 - t) \leq 1$ for real t,

$$\int_\Lambda e^{\frac14 d^2(A,x)}\, d\,\mathbb{P}(x) = \mathbb{P}(A) + e^{1/4}\bigl(1 - \mathbb{P}(A)\bigr) \leq 2 - \mathbb{P}(A) \leq \frac{1}{\mathbb{P}(A)}.$$

Now assume that the result holds for some $N \geq 1$. Let us write $\Lambda^{(k)} = \prod_1^k \Lambda_i$ and $\mathbb{P}_k = \prod_1^k \mu_i$, and denote elements in $\Lambda^{(N+1)}$ by (x, λ), with $x \in \Lambda^{(N)}$ and $\lambda \in \Lambda_{N+1}$.

Let A be a (measurable) subset of $\Lambda^{(N+1)}$, and define, for every $\lambda \in \Lambda_{N+1}$, $A(\lambda) = \{x \in \Lambda^{(N)} : (x, \lambda) \in A\}$ (a section of A). Define also $B = \cup_\lambda A(\lambda)$ (the projection of A on $\Lambda^{(N)}$). Each $A(\lambda)$ is measurable, but, in general, B is not; thus we also select a measurable subset $B_0 \subseteq B$ of maximal $\mathbb{P}_N$ measure. Note that thus $\mathbb{P}_N(B_0) \geq \mathbb{P}_N(A(\lambda))$ for every λ. We may assume that $\mathbb{P}_N(B_0) > 0$, since otherwise $\mathbb{P}_{N+1}(A) = 0$ and the result is trivial.

The basic observation is that, for any $x \in \Lambda^{(N)}$ and $\lambda \in \Lambda_{N+1}$,

$$\begin{aligned} s \in U_{A(\lambda)}(x) &\implies (s, 0) \in U_A\big((x, \lambda)\big), \\ t \in U_B(x) &\implies (t, 1) \in U_A\big((x, \lambda)\big). \end{aligned}$$

It follows that if $s \in V_{A(\lambda)}(x)$, $t \in V_{B_0}(x) \subseteq V_B(x)$ and $0 \leq \tau \leq 1$, then $(s, 0) \in V_A\big((x, \lambda)\big)$ and $(t, 1) \in V_A\big((x, \lambda)\big)$, and thus also $\big((1-\tau)s + \tau t, \tau\big) \in V_A\big((x, \lambda)\big)$, which yields, using the convexity of the function $u \mapsto u^2$,

$$\begin{aligned} d^2\big(A, (x, \lambda)\big) &\leq \|((1-\tau)s + \tau t, \tau)\|_2^2 = \sum_1^N \big((1-\tau)s_i + \tau t_i\big)^2 + \tau^2 \\ &\leq (1-\tau)\|s\|_2^2 + \tau\|t\|_2^2 + \tau^2. \end{aligned}$$

Taking the infimum over s and t, we thus obtain, for every $\lambda \in \Lambda_{N+1}$ and $\tau \in [0, 1]$,

$$d^2\big(A, (x, \lambda)\big) \leq (1-\tau) d^2(A(\lambda), x) + \tau d^2(B_0, x) + \tau^2.$$

Hölder's inequality and the induction hypothesis now yield

$$\begin{aligned} &\int_{\Lambda^{(N)}} e^{\frac{1}{4}d^2(A,(x,\lambda))}\, d\,\mathbb{P}_N(x) \\ &\leq e^{\frac{1}{4}\tau^2} \Big(\int_{\Lambda^{(N)}} e^{\frac{1}{4}d^2(A(\lambda),x)}\, d\,\mathbb{P}_N(x)\Big)^{1-\tau} \Big(\int_{\Lambda^{(N)}} e^{\frac{1}{4}d^2(B_0,x)}\, d\,\mathbb{P}_N(x)\Big)^{\tau} \\ &\leq e^{\frac{1}{4}\tau^2} \Big(\frac{1}{\mathbb{P}_N(A(\lambda))}\Big)^{1-\tau} \Big(\frac{1}{\mathbb{P}_N(B_0)}\Big)^{\tau} \\ &= \frac{1}{\mathbb{P}_N(B_0)} e^{\frac{1}{4}\tau^2} \Big(\frac{\mathbb{P}_N(A(\lambda))}{\mathbb{P}_N(B_0)}\Big)^{\tau-1}. \end{aligned}$$

By Lemma 2.42 with $r = \mathbb{P}_N(A(\lambda)) / \mathbb{P}_N(B_0) \leq 1$, we obtain by taking the infimum over τ

$$\int_{\Lambda^{(N)}} e^{\frac{1}{4}d^2(A,(x,\lambda))}\, d\,\mathbb{P}_N(x) \leq \frac{1}{\mathbb{P}_N(B_0)}\Big(2 - \frac{\mathbb{P}_N(A(\lambda))}{\mathbb{P}_N(B_0)}\Big).$$

Finally, we integrate this over λ, using Fubini's theorem and the inequality $2 - t \leq 1/t$ for $t \geq 0$, to obtain

$$\begin{aligned} \int_{\Lambda^{(N+1)}} e^{\frac{1}{4}d^2(A,z)}\, d\,\mathbb{P}_{N+1}(z) &\leq \frac{1}{\mathbb{P}_N(B_0)}\Big(2 - \frac{\mathbb{P}_{N+1}(A)}{\mathbb{P}_N(B_0)}\Big) \\ &\leq \frac{1}{\mathbb{P}_{N+1}(A)}. \end{aligned}$$

■

2.6 THE UPPER TAIL

As remarked above (Remark 2.17), the upper tail counterpart of Theorem 2.14 is not true, in general. As an exponential bound is often needed also for the upper tail, we present here briefly a few simple ideas on how to cope with this problem in certain situations. For a more thorough account see Janson and Ruciński (2000+).

Recall that in a random set Γ_p, each element of Γ is included with the same probability p. Furthermore, as in Section 2.2, let $\mathcal{S}$ be a family of subsets of Γ. For the sake of clarity, we confine ourselves here to the slightly simplified case in which all members of $\mathcal{S}$ are of the same size s.

One possible idea is to convert an upper tail probability into a lower tail and then to apply Theorem 2.14. This can be done by setting $Z = |\Gamma_p|$, $\bar{\mathcal{S}} = [\Gamma]^s \setminus \mathcal{S}$, and $\bar{X} = \sum_{A \in \bar{\mathcal{S}}} I_A = \binom{Z}{s} - X$. As this approach is limited in applications only to large families $\mathcal{S}$, we will not pursue it any further.

The first result we do present was stimulated by the following problem (Rödl and Ruciński 1994) on a random graph obtained by a random deletion of vertices (*cf.* Section 1.1).

Example 2.43. Let $G = (V, E)$ be a graph with $|V| = n$ and $|E| \le \eta\binom{n}{2}$, $0 < \eta \le 1$, and let $R = V_p$ be a binomial random subset of the vertex set V, $0 < p < 1$. Using Proposition 2.44 below one can show that with probability $1 - e^{-\Omega(np)}$ we have $|[R]^2 \cap E| \le 2\eta\binom{|R|}{2}$ (Exercise!).

The underlying idea is to break the family $\mathcal{S}$ into disjoint subfamilies of disjoint sets, and apply Theorem 2.1 to one subfamily. Set $L = L(\mathcal{S})$ for the standard dependency graph of the family of indicators $\{I_A : A \in \mathcal{S}\}$, where an edge joins A and B if and only if $A \cap B \ne \emptyset$ (see Example 1.5). Note that the maximum degree $\Delta(L)$ can be as large as $|\mathcal{S}| - 1$. The following result has appeared in a slightly more complicated form in Rödl and Ruciński (1994).

Proposition 2.44. *With the notation of Section 2.2, for every $t \ge 0$,*

$$\mathbb{P}(X \ge \mathbb{E}\, X + t) \le (\Delta(L) + 1) \exp\left(-\frac{t^2}{4(\Delta(L) + 1)(\lambda + t/3)}\right).$$

Proof. A matching in $\mathcal{S}$ is a subfamily $\mathcal{M} \subseteq \mathcal{S}$ consisting of pairwise disjoint sets. We claim that the family $\mathcal{S}$ can be partitioned into $\Delta(L) + 1$ disjoint matchings, each of size equal to either $\lceil |\mathcal{S}|/(\Delta(L) + 1) \rceil$ or $\lfloor |\mathcal{S}|/(\Delta(L) + 1) \rfloor$. Indeed, by the well known Hajnal–Szemerédi Theorem (Hajnal and Szemerédi 1970, Bollobás 1978), the vertex set of the graph L can be partitioned into $\Delta(L) + 1$ independent sets of the above order, which correspond to matchings $\mathcal{M}_i$, $i = 1, \ldots, \Delta(L) + 1$, in $\mathcal{S}$. Note that for each i, $|\mathcal{M}_i| > |\mathcal{S}|/2(\Delta(L) + 1)$.

If $X \ge \lambda + t$, then, by simple averaging, there exists a matching $\mathcal{M}_i$ such that $|[\Gamma_p]^s \cap \mathcal{M}_i| \ge p^s|\mathcal{M}_i| + t|\mathcal{M}_i|/|\mathcal{S}|$. Since $|[\Gamma_p]^s \cap \mathcal{M}_i|$ is a random variable with the binomial distribution $\mathrm{Bi}(|\mathcal{M}_i|, p^s)$, we conclude by Theorem 2.1 that

$$\begin{aligned}\mathbb{P}(X \geq \lambda + t) &\leq \sum_{i=1}^{D+1} \exp\left(-\frac{t^2|\mathcal{M}_i|}{2|\mathcal{S}|(\lambda + t/3)}\right) \\ &\leq (\Delta(L) + 1) \exp\left(-\frac{t^2}{4(\Delta(L) + 1)(\lambda + t/3)}\right). \qquad \blacksquare\end{aligned}$$

As our next example shows, Proposition 2.44 can be applied to quite sparse families $\mathcal{S}$.

Example 2.45. Let $\mathcal{S}$ be the family of all edge sets of triangles in the complete graph $[n]^2$. Then $\Delta(L(\mathcal{S})) = 3(n-3)$, and by Proposition 2.44, for fixed $\rho > 0$ and with $X = X_{K_3}$ denoting the number of triangles in $\mathbb{G}(n, p)$,

$$\mathbb{P}\big(X \geq (1 + \rho)\,\mathbb{E}\,X\big) = O(n) \exp\left(-\Theta(n^2p^3)\right),$$

which is a fair bound provided $p \gg n^{-2/3} \log^{1/3} n$.

The next idea also uses disjoint sets to force independence, and is based on Spencer (1990) (see also Alon and Spencer (1992) and Janson (1990b)). Let $\mathcal{S}_p$ be the subfamily of $\mathcal{S}$ consisting of those sets which are entirely contained in Γ_p. Consider also the intersection graph $L_p = L(\mathcal{S}_p)$ of $\mathcal{S}_p$, in which each vertex represents one set and the edges join pairs of vertices representing pairs of intersecting sets. (Clearly, L_p is an induced subgraph of L defined above.) Let X_0 count the largest number of disjoint sets of $\mathcal{S}$ which are present in Γ_p. So, X is the number of vertices and X_0 is the independence number of L_p. Furthermore, set $T = \Delta(L_p)$ and X_1 for the size of the largest induced matching in L_p. Then it follows by an elementary graph theory argument that $X \leq X_0 + 2TX_1$ (Exercise!). Hence, if ad hoc estimates can be found for both T and X_1 then it remains only to bound the upper tail of X_0.

Lemma 2.46. *If $t \geq 0$, then, with $\varphi(x) = (1+x)\log(1+x) - x$ and $\lambda = \mathbb{E}\,X$,*

$$\mathbb{P}(X_0 \geq \mathbb{E}\,X + t) \leq \exp\left(-\lambda\varphi\Big(\frac{t}{\lambda}\Big)\right) \leq \exp\left(-\frac{t^2}{2(\lambda + t/3)}\right).$$

Proof. Let k be an integer ($0 \leq k \leq \lceil \lambda + t \rceil$) and consider the number Z of k-element sequences of disjoint sets of $\mathcal{S}_p$. Clearly, we have

$$\mathbb{E}\,Z \leq |\mathcal{S}|^k p^{sk} = \lambda^k.$$

If $X_0 \geq \lambda + t$, then $Z \geq (\lambda + t)_k = \prod_{i=0}^{k-1} (\lambda + t - i)$, and thus, by Markov's inequality (1.3),

$$\mathbb{P}(X_0 \geq \lambda + t) \leq \mathbb{P}\big(Z \geq (\lambda + t)_k\big) \leq \frac{\lambda^k}{(\lambda + t)_k} = \prod_{i=0}^{k-1} \frac{\lambda}{\lambda + t - i}.$$

If we increase k by 1, the right-hand side is multiplied by $\lambda/(\lambda+t-k)$, which is less than 1 for $k < t$. Hence, the right-hand side is minimized by choosing $k = \lceil t \rceil$. Consequently,

$$\log \mathbb{P}(X_0 \geq \lambda + t) \leq \sum_{i=0}^{\lceil t \rceil - 1} \log(\lambda/(\lambda + t - i)) \leq \int_0^t \log(\lambda/(\lambda + t - x))\, dx,$$

which implies the first inequality of Lemma 2.46 by a straight-forward integration. The second inequality follows by the lower bound on $\varphi(x)$ established at the end of the proof of Theorem 2.1. ∎

Remark 2.47. By choosing $k = \lambda + t$ (assumed to be an integer) in the above proof, we obtain the weaker estimate (Erdős and Tetali 1990, Alon and Spencer 1992), valid for any integer $k \geq 0$,

$$\log \mathbb{P}(X_0 \geq k) \leq \frac{\lambda^k}{k!} \leq \left(\frac{e\lambda}{k}\right)^k = \exp\left(\lambda - \lambda\varphi\left(\frac{t}{\lambda}\right)\right).$$

Remark 2.48. Note that X_0 satisfies the Lipschitz condition (L) of Corollary 2.27 with all $c_i = 1$, as well as the condition (C) of Theorem 2.29 with $\psi(r) = s\lceil r \rceil$. Therefore, similar but weaker versions of Lemma 2.46 can be derived from these results (*cf.* Theorem 3.29 in Section 3.5).

Example 2.49. Let $X = X_{K_3}$ be the number of triangles in $\mathbb{G}(n,p)$, $np \to \infty$. Then X_1 is the maximum number of edge disjoint copies of the diamond graph K_4^- (see Figure 4.7), and Lemma 2.46 can be applied to it. The random variable T can easily be bounded by using Theorem 2.1. For example, if $\rho > 0$ is fixed and $np^2 \to 0$, one can prove (Exercise!) that

$$\mathbb{P}\big(X \geq (1+\rho)\,\mathbb{E}\,X\big) \leq 2e^{-\Theta(n^4p^5)} + n^2 e^{-\Theta(1/np^2)}.$$

This bound is meaningful if $n^4p^5 \to \infty$, and better than that of Example 2.45 if $p = o(n^{-3/5})$ (Exercise!).

Remark 2.50. Using a variant of the martingale approach, very recently Vu (2000+) has proved that for a class of graphs G on k vertices, the inequalities

$$\begin{aligned}\exp\left\{-\Theta\left((\mathbb{E}\,X_G)^{2/k}\log n\right)\right\} &\leq \mathbb{P}(X_G \geq (1+\rho)\,\mathbb{E}\,X_G)\\ &\leq \exp\left\{-\Theta\left((\mathbb{E}\,X_G)^{1/(k-1)}\right)\right\}\end{aligned}$$

hold in a wide range of $p = p(n)$. Here X_G is the number of copies of G in $\mathbb{G}(n,p)$.

The lower bound provides another counterexample to the existence of an upper tail analogue of Theorem 2.14 (*cf.* Remark 2.17). The upper bound competes well against the results of this section. In particular, when $G = K_3$,

Vu's bound is better than that in Example 2.49, but for $p \gg n^{-1/3}$ it is not as good as the bound presented in Example 2.45.

Our final idea for establishing a bound on the upper tail of X incorporates some kind of cheating. We allow ourselves to delete some elements of Γ_p and claim the concentration of X in the remainder. Surprisingly, such results turned out to be useful and even crucial in the context of partition properties of random graphs. This approach was developed in Rödl and Ruciński (1995) and is based on two elementary lemmas. We conclude this section with these technical lemmas. The first of them can easily be proved by a method similar to that used in the proof of Lemma 2.46 (Exercise!).

Lemma 2.51. *Let $\mathcal{S} \subseteq [\Gamma]^s$ and $0 < p < 1$. Then, for every pair of integers k and t, with probability at least $1 - \exp\left(-\frac{kt}{s(\lambda+t)}\right)$, there exists a set $E_0 \subset \Gamma_p$ of size k such that $\Gamma_p \setminus E_0$ contains at most $\lambda + t$ sets from $\mathcal{S}$.* ∎

Hence, substantially exceeding the expectation is exponentially unlikely, provided we are allowed to destroy some of the subsets in the count, by deleting a certain number of elements from the random set. Then, of course, there is the danger of losing other properties held by the random set. It turns out, however, that monotone properties held with exponential probabilities survive the deletion. The next lemma, also from Rödl and Ruciński (1995), makes this precise.

For an increasing family $\mathcal{Q}$ of subsets of a set Γ and for a nonnegative integer k, let

$$\mathcal{Q}_k = \{A \subseteq \Gamma : \forall B \subseteq A, \text{ if } |B| \le k, \text{ then } A \setminus B \in \mathcal{Q}\}.$$

In other words, given a property $\mathcal{Q}$, the property $\mathcal{Q}_k$ assures that $\mathcal{Q}$ holds even after deleting up to k arbitrary elements from the set. For instance, if $\mathcal{Q}$ is the graph property of being connected, then $\mathcal{Q}_k$ is the property of being $(k+1)$-edge-connected.

Lemma 2.52. *Let c and $\delta < 1$ satisfy*

$$\delta(3 - \log \delta) \le c. \tag{2.45}$$

Then, for every increasing family $\mathcal{Q} = \mathcal{Q}(N)$ of subsets of an N-element set Γ, and for every k, $0 \le k \le \delta Np/2$, if $\mathbb{P}(\Gamma_{(1-\delta)p} \notin \mathcal{Q}) < e^{-cNp}$ then $\mathbb{P}(\Gamma_p \notin \mathcal{Q}_k) < 3\sqrt{Np}\,e^{-(c/2)Np} + e^{-(\delta^2/8)Np}$. ∎

Thus, if a random binomial subset has an increasing property with probability extremely close to 1, then a slightly enlarged random subset will enjoy the same property, and with similar probability, even after a small fraction of its elements are arbitrarily destroyed. The elementary proof is left to the reader (Exercise!).

In Chapter 8, in the outline of the proof of Theorem 8.23, we indicate how Lemmas 2.51 and 2.52 were utilized to establish thresholds for Ramsey properties of $\mathbb{G}(n,p)$. In the same proof, Lemma 2.52 alone is also applied.

3

Small Subgraphs

3.1 THE CONTAINMENT PROBLEM

In 1960 Erdős and Rényi published the most fundamental of their random graphs papers (Erdős and Rényi 1960). The first problem studied there was that of the existence in $\mathbb{G}(n, M)$ of at least one copy of a given graph G. Since the graph G is fixed and the random graph $\mathbb{G}(n, M)$ grows with $n \to \infty$, copies of G in $\mathbb{G}(n, M)$ are called small subgraphs, regardless whether G is a triangle or a graph with one billion vertices, as opposed to subgraphs of $\mathbb{G}(n, M)$ which grow with n like, say, a Hamilton cycle.

Erdős and Rényi (1960) found the threshold for the property of containing G only in the special case in which G is a balanced graph (see Section 3.2 for the definition). Twenty-one years later Bollobás (1981b) settled the problem in full generality. Still later a simpler proof was found by Ruciński and Vince (1985) and we will present it here. It is a classical example of an application of the commonly used methods of the first and the second moment. This problem is also instructive in that it shows that the behavior of the expectation alone can be sometimes misleading.

To better comprehend this feature and to have a gentle start, we will consider first the somewhat simpler problem of finding the threshold for the containment of at least one arithmetic progression of length k in a random subset of integers $[n]_p$, where, recall, $[n] = \{1, 2, \ldots, n\}$ and $p = p(n)$ is the probability of including each element of $[n]$, independently of the others, in the random subset $[n]_p$.

The first and second moment methods

Before presenting the applications, let us describe the first and the second moment methods. As a special instance of the Markov inequality (1.3), for every non-negative, integer valued random variable X, the inequality

$$\mathbb{P}(X > 0) \le \mathbb{E}\, X \tag{3.1}$$

holds. *The first moment method* relies on showing that $\mathbb{E}\, X_n = o(1)$, and thus concluding by (3.1) that $X_n = 0$ *a.a.s.* *The second moment method* is based on Chebyshev's inequality (1.2), which implies (Exercise!) that for every random variable X with $\mathbb{E}\, X > 0$

$$\mathbb{P}(X = 0) \le \frac{\operatorname{Var} X}{(\mathbb{E}\, X)^2}. \tag{3.2}$$

Hence, by showing that the right-hand side of (3.2) (with X replaced by X_n) converges to 0, one concludes that $X_n > 0$ *a.a.s.* By the same token one obtains a stronger statement, which also follows from Chebyshev's inequality: If $\operatorname{Var} X_n/(\mathbb{E}\, X_n)^2 = o(1)$ then $X_n = \mathbb{E}\, X_n + o_p(\mathbb{E}\, X_n)$ or, equivalently, $X_n/\mathbb{E}\, X_n \xrightarrow{p} 1$. In particular then, $X_n = \Theta_C(\mathbb{E}\, X_n)$.

Remark 3.1. Inequality (3.2) may be improved. The Cauchy–Schwarz inequality applied to $X = X\mathbf{1}[X \ne 0]$ yields $(\mathbb{E}\, X)^2 \le \mathbb{E}\, X^2\, \mathbb{P}(X \ne 0)$, that is,

$$\mathbb{P}(X \ne 0) \ge \frac{(\mathbb{E}\, X)^2}{\mathbb{E}\, X^2}, \tag{3.3}$$

and thus

$$\mathbb{P}(X = 0) \le 1 - \frac{(\mathbb{E}\, X)^2}{\mathbb{E}\, X^2} = \frac{\operatorname{Var} X}{\mathbb{E}\, X^2} = \frac{\operatorname{Var} X}{(\mathbb{E}\, X)^2 + \operatorname{Var} X}. \tag{3.4}$$

For the purpose of showing $X_n > 0$ *a.a.s.*, (3.2) is just as good as the improvement (3.4), but in Chapter 7 we will see a situation where the improvement is essential.

Example 3.2 (arithmetic progressions). Let X_k be the number of arithmetic progressions of length k in $[n]_p$, where $k \ge 2$ is a fixed integer. (We suppress the subscript n here.) To compute $\mathbb{E}(X_k)$ we need to know the number $f(n,k)$ of all arithmetic progressions of length k in $[n]$. In fact, we only care about the order of magnitude of $f(n,k)$ which equals n^2, since every arithmetic progression is uniquely determined by its first two elements. Let us number the arithmetic progressions of length k in $[n]$ by $1, \dots, f(n,k)$ and, for each $i = 1, \dots, f(n,k)$, define a zero-one random variable (indicator) I_i equal to 1 if the i-th arithmetic progression of length k is entirely present in

$[n]_p$ and equal to 0 otherwise. With this notation, $X_k = \sum_{i=1}^{f(n,k)} I_i$ and, by the linearity of expectation,

$$\mathbb{E}(X_k) = f(n,k)p^k = \Theta(n^2p^k).$$

Hence, if $p \ll n^{-2/k}$ then $\mathbb{E}(X_k) \to 0$ as $n \to \infty$ and, by the first moment method, (i.e. by (3.1)), $\mathbb{P}(X_k > 0) = o(1)$.

If, on the other hand, $p \gg n^{-2/k}$, then $\mathbb{E}(X_k) \to \infty$, but this fact alone is not sufficient to claim that $\mathbb{P}(X_k > 0) \to 1$. One has to work for it, using the second moment method. Observe that I_i and I_j are independent if the i-th and j-th arithmetic progressions have no element in common; in that case the covariance $\mathrm{Cov}(I_i, I_j)$ equals zero. In the remaining cases, we use the inequality $\mathrm{Cov}(I_i, I_j) \leq \mathbb{E}(I_iI_j)$. There are $O(n^3)$ pairs (I_i, I_j) which share one element and then $\mathbb{E}(I_iI_j) = p^{2k-1}$, and only $O(n^2)$ pairs which share two or more elements, in which case $\mathbb{E}(I_iI_j) \leq p^k$. We thus can estimate the variance of X_k as follows:

$$\mathrm{Var}(X_k) = \sum_{i=1}^{f(n,k)} \sum_{j=1}^{f(n,k)} \mathrm{Cov}(I_i, I_j) = O(n^3p^{2k-1} + n^2p^k).$$

Consequently, by the second moment method, (i.e., by (3.2)), if $p \gg n^{-2/k}$, then

$$\mathbb{P}(X_k = 0) = O\left(\frac{1}{np} + \frac{1}{n^2p^k}\right) = o(1).$$

Together, these results show that the threshold for existence of a k-term arithmetic progression in $[n]_p$ is $n^{-2/k}$.

Thresholds for subgraph containment

Returning to small subgraphs of random graphs, we let X_G stand for the number of copies of a given graph G that can be found in the binomial random graph $\mathbb{G}(n,p)$. Let $v = v_G$ and $e = e_G$ stand for the number of vertices and edges of G, respectively. There are exactly $f(n,G) = \binom{n}{v}v!/\mathrm{aut}(G)$ copies of G in the complete graph K_n, where, recall, $\mathrm{aut}(G)$ denotes the number of automorphisms of G. For each copy G' of G in K_n define the indicator random variable $I_{G'} = \mathbf{1}[\mathbb{G}(n,p) \supseteq G']$. Then

$$\mathbb{E}(X_G) = f(n,G)p^e = \Theta(n^vp^e) \to \begin{cases} 0 & \text{if } p \ll n^{-v/e} \\ \infty & \text{if } p \gg n^{-v/e} \end{cases}$$

and, by the first moment method,

$$\mathbb{P}(X_G > 0) = o(1) \qquad \text{if } p \ll n^{-v/e}. \tag{3.5}$$

Is it then true that $\mathbb{P}(X_G > 0) = 1 - o(1)$ if $p \gg n^{-v/e}$? Consider first an example.

Example 3.3. Let H_0 be the graph with 4 vertices and 5 edges and let G_0 be a graph obtained by adding one vertex to H_0 and connecting it to just one vertex of H_0. (There are two nonisomorphic ways to do so, and it does not matter which one we choose – see Figure 3.1 for one version of G_0.) Take any sequence $p = p(n)$ such that $n^{-5/6} \ll p \ll n^{-4/5}$, say, $p = n^{-9/11}$. Then $\mathbb{E} X_{G_0} = \Theta(n^5 p^6) \to \infty$, but by (3.5) applied to H_0, *a.a.s.* there is no copy of H_0 in $\mathbb{G}(n,p)$, and therefore there is no copy of G_0 either.

Hence, things are more complicated for graphs than for arithmetic progressions. It should be clear at this point that the behavior of the expectation is deceptive in case of G_0, because G_0 contains a subgraph (*viz.* H_0) denser than itself, and that the right threshold should be $n^{-4/5}$. Indeed, this was confirmed by Bollobás (1981b) in the following, general result.

Recall that $m(G)$ is the ratio of the number of edges to the number of vertices in the densest subgraph of G, that is,

$$m(G) = \max\left\{\frac{e_H}{v_H} : H \subseteq G,\, v_H > 0\right\}. \tag{3.6}$$

Theorem 3.4. *For an arbitrary graph G with at least one edge,*

$$\lim_{n\to\infty} \mathbb{P}(\mathbb{G}(n,p) \supset G) = \begin{cases} 0 & \text{if } p \ll n^{-1/m(G)}, \\ 1 & \text{if } p \gg n^{-1/m(G)}. \end{cases}$$

Proof. There are two statements to be proved, the 0-statement, and the 1-statement. To prove the former one, assume that $p \ll n^{-1/m(G)}$ and let H' be a subgraph of G for which $e(H')/v(H') = m(G)$. Then, by (3.5), *a.a.s.* there is no copy of H', and thus, no copy of G in $\mathbb{G}(n,p)$.

To prove the 1-statement we use the second moment method; we then need to bound the variance of X_G from above. For future reference, we state the result as a lemma. We define

$$\Phi_G = \Phi_G(n,p) = \min\{\mathbb{E}(X_H) : H \subseteq G,\, e_H > 0\} \tag{3.7}$$

and note that

$$\Phi_G \asymp \min_{H\subseteq G,\, e_H>0} n^{v_H} p^{e_H}; \tag{3.8}$$

this quantity will be useful on several occasions in the sequel.

Lemma 3.5. *Let G be a graph with at least one edge. Then*

$$\begin{aligned} \mathrm{Var}(X_G) &\asymp (1-p) \sum_{H\subseteq G,\, e_H>0} n^{2v_G - v_H} p^{2e_G - e_H} \\ &\asymp (1-p) \max_{H\subseteq G,\, e_H>0} \frac{(\mathbb{E} X_G)^2}{\mathbb{E} X_H} = (1-p)\frac{(\mathbb{E} X_G)^2}{\Phi_G}, \end{aligned} \tag{3.9}$$

where the implicit constants depend on G but not on n or p. In particular, $\operatorname{Var} X_G = O((\mathbb{E} X_G)^2/\Phi_G)$, *and if* $p = p(n)$ *is bounded away from* 1, *then* $\operatorname{Var} X_G \asymp (\mathbb{E} X_G)^2/\Phi_G$.

Proof. Using the fact that $I_{G'}$ and $I_{G''}$ are independent if $E(G') \cap E(G'') = \emptyset$, and that for each $H \subseteq G$ there are $\Theta(n^{v_H} n^{2(v_G - v_H)}) = \Theta(n^{2v_G - v_H})$ pairs (G', G'') of copies of G in the complete graph K_n with $G' \cap G''$ isomorphic to H, we have

$$\begin{aligned}\operatorname{Var}(X_G) &= \sum_{G', G''} \operatorname{Cov}(I_{G'}, I_{G''}) = \sum_{E(G') \cap E(G'') \neq \emptyset} [\mathbb{E}(I_{G'} I_{G''}) - \mathbb{E}(I_{G'}) \mathbb{E}(I_{G''})] \\ &\asymp \sum_{H \subseteq G,\, e_H > 0} n^{2v_G - v_H} (p^{2e_G - e_H} - p^{2e_G}) \\ &\asymp \sum_{H \subseteq G,\, e_H > 0} n^{2v_G - v_H} p^{2e_G - e_H} (1 - p). \qquad (3.10)\end{aligned}$$

■

The simple observation below is often useful.

Lemma 3.6. *The following are equivalent, for any graph G with $e_G > 0$.*

(i) $np^{m(G)} \to \infty$.

(ii) $n^{v_H} p^{e_H} \to \infty$ *for every* $H \subseteq G$ *with* $v_H > 0$.

(iii) $\mathbb{E}(X_H) \to \infty$ *for every* $H \subseteq G$ *with* $v_H > 0$.

(iv) $\Phi_G \to \infty$.

Proof. By (3.6) (and $p \leq 1$), (i) holds if and only if $np^{e_H/v_H} \to \infty$ for every $H \subseteq G$ with $v_H > 0$; since

$$\mathbb{E} X_H \asymp n^{v_H} p^{e_H} = (np^{e_H/v_H})^{v_H},$$

this is equivalent to both (ii) and (iii). Finally, by the definition (3.7), Condition (iv) is equivalent to $\mathbb{E}(X_H) \to \infty$ for every $H \subseteq G$ with $e_H > 0$; this is equivalent to (iii) since the case $v_H > 0$ and $e_H = 0$ is trivial. ■

To complete the proof of Theorem 3.4, we observe that if $p \gg n^{-1/m(G)}$, then by Lemma 3.6 $\Phi_G \to \infty$. Consequently, (3.2) and Lemma 3.5 yield

$$\mathbb{P}(\mathbb{G}(n,p) \not\supset G) = \mathbb{P}(X_G = 0) \leq \frac{\operatorname{Var}(X_G)}{(\mathbb{E} X_G)^2} = O(1/\Phi_G) = o(1). \qquad ■$$

Remark 3.7. It follows from the above proof that if $\Phi_G(n,p) \to \infty$, then not only $\mathbb{P}(\mathbb{G}(n,p) \not\supset G) \to 1$ but further $X_G / \mathbb{E}(X_G) \xrightarrow{p} 1$.

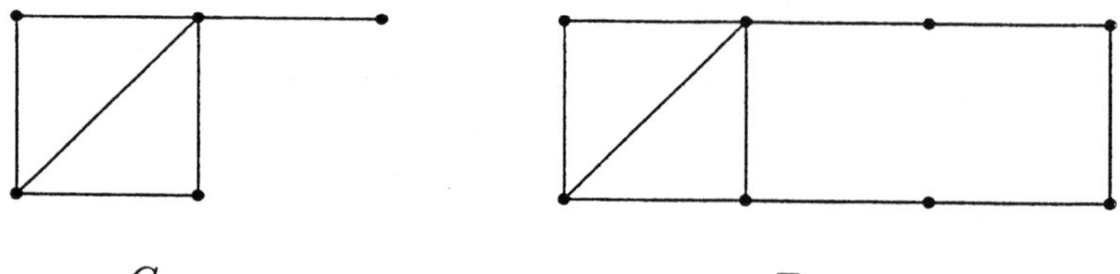

Fig. 3.1 A non-balanced graph and a balanced extension of it.

Remark 3.8. As we mentioned before, Erdős and Rényi proved Theorem 3.4 already in 1960, but only in the special case of balanced graphs, that is, graphs G with $m(G) = e_G/v_G$, while the general case was proved much later by Bollobás (1981b). Another approach to the general case was suggested by Karoński and Ruciński (1983b), who proposed a "deterministic" argument to derive Theorem 3.4 from the Erdős and Rényi result for balanced graphs. It was based on a conjecture proved a few years later (Győri, Rothschild and Ruciński 1985, Payan 1986) that every graph G is a subgraph of a graph F, called a *balanced extension* of G, which is balanced and, what is crucial here, is not denser than the densest subgraph of G, that is, $m(F) = e_F/v_F = m(G)$. The validity of the 1-statement of Theorem 3.4 for F implies its validity for G, since, trivially, $\mathbb{P}(X_G > 0) \geq \mathbb{P}(X_F > 0)$.

An example of a balanced extension F_0 of the graph G_0 is presented in Figure 3.1. (The smallest balanced extension of a graph G can be much larger than G; see Ruciński and Vince (1993).)

An exponential rate of decay

Analyzing the proof of Theorem 3.4, one comes to the conclusion that a decisive role is played by the quantity Φ_G. Indeed, the two parts of the above proof can be combined into one pair of inequalities

$$1 - \Phi_G \leq \mathbb{P}(\mathbb{G}(n,p) \not\supset G) \leq O(1/\Phi_G), \tag{3.11}$$

which together imply Theorem 3.4, since $\Phi_G \to \infty$ if and only if $np^{m(G)} \to \infty$ by Lemma 3.6, and by a similar argument $\Phi_G \to 0$ if and only if $np^{m(G)} \to 0$. What is especially nice about the inequalities (3.11) is that both sides can be strengthened to an exponential rate of decay.

Theorem 3.9. *Let G be a graph with at least one edge. Then, for every sequence $p = p(n) < 1$,*

$$\exp\left\{-\frac{1}{1-p}\Phi_G\right\} \leq \mathbb{P}(\mathbb{G}(n,p) \not\supset G) \leq \exp\{-\Theta(\Phi_G)\}.$$

Proof. The left-hand inequality follows immediately from Corollary 2.13, a consequence of the FKG inequality, with X replaced by $X_{H'}$, where $\mathbb{E}(X_{H'}) =$

Φ_G. The other inequality is implied by Theorem 2.18(ii) with $S = X_G$ and the I_A's replaced by I_G's. Indeed, the denominator of the exponent there becomes

$$\sum_{H \subseteq G, e_H > 0} \sum_{G'} \sum_{G'' \cap G' = H} p^{2e_G - e_H} = \Theta\big((\mathbb{E} X_G)^2 / \Phi_G\big) \tag{3.12}$$

and the right-hand inequality in Theorem 3.9 follows. ∎

Note that Theorem 3.9 implies Theorem 3.4.

Remark 3.10 (Martingale approach). There are at least two other ways to deduce the right-hand inequality of Theorem 3.9; they are based, respectively, on the martingale and Talagrand inequalities given in Chapter 2. Here we present the martingale approach and the other one will be given in Section 3.5.

We confine ourselves to the special case in which for every proper subgraph H of G with $e_H > 0$, $\mathbb{E}(X_H) \gg \Phi_G$. In particular, $\Phi_G = \mathbb{E}(X_G)$. (The general case is quite involved and we refer the reader to Janson, Łuczak and Ruciński (1990).) Let $f = c\Phi_G$ for a suitably chosen constant $c > 0$, and let $\pi = (A_1, \ldots, A_f)$ be an arbitrary partition of the set $[n]^2$ into sets of size $|A_i| < n^2/f$, $i = 1, \ldots, f$. Two copies of G are called π-disjoint if for each index i at most one of them has an edge in A_i. Let $D_{\pi,G}$ be the maximum number of π-disjoint copies of G in $\mathbb{G}(n,p)$. Then, by Corollary 2.27,

$$\mathbb{P}(X_G = 0) = \mathbb{P}(D_{\pi,G} = 0) \le \mathbb{P}(|D_{\pi,G} - t| \ge t) \le 2\exp\{-t^2/2f\},$$

where $t = \mathbb{E}(D_{\pi,G}) \le \Phi_G$. Now we need to bound t from below. Let $Y_{\pi,G}$ be the number of non-π-disjoint pairs of copies of G. Clearly, $D_{\pi,G} \ge X_G - Y_{\pi,G}$, and so $t \ge \Phi_G - \mathbb{E}(Y_{\pi,G})$. When computing $\mathbb{E}(Y_{\pi,G})$ asymptotically, we may ignore all pairs of G sharing at least one edge, as their expectation is $o(\Phi_G)$. The expected number of the other non-π-disjoint pairs is

$$O\left(f\binom{n^2/f}{2} n^{2(v_G-2)} p^{2e_G}\right) = O(\Phi_G^2/f).$$

Hence, for sufficiently large c, $t = \Omega(\Phi_G)$ and the right-hand inequality of Theorem 3.9 follows.

The uniform model $\mathbb{G}(n, M)$

In this section we consider the containment problem for the uniform random graph $\mathbb{G}(n,M)$. We note first that by Corollary 1.16 (or Remark 1.18), Theorem 3.4 immediately implies the corresponding result for $\mathbb{G}(n,M)$: the threshold is $n^{2-1/m(G)}$. (This can also be shown directly by the first and second moment methods.)

For the exponential bounds in Theorem 3.9, the situation is somewhat more complicated. Not only do neither part of the proof of Theorem 3.9 carry over to $\mathbb{G}(n, M)$, but the result cannot be true in general. Indeed, for dense graphs Turán's theorem (see, e.g., Bollobás (1998)) shows, for example, that if $G = K_3$ and $M > \frac{1}{4}n^2 \sim \frac{1}{2}\binom{n}{2}$, then $\mathbb{G}(n, M)$ *always* contains a copy of G, so $\mathbb{P}(\mathbb{G}(n, M) \not\supset G) = 0$. More generally, by the Erdős–Stone–Simonovits Theorem (Erdős and Stone 1946, Erdős and Simonovits 1966, Diestel 1996, Bollobás 1998), the same holds for any graph G and for $M \geq c\binom{n}{2}$, provided $c > 1 - 1/(\chi(G) - 1)$ is fixed and n is large enough. However, if M is not too large, both inequalities in Theorem 3.9 have counterparts for $\mathbb{G}(n, M)$.

For a sequence $M = M(n) \leq \binom{n}{2}$, define Φ_G by (3.7) with $p = M/\binom{n}{2}$, and note that (if G is non-empty) $\Phi_G \leq \mathbb{E}(X_{K_2}) = \binom{n}{2}p = M$.

Theorem 3.11. *Let G be a graph with at least one edge.*

(i) *If $M \geq e_G$, then*

$$\mathbb{P}(\mathbb{G}(n, M) \not\supset G) \leq \exp\{-\Theta(\Phi_G)\}.$$

(ii) *If, in addition, either $\Phi_G \leq cM$, where c is some small positive constant depending on G, or G is not bipartite and $M \leq c\binom{n}{2}$, where $c < 1 - 1/(\chi(G) - 1)$ is fixed, then*

$$\mathbb{P}(\mathbb{G}(n, M) \not\supset G) = \exp\{-\Theta(\Phi_G)\}.$$

Proof. We will give several arguments which are valid for different ranges of M and together yield the results. We let $c_1, c_2, \ldots$ denote some positive constants depending on G only. Note that we may assume that n is large, since the results are trivial for any finite number of small n.

(i) For $\Phi_G \gg \log n$, the estimate in (i) follows immediately from the upper bound in Theorem 3.9 and Pittel's inequality (1.6).

Alternatively and more generally, we find by monotonicity and the law of total probability, as in (1.5),

$$\mathbb{P}(\mathbb{G}(n, p/2) \supset G) \leq \mathbb{P}(\mathbb{G}(n, M) \supset G) + \mathbb{P}(e(\mathbb{G}(n, p/2)) > M)$$

and thus, using the Chernoff bound (2.7) (or (2.5)), Theorem 3.9 and the fact that $\Phi_G(n, p/2) \geq 2^{-e(G)}\Phi_G(n, p)$, we get

$$\begin{aligned} \mathbb{P}(\mathbb{G}(n, M) \not\supset G) &\leq \mathbb{P}(\mathbb{G}(n, p/2) \not\supset G) + e^{-M/6} \\ &\leq e^{-c_1\Phi_G(n,p/2)} + e^{-M/6} \\ &\leq e^{-c_2\Phi_G} + e^{-M/6}. \end{aligned} \tag{3.13}$$

Since $M \geq \Phi_G$ as remarked above, (3.13) yields the upper bound $2e^{-c_3\Phi_G}$, which yields the sought bound $e^{-\Theta(\Phi_G)}$, provided $\Phi_G \geq c_4 = 1/c_3$, say.

For $\Phi_G \leq c_4$, (3.13) implies further

$$\mathbb{P}(\mathbb{G}(n, M) \not\supset G) \leq 1 - c_5\Phi_G + e^{-M/6},$$

which implies the result provided $M \geq \log^2 n$ and thus $e^{-M/6} \leq \frac{1}{2}c_5\Phi_G$ for large n; note that $\Phi_G > n^{-2e_G}$ for $M \geq 1$.

Finally, in the rather uninteresting case in which $e_G \leq M < \log^2 n$, we assume for simplicity that every component of G has at least three vertices. (The general case follows easily by treating isolated edges and vertices separately. – Exercise!) It is then easy to see that if we let $\widetilde{X}_G$ be the number of copies of G in $\mathbb{G}(n, M)$, then $\mathbb{E}(\widetilde{X}_G) \asymp n^{v_G}p^{e_G} = \Phi_G$ and $\mathbb{E}(\widetilde{X}_G^2) \sim \mathbb{E}(\widetilde{X}_G)$, and thus, by (3.3),

$$\mathbb{P}(\mathbb{G}(n,M) \not\supset G) = \mathbb{P}(\widetilde{X}_G = 0) \leq 1 - c_6\Phi_G \leq \exp\{-c_6\Phi_G\}.$$

(ii) To obtain a lower bound, we argue similarly. By monotonicity,

$$\mathbb{P}(\mathbb{G}(n,2p) \not\supset G) \leq \mathbb{P}(\mathbb{G}(n,M) \not\supset G) + \mathbb{P}(e(\mathbb{G}(n,2p)) < M)$$

and thus, using Theorem 3.9 and the Chernoff bound (2.6),

$$\mathbb{P}(\mathbb{G}(n,M) \not\supset G) \geq e^{-c_7\Phi_G} - e^{-M/4}.$$

If $1/2 \leq \Phi_G \leq c_8 M$, say, this yields the desired lower bound $e^{-c_9\Phi_G}$.

If $\Phi_G \leq 1/2$, let H_0 be a subgraph of G with $\mathbb{E}\,X_{H_0} = \Phi_G$ (in $\mathbb{G}(n,p)$), and observe that then $\mathbb{E}\,\widetilde{X}_{H_0} \leq \Phi_G$ (in $\mathbb{G}(n,M)$). Hence,

$$\mathbb{P}(\mathbb{G}(n,M) \supset G) \leq \mathbb{P}(\mathbb{G}(n,M) \supset H_0) \leq \mathbb{E}\,\widetilde{X}_{H_0} \leq \Phi_G \leq 1 - e^{c_{10}\Phi_G}.$$

For the remaining case in which $\Phi_G > c_8 M$, and thus $\Phi_G = \Theta(M)$, the above approach may be useless. Note that in this case, the lower bound $e^{-\Theta(n^2p)}$ in Theorem 3.9 can be obtained by just considering the event that $\mathbb{G}(n,p)$ is empty, which, of course, does not happen in $\mathbb{G}(n,M)$, $M \geq 1$. Fortunately, a simple and entirely different argument still yields the desired lower bound for graphs which are not bipartite. Let $k = \chi(G) \geq 3$, where $\chi(G)$ is the chromatic number of G. Clearly, if $\mathbb{G}(n,M)$ is $(k-1)$-partite, then there is no room for a copy of G. Let us fix a partition of the vertex set $[n]$ into $k-1$ sets of size $\lfloor n/(k-1) \rfloor$ or $\lceil n/(k-1) \rceil$. It is easy to show (Exercise!) that, as long as $M \leq c\binom{n}{2}$, the probability of no edge of $\mathbb{G}(n,M)$ falling within any of the sets is at least $(1 - 1/(k-1) - c)^M$. ■

For non-bipartite graphs G, we thus have an almost complete description: $\mathbb{P}(\mathbb{G}(n,M) \not\supset G) = \exp(-\Theta(\Phi_G))$ almost all the way up to the point where the probability becomes zero by the Erdős–Stone–Simonovits Theorem.

For bipartite graphs, the result is less satisfactory. Clearly, the final argument in the proof above does not work. The condition $\Phi_G \leq cM$ in the theorem is equivalent (Exercise!) to $M \leq c'n^{2-1/m^{(2)}(G)}$, where $m^{(2)}(G)$ is defined in (3.18) in the next section, and for larger M we have no precise description of $\mathbb{P}(G(n,M) \not\supset G)$. Indeed, for a general bipartite G, it is not even known when this probability vanishes.

In fact, it is conjectured that if G is bipartite and $M \gg n^{2-1/m^{(2)}(G)}$, then the probability that $\mathbb{G}(n, M)$ contains no copy of G tends to 0 with n faster than in the binomial case, and we have

$$-\log \mathbb{P}(\mathbb{G}(n, M) \not\supset G) = o(M). \tag{3.14}$$

This conjecture has been verified for cycles C_4 by Füredi (1994) (see also Kleitman and Winston (1982)) and for all even cycles C_{2k}, $k \geq 2$, by Haxell, Kohayakawa and Łuczak (1995). For further results in this direction see Kohayakawa, Kreuter and Steger (1998) and Łuczak (2000), where it is shown, using a slightly generalized version of the Szemerédi Regularity Lemma (*cf.* Section 8.3), that (3.14) holds for all bipartite graphs G for which Conjecture 8.35 of Chapter 8 holds.

3.2 LEADING OVERLAPS AND THE SUBGRAPH PLOT

Leading overlaps

When $p = p(n) \to 0$, the logarithm of the lower bound in Theorem 3.9 becomes asymptotically equal to $-\Phi_G$. When can the same be concluded about the upper bound? To answer this question we introduce a related concept.

A subgraph H' of G with $e_{H'} > 0$ is called a *leading overlap* of G (for a given sequence $p(n)$) if $\liminf \mathbb{E}(X_{H'})/\Phi_G < \infty$. In other words, H' is a leading overlap if and only if $\mathbb{E}(X_{H'}) \gg \Phi_G$ does *not* hold. If we, for simplicity, assume that the sequence $p(n)$ is sufficiently regular, so that $\lim \mathbb{E}(X_H)/\Phi_G$ exists in $[1, \infty]$ for every $H \subseteq G$, this is equivalent to $\mathbb{E}(X_{H'}) = \Theta(\Phi_G)$; in other words, a leading overlap is a subgraph of G, which, up to a constant, achieves the minimum in $\Phi_G = \min_H \mathbb{E}\, X_H$. (For general $p(n)$, this holds at least along a suitable subsequence.)

Leading overlaps owe their name to the fact that they correspond to the leading terms of the asymptotic expression (3.9) for the variance of X_G as well as in (3.12).

Returning to Theorem 3.9, a detailed analysis of expression (3.12) reveals that the coefficient hidden in the Θ term in the upper bound in Theorem 3.9 becomes $1-o(1)$ if there is just one leading overlap H' of G, and the uniqueness holds in the strong form, that is, there is just one copy of H' in G. (The converse holds if we assume that $p(n)$ is regular as above.)

Thus, we arrive at the following corollary.

Corollary 3.12. *If $p = p(n) \to 0$ in such a way that H is a unique leading overlap of G then*

$$\log \mathbb{P}(\mathbb{G}(n,p) \not\supset G) \sim -\mathbb{E}(H) . \tag{3.15}$$

■

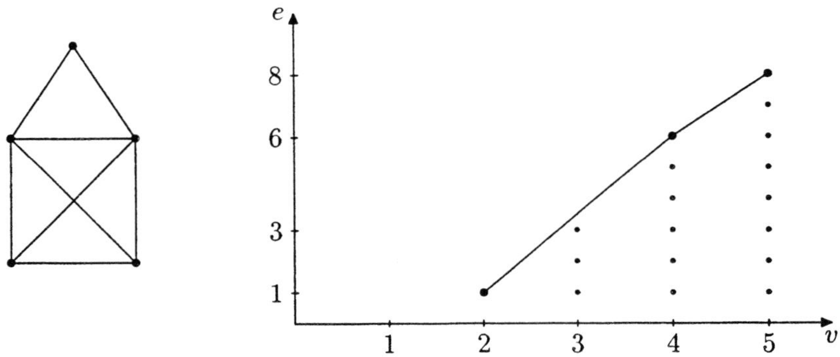

Fig. 3.2 A graph and its subgraph plot.

Subgraph plot

The leading overlaps vary with $p = p(n)$, but there is an easy geometrical way to detect them all at once, by plotting in the xy-plane points that represent subgraphs of G and focusing on the upper boundary of the convex hull of the obtained set of points.

Formally, the *subgraph plot* of a graph G is defined as the set of points in the xy-plane

$$\Sigma(G) = \{(v_H, e_H) : H \subseteq G,\ v_H \geq 2\}.$$

Remark 3.13. Note that we do not include $(0,0)$ or $(1,0)$; for other purposes it might be convenient to define versions of the subgraph plot containing one or both these points. Similarly, it may be natural to consider only induced subgraphs H in the definition; for our purposes this makes no difference.

We call the upper boundary $\widehat{\Sigma}(G)$ of the convex hull of $\Sigma(G)$ the *roof*, and we say that a subgraph H *lies on the roof* if the point (v_H, e_H) does. Observe that $\widehat{\Sigma}(G)$ is a piecewise linear curve, with endpoints in $(2,1)$ and (v_G, e_G). An example of a graph and its subgraph plot is presented in Figure 3.2.

Elementary calculations yield a full description of the entire spectrum of leading overlaps of G. If, for simplicity, G lacks isolated vertices, then a subgraph H of G is a leading overlap of G for some range of $p = p(n)$ if and only if it lies on the roof $\widehat{\Sigma}(G)$ of the subgraph plot $\Sigma(G)$. Moreover, the range of $p = p(n)$ in which H is a leading overlap is determined by the slopes a_H^- and a_H^+ of the straight line segments to the left and to the right from (v_H, e_H). (Note that $a_H^- \geq a_H^+$; we set $a_{K_2}^- = \infty$ and $a_G^+ = 0$ for convenience.) Indeed, H is a leading overlap as long as $np^{a_H^-} = O(1)$ and, at the same time, $np^{a_H^+} = \Omega(1)$; this condition is necessary too if $p(n)$ is regular as above. (If G has isolated vertices, and G_0 is the subgraph obtained by removing them, then the possible leading overlaps are the subgraphs on the roof $\widehat{\Sigma}(G_0)$, which equals $\widehat{\Sigma}(G)$ without its final, flat part where $a_H^- = 0$.)

As one can see in Figure 3.2, the points $(s, \max_{v_H=s} e_H)$, $s = 3, \ldots, v_G - 1$, do not necessarily lie on the roof. In fact, there are graphs with only two roof subgraphs, K_2 and G. They are easily characterized, assuming $v_G \geq 3$, by the condition that for all $H \subset G$ with $2 < v_H < v_G$ the inequality

$$\frac{e_H - 1}{v_H - 2} < \frac{e_G - 1}{v_G - 2} \tag{3.16}$$

holds. On the other extreme, there are graphs G with as many as about $\frac{2}{5}v_G$ of their subgraphs being leading overlaps for various (mutually distinct) ranges of p. For this and other related results, see Ruciński (1991) and Łuczak and Ruciński (1992).

Measures of graph density

The subgraph plot can also be used to visualize several other useful concepts.

First, the *density* $d(G) = e_G/v_G$ of G (with $v_G > 0$) equals the slope of the line L_d from $(0,0)$ to the top point (v_G, e_G).

The *maximum density* $m(G) = \max\{d(H) : H \subseteq G, v_H > 0\}$ equals the slope of the least steep line L_m from $(0,0)$, such that the entire subgraph plot lies below or on L_m; in other words, L_m is the tangent from $(0,0)$ to the roof.

A graph G is called *balanced* if $m(G) = d(G)$, that is, if $d(H) \leq d(G)$ for every $H \subset G$. (In words: G does not contain a subgraph denser than itself.) This is equivalent to $L_d = L_m$, and thus G is balanced if and only if the subgraph plot lies below or on L_d. In Example 3.3, H_0 is balanced and G_0 is not.

A graph G is called *strictly balanced* if $d(H) < d(G)$ whenever $H \subsetneq G$, which is to say that every proper subgraph of G is strictly less dense than the graph itself; equivalently, the subgraph plot lies strictly below L_d, except for the top point. Trees, regular connected graphs as well as the graph H_0 from Example 3.3 are all strictly balanced. An example of a balanced graph that is not strictly balanced is the union of a cycle and a path (of length ≥ 1) which are disjoint except that one endpoint of the path lies on the cycle. Another example is given by the disjoint union of two copies of any balanced graph, or by any balanced extension of a non-balanced graph (*cf.* the graph F_0 in Figure 3.1).

We will further use some related notions, which are natural, for example, when considering graphs with a distinguished vertex or edge. For a graph G with $v_G \geq 2$, define $d^{(1)}(G) = e_G/(v_G - 1)$; let $d^{(1)}(K_1) = 0$. Then define

$$m^{(1)}(G) = \max\{d^{(1)}(H) : H \subseteq G\}. \tag{3.17}$$

When $v_G \geq 2$, $d^{(1)}(G)$ and $m^{(1)}(G)$ are the slopes of the line $L_d^{(1)}$ from $(1,0)$ to the top point, and of $L_m^{(1)}$, the tangent from $(1,0)$ to the roof.

Similarly, for a graph G with $v_G \geq 3$, define $d^{(2)}(G) = (e_G - 1)/(v_G - 2)$; let $d^{(2)}(K_1) = d^{(2)}(2K_1) = 0$ and $d^{(2)}(K_2) = 1/2$. Then define

$$m^{(2)}(G) = \max\{d^{(2)}(H) : H \subseteq G\}. \tag{3.18}$$

The definition of $d^{(2)}(K_2)$ may look artificial but turns out to be convenient (*cf.* Remark 3.14 below). Note that if $e_G \geq 2$, then $m^{(2)}(G) = \max\{d^{(2)}(H) : H \subseteq G,\ v_H \geq 3\}$, so the special case does not matter.

When $e_G \geq 2$, $d^{(2)}(G)$ and $m^{(2)}(G)$ are the slopes of the line $L_d^{(2)}$ from $(2,1)$ to the top point, and of $L_m^{(2)}$, the tangent from $(2,1)$ to the roof.

In analogy with the above, a graph G is called *K_1-balanced* if $m^{(1)}(G) = d^{(1)}(G)$, or equivalently, $d^{(1)}(H) \leq d^{(1)}(G)$ for all $H \subseteq G$; furthermore, graphs with $d^{(1)}(H) < d^{(1)}(G)$ for all $H \subsetneq G$ are called *strictly K_1-balanced.* Analogously, we define *K_2-balanced* and *strictly K_2-balanced* graphs. These notions have applications in the study of solitary subgraphs (see Section 3.6), G-factors (see Chapter 4), and Ramsey properties of random graphs (see Section 7.6 and Chapter 8).

Remark 3.14. Below we collect some simple but useful facts about the parameters $m(G)$, $m^{(1)}(G)$ and $m^{(2)}(G)$. The proofs are left to the reader (Exercise!).

For convenience, let $m^{(0)}(G) = m(G)$. Let $G_1, \dots, G_k$ be the connected components of G. Then $m^{(i)}(G) = \max_j m^{(i)}(G_j)$ for $i = 0, 1, 2$. This implies that strictly balanced, strictly K_1-balanced, and strictly K_2-balanced graphs are all connected.

We have $m^{(i)}(G) = 0$ if and only if G is empty, that is, $e_G = 0$, for every i. Moreover, $m(G) < 1$ if and only if G is a forest (and then $m(G) = 1 - 1/s$, where s is the order of the largest component), and $m(G) = 1$ if and only if the densest component of G is unicyclic. For all other graphs, $m(G) > 1$.

As far as $m^{(1)}(G)$ is concerned, $m^{(1)}(G) = 1$ if G is a non-empty forest, and $m^{(1)}(G) > 1$ if G is not a forest.

Finally, $m^{(2)}(G) = 1/2$ when the maximum degree $\Delta(G) = 1$ (i.e., when G consists of isolated edges and possibly some isolated vertices), $m^{(2)}(G) = 1$ if G is a forest with $\Delta(G) \geq 2$, and $m^{(2)}(G) > 1$ if G is not a forest.

In Chapter 6, we will use the following observation.

Lemma 3.15. *If $np^{m(G)} \to \infty$, then every leading overlap is connected.*

Proof. Suppose that $H \subseteq G$ with $e_H > 0$. If H is the disjoint union of two proper subgraphs H_1 and H_2, where, say, $e(H_1) > 0$, then Lemma 3.6 yields $n^{v(H_2)}p^{e(H_2)} \to \infty$ and thus

$$n^{v_H}p^{e_H} = n^{v(H_1)}p^{e(H_1)}n^{v(H_2)}p^{e(H_2)} \gg n^{v(H_1)}p^{e(H_1)} \asymp \mathbb{E}(X_{H_1}) \geq \Phi_G.$$

Consequently, H is not a leading overlap. ∎

Remark 3.16. For $e_G \geq 2$, the slope $a^+_{K_2}$ defined above equals $m^{(2)}(G)$. Hence, under this condition, K_2 is a leading overlap when $np^{m^{(2)}(G)} = \Omega(1)$, and the only leading overlap when $np^{m^{(2)}(G)} \to \infty$.

Remark 3.17. Assume $v_G \geq 3$. Then the condition (3.16) characterizing graphs G with only two roof subgraphs (*viz.* K_2 and G) may be expressed as $d^{(2)}(H) < d^{(2)}(G)$ for all $H \subsetneq G$ with $v_H \geq 3$; this is equivalent (Exercise!) to G being strictly K_2-balanced, except for the two cases $G = 2K_2$ and G being a union of an edge and an isolated vertex. Consequently, if G lacks isolated vertices, then G has only two possible leading overlaps if and only if G is strictly K_2-balanced or $G = 2K_2$.

Remark 3.18. The *arboricity* of a graph is defined as the least number of forests that together cover the edge set of the graph. This seemingly unrelated notion is, in fact, closely connected to the quantities just defined; by a theorem of Nash-Williams (1964) (see e.g. Diestel (1996)), the arboricity of G equals $\lceil m^{(1)}(G) \rceil$.

3.3 SUBGRAPH COUNT AT THE THRESHOLD

When $p = \Theta(n^{-1/m(G)})$, we have $\Phi_G = \Theta(1)$ and, by Theorem 3.9,

$$0 < \liminf_{n\to\infty} \mathbb{P}(\mathbb{G}(n,p) \supset G) \leq \limsup_{n\to\infty} \mathbb{P}(\mathbb{G}(n,p) \supset G) < 1.$$

This ensures that the threshold in Theorem 3.4 cannot be sharpened. For this range of $p = p(n)$ the derivation of $\lim_{n\to\infty} \mathbb{P}(\mathbb{G}(n,p) \supset G)$ may not be easy. However, for the class of strictly balanced graphs defined in the preceding section, not only the precise value of $\lim_{n\to\infty} \mathbb{P}(\mathbb{G}(n,p) \supset G)$, but the entire limiting distribution of X_G can be computed.

The following result was proved independently in Bollobás (1981b) and Karoński and Ruciński (1983a), and generalizes earlier results about trees (Erdős and Rényi 1960) and complete graphs (Schürger 1979).

Theorem 3.19. *If G is a strictly balanced graph and $np^{m(G)} \to c > 0$, then $X_G \xrightarrow{d} \mathrm{Po}(\lambda)$, the Poisson distribution with expectation $\lambda = c^{v_G} / \mathrm{aut}(G)$.*

Proof. This proof exemplifies the technique called *the method of moments*, which is presented in detail in Chapter 6; we use here the version given in Corollary 6.8.

Consider the factorial moments of X_G, defined as $\mathbb{E}(X_G)_k = \mathbb{E}[X_G(X_G - 1)\cdots(X_G - k + 1)]$. We have, for $k = 1, 2, \ldots,$

$$\mathbb{E}(X_G)_k = \sum_{G_1,\ldots,G_k} \mathbb{P}(I_{G_1}\cdots I_{G_k} = 1) = E'_k + E''_k,$$

where the summation extends over all ordered k-tuples of distinct copies of G in K_n, and E'_k is the partial sum where the copies in a k-tuple are mutually vertex disjoint. It is easy to verify (Exercise!) that

$$E'_k \sim (\mathbb{E}\, X_G)^k \sim \left(c^{v_G} / \operatorname{aut}(G)\right)^k.$$

This implies that X_G is asymptotically Poisson if $E''_k = o(1)$, and it remains to be proved that $E''_k = o(1)$. Let e_t be the minimum number of edges in a t-vertex union of k not mutually vertex disjoint copies of G.

Claim. *For every $k \geq 2$ and $k \leq t < kv_G$, we have $e_t > tm(G)$.*

Proof of Claim. For a graph F define $f_F = m(G)v_F - e_F$. Note that $f_G = 0$ and, since G is strictly balanced, $f_H > 0$ for every proper subgraph H of G. We are to prove that for every graph F which is a union of k not mutually vertex disjoint copies of G, $f_F < 0$. We will do it by induction on k, relying heavily on the modularity of f, that is, on the equality

$$f_{F_1 \cup F_2} = f_{F_1} + f_{F_2} - f_{F_1 \cap F_2} \tag{3.19}$$

valid for any two graphs F_1 and F_2. Let $F = \bigcup_{i=1}^k G_i$, where each G_i is a copy of G, and the copies are numbered so that $G_1 \cap G_2 \neq \emptyset$. For $k = 2$, (3.19) yields $f_{G_1 \cup G_2} = -f_{G_1 \cap G_2} < 0$, because $G_1 \cap G_2$ is a proper subgraph of G. For arbitrary $k \geq 3$ we let $F' = \bigcup_{i=1}^{k-1} G_i$ and $H = F' \cap G_k$. Then H may be any subgraph of G including G itself and the null graph, but in any case $f_H \geq 0$. Moreover, $f_{F'} < 0$ by the induction assumption. Thus

$$f_F = f_{F'} + f_{G_k} - f_H < 0. \qquad \blacksquare$$

Having proven the claim, we easily complete the proof of Theorem 3.19 with one line. Indeed

$$E''_k = \sum_{t=k}^{kv-1} O(n^t p^{e_t}) = o(1). \qquad \blacksquare$$

Remark 3.20. Using Theorem 6.10, Theorem 3.19 can be extended to joint convergence of several subgraph counts, with the limit variables independent (Exercise!).

Still assuming that $p = \Theta(n^{-1/m(G)})$, consider graphs other than strictly balanced graphs. If G is nonbalanced, then the expectation of X_G tends to infinity. It turns out that there is a nonrandom sequence $a_n(G) \to \infty$, such that the asymptotic distribution of $X_G/a_n(G)$ coincides with that of X_H, where H is the largest subgraph of G for which $d(H) = m(G)$. Clearly, H is balanced and we are back to the balanced case. The sequence $a_n(G)$ is equal to the expected number of extensions of a given copy of H to a copy of G in the random graph $\mathbb{G}(n,p)$. For details, see Ruciński (1990).

If a graph G is balanced but not strictly balanced, then the limiting distribution of X_G is no longer Poisson. Although, in principle, as shown by Bollobás and Wierman (1989), the limiting distribution can be computed, there is no compact formula. We give only three simple examples, illustrating typical phenomena.

Example 3.21. We consider three balanced but not strictly balanced graphs. All three have $m(G) = 1$, and thus we assume $p = c/n$ for some $c > 0$.

First, if $G = 2C_3$, a union of two disjoint triangles, then *a.a.s.* $X_G = \frac{1}{2}X_{C_3}(X_{C_3}-1)$ (Exercise!). Since $X_{C_3} \xrightarrow{d} Z_3 \in \mathrm{Po}(c^3/6)$ by Theorem 3.19, and continuous functions preserve convergence of distribution (Billingsley 1968, Section 5), we obtain $X_G \xrightarrow{d} \frac{1}{2}Z_3(Z_3 - 1)$. In particular, for the probability of no copy of G, $\mathbb{P}(X_G = 0) \to (1 + c^3/6)\exp(-c^3/6)$.

Second, if G is a disjoint union of a C_3 and a C_4, then *a.a.s.* $X_G = X_{C_3}X_{C_4}$. By Remark 3.20, $(X_{C_3}, X_{C_4}) \xrightarrow{d} (Z_3, Z_4)$, with $Z_3 \in \mathrm{Po}(c^3/6)$ and $Z_4 \in \mathrm{Po}(c^4/8)$ independent. Consequently, $X_G \xrightarrow{d} Z_3Z_4$. In particular, $\mathbb{P}(X_G = 0) \to 1 - \big(1 - \exp(-c^3/6)\big)\big(1 - \exp(-c^4/8)\big)$.

Third, if G is the *whisk graph* K_3^+, that is, a triangle with a pendant edge (see Figure 3.3), then $X_G \xrightarrow{d} \sum_{i=1}^{Z_3} W_i$, where $W_i \in \mathrm{Po}(3c)$ are independent of $Z_3 \in \mathrm{Po}(c^3/6)$ and of each other. In particular, $\mathbb{P}(X_G = 0) \to \exp\big(-(1 - e^{-3c})c^3/6\big)$. The idea behind this is that, asymptotically, there is a $\mathrm{Po}(c^3/6)$ distributed number of triangles, and each triangle has a $\mathrm{Po}(3c)$ distributed number of pendant edges, each creating one copy of K_3^+. For details, see Bollobás and Wierman (1989) or Janson (1987).

Finally, let us mention that for $p \gg n^{-1/m(G)}$, X_G has an asymptotic normal distribution (Theorem 6.5).

3.4 THE COVERING PROBLEM

The next topic covered in this chapter deals with covering every vertex of a random graph by a copy of a given graph G. The graph property that every vertex belongs to a copy of G will be denoted throughout this chapter and Chapter 4 by COV_G. If G contains an isolated vertex (and $n \geq v_G$), then, trivially, the property COV_G coincides with the presence of a copy of H, where H is obtained from G by removing one isolated vertex. Since this property has been discussed before, throughout this section we will be assuming that the minimum degree of G is at least 1.

For a particular vertex $i \in [n]$, there are possibly several positions it may take in a copy of G which covers it. For the purpose of classifying them, let us introduce the notion of a *rooted graph* (v, G), where G is a graph and $v \in V(G)$ is the *root*. For example, there is only one (up to isomorphism) rooted version of K_3, while the whisk graph K_3^+ enjoys three nonisomorphic rooted versions (see Figure 3.3).

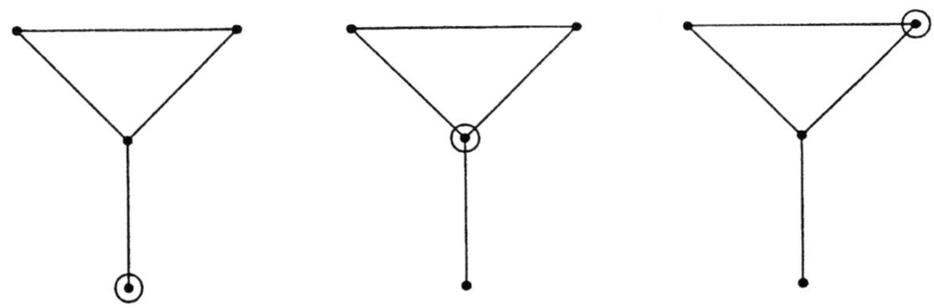

Fig. 3.3 Three rooted versions of the whisk graph; the roots are indicated by open circles.

For a rooted graph (v, G), with $v_G > 1$, let $d(v, G) = e_G/(v_G - 1)$ and let

$$m(v, G) = \max_{H: v \in H \subseteq G} d(v, H).$$

(Thus, $d(v, G) = d^{(1)}(G)$ does not depend on v, but $m(v, G)$ does, in general.)

A rooted graph (v, G) is called *balanced* if $d(v, G) = m(v, G)$ and *strictly balanced* if $d(v, H) < d(v, G)$ for every proper subgraph H of G containing the vertex v. For instance, among the three rooted versions of K_3^+ only one is strictly balanced, while the other two are not balanced (Exercise!). Note that a graph is strictly K_1-balanced if and only if all its rooted versions are strictly balanced (Exercise!). In particular, all cycles and complete graphs have only one rooted version, and that is strictly balanced.

For $i \in [n]$ and $v \in V(G)$, let $U_i(v)$ be the number of copies of (v, G) contained in the random graph $\mathbb{G}(n, p)$, in which vertex i takes the role of the root v, and let U_i be the total number of copies of G containing i. Then, similarly to the problem of containment of ordinary subgraphs, $p = n^{-1/m(v,G)}$ is the threshold for the property "$U_i(v) > 0$" (Ruciński and Vince 1986) and consequently, $p = n^{-1/\min_{v \in G} m(v,G)}$ is the threshold for the property "$U_i > 0$" (Exercise!).

For instance, as soon as $p \gg n^{-3/4}$, a fixed vertex, say vertex 1, *a.a.s.* belongs to a copy of K_3^+, but only when $p \gg n^{-2/3}$, does it belong to a triangle.

However, we are mainly interested in the random variable

$$W_G = |\{i \in [n] : U_i = 0\}| = \sum_1^n \mathbf{1}[U_i = 0],$$

which counts the vertices of $\mathbb{G}(n, p)$ not covered by copies of G; hence COV_G is equivalent to "$W_G = 0$". Theorem 3.22 below provides thresholds for the events COV_G which, of course, depend on the structure of G.

For a graph G, let $m_* = \min_{v \in G} m(v, G)$ and $M(G) = \{v \in V(G) : m(v, G) = m_*\}$. For a vertex $v \in M(G)$ let $\mathcal{C}_v$ be the family of all subgraphs H of G which contain v and satisfy the conditions $d(v, H) = m_*$ and $N_H(v) \neq \emptyset$, the latter condition just saying that v is not an isolated vertex in H. Further, let $s_v = \min_{H \in \mathcal{C}_v} e(H)$, with $s_v = \infty$ if $\mathcal{C}_v = \emptyset$, and $s = \max_{v \in M(G)} s_v$. Finally, set $a = |M(G)|/\operatorname{aut}(G)$.

Theorem 3.22. *Let G be a graph with minimum degree at least 1.*

(i) *If for every $v \in M(G)$ the rooted graph (v, G) is strictly balanced, then*

$$\lim_{n\to\infty} \mathbb{P}(\mathbb{G}(n,p) \in \mathsf{COV}_G) = \begin{cases} 0 & \text{if } an^{v_G-1}p^{e_G} - \log n \to -\infty \\ 1 & \text{if } an^{v_G-1}p^{e_G} - \log n \to \infty. \end{cases}$$

Moreover, if $an^{v_G-1}p^{e_G} - \log n \to c$, $-\infty < c < \infty$, then $W_G \xrightarrow{d} \operatorname{Po}(e^{-c})$, and hence $\mathbb{P}(\mathbb{G}(n,p) \in \mathsf{COV}_G) \to \exp(-e^{-c})$.

(ii) *If $s < \infty$, then there exist constants $c, C > 0$ such that*

$$\lim_{n\to\infty} \mathbb{P}(\mathbb{G}(n,p) \in \mathsf{COV}_G) = \begin{cases} 0 & \text{if } p \le c(\log n)^{1/s} n^{-1/m_*}, \\ 1 & \text{if } p \ge C(\log n)^{1/s} n^{-1/m_*}. \end{cases}$$

(iii) *If $s = \infty$, then*

$$\lim_{n\to\infty} \mathbb{P}(\mathbb{G}(n,p) \in \mathsf{COV}_G) = \begin{cases} 0 & \text{if } p \ll n^{-1/m_*} \\ 1 & \text{if } p \gg n^{-1/m_*}. \end{cases}$$

It is easy to check that the assumption in (i) is a special case of that in (ii), with $m_* = d^{(1)}(G)$ and $s = e_G$ (Exercise!). In Case (iii), which is the complement of (ii), the parameter m_* coincides with $m(G)$ appearing in Theorem 3.4 (Exercise!); hence the threshold for covering by copies of G coincides with the threshold for existence of any copy at all.

Remark 3.23. Note that the nicer the structure of G, the sharper threshold one can prove. Indeed, in Case (i), $a^{-1/e_G}(\log n)^{1/e_G} n^{-1/m_*}$ is a sharp threshold. In Case (ii), it follows from Theorem 1.31 that there exists a sharp threshold, although we do not know it exactly. By (ii) above, the sharp threshold is of the form $b(n)(\log n)^{1/s} n^{-1/m_*}$ for some $b(n)$ with $c \le b(n) \le C$; it is reasonable to conjecture that $b(n)$ is a constant, but at present we cannot rule out the possibility that it oscillates somehow.

In Case (iii), in contrast, the threshold is coarse. In fact, if $p = cn^{-1/m_*}$ for any fixed $c > 0$, and H is a minimal subgraph of G such that $d(H) = m(G)$, then Theorem 3.19 shows that $X_H \xrightarrow{d} \operatorname{Po}(\lambda)$ for some $\lambda < \infty$, and thus $\mathbb{P}(X_G = 0) \ge \mathbb{P}(X_H = 0) \to e^{-\lambda} > 0$, so $\mathbb{P}(\mathbb{G}(n,p) \in \mathsf{COV}_G) \not\to 1$.

Part (i) was proved independently by Ruciński (1992a) and, in a slightly disguised form, by Spencer (1990). We will present the proof of part (i) only. The proofs of the other two parts follow from more general results by Spencer

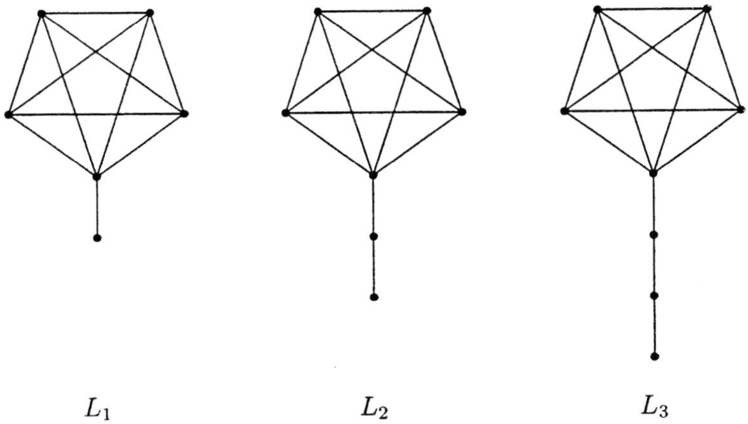

Fig. 3.4 The lollipop graphs.

(1990) on extension statements; see the end of this section. Before presenting the proof of (i), we give a few examples.

Example 3.24. The graphs K_3 and K_3^+ have strictly balanced rooted versions and, by Case (i) of Theorem 3.22, the respective thresholds for the properties COV_{K_3} and $\mathsf{COV}_{K_3^+}$ are $(\log n)^{1/3}n^{-2/3}$ and $(\log n)^{1/4}n^{-3/4}$, respectively. In particular, for $(\log n)^{1/4}n^{-3/4} \ll p(n) \ll n^{-2/3}$ *a.a.s.* every vertex belongs to a copy of K_3^+, but since there are only $o(n)$ triangles, most vertices take the "off-triangle" position.

Example 3.25. The threshold for COV_{K_5} equals $n^{-2/5}\log^{1/10} n$ by Theorem 3.22(i). Consider now the lollipop graphs L_r obtained from a clique K_5 by attaching to it a path P_r (see Figure 3.4).

Let t denote the vertex of degree one (the tail vertex). The lollipop L_1 has $m = 11/5$ and $M = \{t\}$, and the rooted graph (t, L_1) is strictly balanced; thus Case (i) applies. For L_2, we have $m = 2$ and again $M = \{t\}$, but this time the rooted graph (t, L_2) is balanced, but not strictly balanced. Moreover, $\mathcal{C}_t = \{L_2\}$, so $s = s_t = e(L_2)$, and thus Case (ii) applies. Hence, the thresholds for covering every vertex of $\mathbb{G}(n,p)$ by copies of L_1 and L_2 are, respectively, $n^{-5/11}(\log n)^{1/11}$ and $n^{-1/2}(\log n)^{1/12}$.

Finally, consider lollipops with $r \geq 3$. We have $m = 2$ and $t \in M$, but this time the only pair (t, H) which achieves $d(v, H) = m$ is such that H is the clique K_5 together with the tail vertex t, which is isolated in H. Thus $\mathcal{C}_t = \emptyset$, $s_t = \infty$, Case (iii) applies, and the threshold for COV_{L_r} coincides with that for the existence of K_5, that is, $n^{-1/2}$. In other words, as soon as copies of K_5 begin to appear in $\mathbb{G}(n,p)$, every vertex is at distance at most three from one of them. This particular observation follows also from the known fact that the threshold for diameter three is $n^{-2/3}\log^{1/3} n$ (Bollobás 1985, Chapter X), which is well below the threshold for existence of K_5.

Proof of Theorem 3.22(i). We use a mixture of the second moment method and the correlation inequality of Theorem 2.18(i). By the monotonicity of the property COV_G we may assume that $n^{v_G-1}p^{e_G} = \Theta(\log n)$. Since for every $v \notin M(G)$, $\mathbb{P}(U_i(v) > 0) = o(1)$, the decisive role in covering the vertices of $\mathbb{G}(n,p)$ is played by the rooted versions (v, G), where $v \in M(G)$. Let $\{v_j\}$, $i = 1, \ldots, l$, be a maximal collection of vertices of $M(G)$ for which the rooted graphs (v_j, G) are pairwise nonisomorphic. Set $\bar{U}_i = \sum_{j=1}^{l} U_i(v_j)$ and observe that

$$\sum_{i=1}^{n} \bar{U}_i = |M| X_G.$$

Hence, by symmetry, $\mathbb{E}(\bar{U}_1) = \frac{|M|}{n}\,\mathbb{E}(X_G) = an^{v_G-1}p^{e_G} + o(1)$.

Further, for each $v \notin M(G)$, choose a minimal subgraph $H_v \subseteq G$ containing v such that $d(v, H_v) = m(v, G) > m_*$, let $U_i^*(v)$ be the number of copies of (v, H_v) in $\mathbb{G}(n,p)$ rooted at i, and let $U_i^* = \sum_{v \notin M(G)} U_i^*(v)$. Then $\mathbb{E}\, U_1^* = o(1)$.

Since $\mathbb{E}(W_G) = n\,\mathbb{P}(U_1 = 0)$, we need a sensitive asymptotic for $\mathbb{P}(U_1 = 0)$. Note that $\bar{U}_1 + U_1^* = 0$ implies $U_1 = 0$. Thus, by Corollary 2.13 we have

$$\mathbb{P}(U_1 = 0) \geq \mathbb{P}(\bar{U}_1 + U_1^* = 0) \geq e^{-\mathbb{E}(\bar{U}_1 + U_1^*)/(1-p)} = e^{-\mathbb{E}(\bar{U}_1) + o(1)}.$$

For an upper bound, let $\mathcal{S}$ be the family of all edge sets of rooted copies of (v_j, G), $j = 1, \ldots, l$, in the complete graph K_n with vertex 1 as the root. For each $A \in \mathcal{S}$, we set

$$I_A = \mathbf{1}[A \subseteq \mathbb{G}(n,p)].$$

We then have

$$\sum_{A} \sum_{B \neq A,\, B \cap A \neq \emptyset} \mathbb{E}(I_A I_B) = O\left(\sum_{(s,t)} n^{2v_G - s - 1} p^{2e_G - t} \right)$$

$$= O\left(n^{2v_G-2} p^{2e_G} \sum_{(s,t)} n^{-(s-1)} p^{-t} \right),$$

where s and t represent, respectively, the number of common vertices and edges of a pair of two copies of G, each rooted at a vertex of $M(G)$ and both containing vertex 1 as the root. Such an intersection is a proper subgraph of G containing a vertex $v \in M(G)$ and hence, by the fact that (v, G) is strictly balanced, we always have

$$\frac{t}{s-1} < \frac{e_G}{v_G - 1}.$$

Thus, in our range of $p(n)$, $n^{s-1}p^t \geq n^{\varepsilon}$ for some $\varepsilon > 0$. This together with Theorem 2.18(i) implies that

$$\mathbb{P}(U_1 = 0) \leq \mathbb{P}(\bar{U}_1 = 0) \leq e^{-\mathbb{E}(\bar{U}_1) + o(1)}.$$

Hence,

$$\mathbb{E}(W_G) = n\,\mathbb{P}(U_1 = 0) = ne^{-\,\mathbb{E}(\bar{U}_1)+o(1)} = ne^{-an^{v_G-1}p^{e_G}+o(1)}. \tag{3.20}$$

If $an^{v_G-1}p^{e_G} - \log n \to \infty$, then $\mathbb{E}(W_G) = o(1)$ and, by the first moment method, $\mathbb{P}(W_G > 0) = o(1)$. On the other hand, if $an^{v_G-1}p^{e_G} - \log n \to -\infty$, then $\mathbb{E}(W_G) \to \infty$ and we apply the second moment method to W_G. To this end, as W_G is a sum of mutually dependent indicators, it is convenient to express the variance of W_G in the form

$$\mathrm{Var}(W_G) = \mathbb{E}(W_G(W_G-1)) + \mathbb{E}(W_G) - (\mathbb{E}(W_G))^2.$$

We have

$$\mathbb{E}(W_G(W_G-1)) = n(n-1)\,\mathbb{P}(U_1 = U_2 = 0) \le n(n-1)\,\mathbb{P}(\bar{U}_1 = \bar{U}_2 = 0)$$

and, by another application of Theorem 2.18(i) (this time to the family of all edge sets of copies of (v_j, G), $j = 1, \dots, l$, rooted at 1 or 2),

$$\mathbb{P}(\bar{U}_1 = \bar{U}_2 = 0) \le e^{-2\,\mathbb{E}(\bar{U}_1)+o(1)}.$$

Altogether,

$$\mathbb{P}(W_G = 0) \le \frac{\mathrm{Var}(W_G)}{(\mathbb{E}(W_G))^2} = \frac{\mathbb{E}(W_G(W_G-1))}{(\mathbb{E}(W_G))^2} + \frac{1}{\mathbb{E}(W_G)} - 1 \to 0.$$

Similarly one can prove that when $an^{v_G-1}p^{e_G} - \log n \to c$, and thus $\mathbb{E}(W_G) \to e^{-c}$ by (3.20), the k-th factorial moment $\mathbb{E}[W_G(W_G-1)\cdots(W_G-k+1)]$ of W_G converges to e^{-ck} for every $k \ge 1$. This proves, by Corollary 6.8, that W_G converges to the Poisson distribution with expectation e^{-c}. Alternatively, one could apply here Stein's method (*cf.* Theorem 6.24). ■

Extension statements

Spencer (1990) considers a related problem with some applications to the zero-one laws for random graphs discussed in Chapter 10.

Let $R = \{v_1, \dots, v_r\}$ be an independent set of vertices in a graph G. The pair (R, G) will be dubbed a *rooted graph*. For $|R| = 1$ this is the notion introduced at the beginning of this section. We say that a graph F satisfies the extension statement $\mathsf{Ext}(R, G)$, or briefly, $F \in \mathsf{Ext}(R, G)$, if for every r-tuple $R' = \{v'_1, \dots, v'_r\}$ of vertices of F there is a copy of G in F with v'_j mapped onto v_j, $j = 1, \dots, r$.

Example 3.26. If $G = K_2$ and $|R| = 1$, then $F \in \mathsf{Ext}(R, G)$ means that there are no isolated vertices in F. If $G = K_3$ and $|R| = 1$, then $F \in \mathsf{Ext}(R, G)$ is equivalent to $F \in \mathsf{COV}_G$. The same is true for every vertex-transitive graph G; more generally, if $R = \{v\}$, then $F \in \mathsf{Ext}(R, G)$ means that every vertex in F belongs to a copy of G where it corresponds to v. If $G = P_k$ and R is the set of the endpoints of P_k, then $F \in \mathsf{Ext}(R, G)$ says that every pair of vertices of F is connected by a path of length k.

We will now define notions which are straightforward generalizations of the case (v, G) treated above. For a rooted graph (R, G), with $r = |R|$, let $d(R, G) = e_G/(v_G - r)$ and let

$$m(R,G) = \max_{H:R\subseteq H\subseteq G} d(R,H).$$

The rooted graph (R, G) is called *balanced* if for every subgraph H of G, such that $V(H) \supseteq R$, we have $d(R, H) \le d(R, G)$, and *strictly balanced* if this inequality is strict for all proper subgraphs H of G, such that $V(H) \supsetneq R$.

As a generalization of the families C_v appearing before Theorem 3.22 we now define a subgraph H containing R to be *primal* if $d(R, H) = m(R, G)$ and *grounded* if at least one of $v_1, \dots, v_r$ is not isolated in H. We let s_R be the smallest number of edges in a grounded primal subgraph H, with $s_R = \infty$ if no such subgraph exists. Finally, let b_1 be the number of automorphisms of G that fix every element of R, and let b_2 be the number of permutations of R that can be extended to some automorphism of G. Then, the following results hold; for the proof we refer to Spencer (1990).

Theorem 3.27. *Let G be a graph with minimum degree at least 1, and let $R \neq \emptyset$ be an independent set of vertices in G.*

(i) *If the rooted graph (R, G) is strictly balanced, then*

$$\lim_{n\to\infty} \mathbb{P}(\mathbb{G}(n,p) \in \mathsf{Ext}(R,G)) = \begin{cases} 0 & \text{if } n^{v_G - r}p^{e_G} - b_1 r \log n \to -\infty \\ 1 & \text{if } n^{v_G - r}p^{e_G} - b_1 r \log n \to \infty. \end{cases}$$

Moreover, if $n^{v_G-r}p^{e_G} - b_1 r\log n \to c$, $-\infty < c < \infty$, then $\mathbb{P}(\mathbb{G}(n,p) \in \mathsf{Ext}(R,G)) \to \exp(-e^{-c/b_1}/b_2)$.

(ii) *If $s_R < \infty$, then there exist constants $c, C > 0$ such that*

$$\lim_{n\to\infty} \mathbb{P}(\mathbb{G}(n,p) \in \mathsf{Ext}(R,G)) = \begin{cases} 0 & \text{if } p \le c(\log n)^{1/s_R} n^{-1/m(R,G)}, \\ 1 & \text{if } p \ge C(\log n)^{1/s_R} n^{-1/m(R,G)}. \end{cases}$$

(iii) *If $s_R = \infty$, in which case $m(R, G) = m(G)$, then*

$$\lim_{n\to\infty} \mathbb{P}(\mathbb{G}(n,p) \in \mathsf{Ext}(R,G)) = \begin{cases} 0 & \text{if } p \ll n^{-1/m(G)} \\ 1 & \text{if } p \gg n^{-1/m(G)}. \end{cases}$$ ∎

The 1-statements in parts (ii) and (iii) (with $|R| = 1$) immediately imply the corresponding statements in Theorem 3.22. This is not so for the 0-statements, but the 0-statements in Theorem 3.22(ii, iii) follow by the same proofs as for the corresponding 0-statements in Theorem 3.27 given in Spencer (1990). (For the 0-statement in (iii), this is just applying Theorem 3.4.)

Example 3.28. For a graph G, one may ask what the threshold is for the property that for every vertex of $\mathbb{G}(n,p)$ the subgraph induced by its neighborhood contains a copy of G. Let $G + K_1$ be the graph obtained by joining a new vertex w to every vertex of G. Then this property is equivalent to $\mathsf{Ext}(\{w\}, G + K_1)$. For a strictly balanced (in the ordinary, unrooted sense) graph G, the rooted graph $(w, G + K_1)$ is strictly balanced and, by Theorem 3.27(i), the desired threshold is $(\log n)^{1/(v_G+e_G)} n^{-v_G/(v_G+e_G)}$ (which coincides with the threshold for COV_{G+K_1}) (Exercise!).

3.5 DISJOINT COPIES

In this section we consider a problem which will be further developed in Section 4.2. The question we address here is: How many disjoint copies of a given graph G are there in a random graph $\mathbb{G}(n,p)$? As the disjointness may be meant with respect to vertices or with respect to edges, we define two random variables D^v_G and D^e_G equal to the cardinality of the largest collection of vertex- and edge-disjoint copies of G, respectively. Trivially, $D^v_G \le D^e_G \le X_G$, but also $D^e_G \le D^e_H$ for every non-empty subgraph H of G and $D^v_G \le D^v_H$ for every non-null subgraph H of G. Define

$$\Phi^v_G(n,p) = \Phi^v_G \stackrel{\text{def}}{=} \min\{\mathbb{E}(X_H) : H \subseteq G,\ v_H > 0\} = \min(\Phi_G, n),$$

denote $\Phi^e_G = \Phi_G$, where Φ_G was defined in Section 3.1, and observe that $\Phi^v_G \to \infty$ if and only if $\Phi^e_G \to \infty$ (*cf.* Lemma 3.6). We know from Section 3.1 that when $\Phi^e_G \to \infty$, then $X_H = \Theta_C(\mathbb{E} X_H)$ for $H \subseteq G$ (since $\Phi^e_G \to \infty$ implies $\Phi^e_H \to \infty$) and thus $D^v_G = O_C(\Phi^v_G)$ and $D^e_G = O_C(\Phi^e_G)$. In fact, the above quantities provide the correct orders of magnitude for the two random variables in question.

Theorem 3.29. *If $\Phi_G \to \infty$, then $D^v_G = \Theta_C(\Phi^v_G)$ and $D^e_G = \Theta_C(\Phi^e_G)$.*

Proof. The proof below is a slight modification of that from Kreuter (1996) and relies on the second moment method.

Consider first the vertex case and define an auxiliary graph Γ with vertices being the copies of G in $\mathbb{G}(n,p)$ and edges connecting pairs of copies with at least one vertex in common. Thus, $v_\Gamma = X_G$ and $e_\Gamma = \sum_F X_F$, where the sum is taken over all unions $F = G_1 \cup G_2$ of two copies of G sharing at least one vertex. Also, any independent set of vertices in Γ corresponds to a vertex-disjoint collection of copies of G in $\mathbb{G}(n,p)$. Hence, it follows from the

Turán Theorem (see, e.g., Berge (1973, p. 282)) that

$$D_G^v \geq \frac{X_G^2}{X_G + 2\sum_F X_F}. \tag{3.21}$$

In view of this, all we need to show is that

$$X_F = O_C\left(\frac{(\mathbb{E}\, X_G)^2}{\Phi_G^v}\right) \tag{3.22}$$

for every union F of two vertex-intersecting copies of G.

For convenience, set $\Psi_F = n^{v_F}p^{e_F}$ and note that $\mathbb{E}\, X_F = \Theta(\Psi_F)$. The reason we prefer to use Ψ_F rather than $\mathbb{E}\, X_H$ is the log-modularity property

$$\Psi_{F_1 \cup F_2}\Psi_{F_1 \cap F_2} = \Psi_{F_1}\Psi_{F_2}$$

holding for arbitrary graphs F_1 and F_2. Note also that $\Psi_H = \Theta(\mathbb{E}\, X_H) = \Omega(\Phi_G^v)$ if $H \subseteq G$ and $v_H > 0$.

Now assume that $F = G_1 \cup G_2$, where G_1 and G_2 are two copies of G, and let $H = G_1 \cap G_2$ have $v_H > 0$. Then

$$\mathbb{E}\, X_F = \Theta(\Psi_F) = \Theta\left(\frac{\Psi_G^2}{\Psi_H}\right) = O\left(\frac{\Psi_G^2}{\Phi_G^v}\right).$$

In order to bound X_F by the same quantity as $\mathbb{E}\, X_F$ we will apply Chebyshev's inequality. For this we need to estimate the variance of X_F, which, by Lemma 3.5, is of the order Ψ_F^2/Φ_F^e. To bound, in turn, Φ_F^e from below, assume that $L \subseteq F$ with $e_L > 0$ and let $L_i = L \cap G_i$, $i = 1, 2$. Then $L = L_1 \cup L_2$, $L \cup H = (L_1 \cup H) \cup (L_2 \cup H)$, and $(L_1 \cup H) \cap (L_2 \cup H) = H$. Two applications of the log-modularity of Ψ yield

$$\Psi_L = \frac{\Psi_{L\cup H}\Psi_{L\cap H}}{\Psi_H} = \frac{\Psi_{L_1\cup H}\Psi_{L_2\cup H}\Psi_{L\cap H}}{\Psi_H^2}.$$

Here $L_1 \cup H$, $L_2 \cup H$ and $L \cap H$ are all subgraphs of G. Thus, if $v_{L\cap H} > 0$, then $\Psi_L = \Omega\left((\Phi_G^v)^3/\Psi_H^2\right)$.

In the special case $v_{L\cap H} = 0$, the graphs L_1 and L_2 are disjoint and at least one of them is non-empty. Assume that $e_{L_1} > 0$. Then, taking into account that $\Psi_{L_2} \geq \Phi_G^v \to \infty$ if $v_{L_2} > 0$ and $\Psi_{L_2} = 1$ otherwise, we obtain

$$\Psi_L = \Psi_{L_1}\Psi_{L_2} \geq \Psi_{L_1} = \Omega\left(\frac{(\Phi_G^v)^3}{\Psi_H^2}\right).$$

Consequently,

$$\Phi_F^e = \min_{e_L>0} \mathbb{E}\, X_L = \Omega\left(\frac{(\Phi_G^v)^3}{\Psi_H^2}\right)$$

and, using the log-modularity of Ψ again,

$$\operatorname{Var}(X_F) = O\left(\frac{\Psi_F^2}{\Phi_F^e}\right) = O\left(\frac{\Psi_F^2\Psi_H^2}{(\Phi_G^v)^3}\right) = O\left(\frac{\Psi_G^4}{(\Phi_G^v)^3}\right).$$

Hence, by Chebyshev's inequality and the assumption that $\Phi_G^v \to \infty$,

$$\mathbb{P}\left(X_F \geq \mathbb{E}\,X_F + \frac{\Psi_G^2}{\Phi_G^v}\right) = O(1/\Phi_G^v) = o(1),$$

which proves (3.22) and completes the proof of Theorem 3.29 in the vertex case.

The proof for D_G^e follows along the same lines. But instead of repeating the same argument for the edge case, we present an alternative approach involving Markov's and Talagrand's inequalities.

Building the auxiliary graph Γ in a similar way with the obvious modification that now the edges of Γ join edge-intersecting copies of G in $\mathbb{G}(n,p)$, we have, by Markov's inequality (1.3), that with probability at least $\frac{3}{4}$, $e_\Gamma \leq 4\,\mathbb{E}(e_\Gamma)$. We have $4\,\mathbb{E}(e_\Gamma) \leq c_1(\mathbb{E}\,X_G)^2/\Phi_G^e$, for some $c_1 > 0$, and thus, by (3.21) modified to the edge case, and by the fact that $X_G = \Theta_C(\mathbb{E}\,X_G)$, there is another constant $c_2 > 0$ such that $D_G^e \geq c_2\Phi_G^e$ with probability at least, say, $\frac{1}{2}$.

Now, using Talagrand's inequality, we will convert $\frac{1}{2}$ to $1-o(1)$ as required. As we are heading toward an application of Theorem 2.29, let us define Z_i to be the indicator of the presence of the i-th edge of the random graph $\mathbb{G}(n,p)$, $i = 1, 2, \ldots, N = \binom{n}{2}$. Then $D_G^e = f(Z_1, \ldots, Z_N)$, where the function f clearly satisfies the Lipschitz condition (L) with all $c_i = 1$. The other assumption of Theorem 2.29, Condition (C), holds with the function $\psi(r) = e_G r$ for integer $r \geq 0$ (and thus with $\psi(r) = e_G\lceil r\rceil$ for any real $r \geq 0$). Indeed, for any integer r, and for any graph F containing r edge-disjoint copies of G, choose J to be the index set of all $e_G r$ edges belonging to these copies. Then any other graph coinciding with F on the given edges contains r edge-disjoint copies of G too. Therefore, by (2.35) (*cf.* Example 2.33), for $0 < c_3 < c_2$ and with $t = (c_2 - c_3)\Phi_G^e$,

$$\begin{aligned}\mathbb{P}(D_G^e < c_3\Phi_G^e) &\leq 2\,\mathbb{P}(D_G^e < c_3\Phi_G^e)\,\mathbb{P}(D_G^e \geq \lceil c_2\Phi_G^e\rceil) \\ &\leq 2\exp\Big\{-\frac{(c_2-c_3)^2(\Phi_G^e)^2}{4e_G c_2\Phi_G^e}\Big\} = \exp\{-\Theta(\Phi_G^e)\} = o(1). \qquad \blacksquare\end{aligned}$$

Note that the last bound in the proof provides yet another proof of the right-hand inequality of Theorem 3.9. Indeed,

$$\mathbb{P}(X_G = 0) = \mathbb{P}(D_G^e = 0) \leq \mathbb{P}(D_G^e < c_3\Phi_G^e).$$

3.6 VARIATIONS ON THE THEME

There are several other properties related to that of containing a copy of G. One such property is the containment of at least one *induced* copy of G in the random graph $\mathbb{G}(n,p)$. Another variation is counting only those copies of G which are vertex disjoint from all other copies of G contained in $\mathbb{G}(n,p)$.

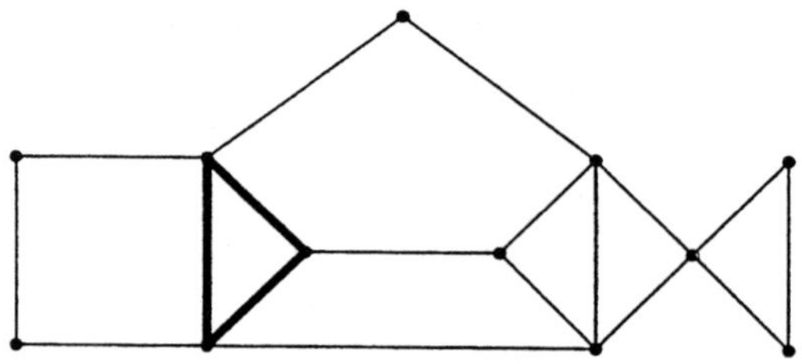

Fig. 3.5 The only solitary triangle in this graph is drawn in bold.

Below we call them *solitary* (see Figure 3.5). Finally, we consider a special case of solitary subgraphs: the *isolated* copies of G.

Induced subgraphs

Let us denote by Y_G the number of induced copies of G in $\mathbb{G}(n,p)$. As we will see in Chapter 6, for p constant, the behavior of Y_G may significantly differ from that of X_G. However, for $p = o(1)$ they are asymptotically the same. Here we only explain why the event "$Y_G > 0$" has the same threshold as the event "$X_G > 0$". The 0-statement of Theorem 3.4 holds for induced copies simply because $Y_G \le X_G$. Let $J_{G'}$ be a zero–one random variable equal to 1 if G', a copy of G in K_n, actually becomes an induced copy of G in $\mathbb{G}(n,p)$. For an application of the second moment method, observe that, assuming $p = o(1)$,

$$\mathbb{E}(J_{G'}) = p^{e_G}(1-p)^{\binom{v_G}{2}-e_G} \sim \mathbb{E}(I_{G'})$$

and, consequently,

$$\mathbb{E}(Y_G) \sim \mathbb{E}(X_G).$$

Moreover, for any two copies G' and G'' of G which share at least one edge

$$\mathrm{Cov}(J_{G'}, J_{G''}) < \mathbb{E}(J_{G'}J_{G''}) \le \mathbb{E}(I_{G'}I_{G''}) \sim \mathrm{Cov}(I_{G'}, I_{G''}),$$

while for any two copies with at most one vertex in common, $\mathrm{Cov}(J_{G'}, J_{G''}) = 0$. Finally, for any two edge-disjoint copies sharing t vertices, where $t \ge 2$, $\mathrm{Cov}(J_{G'}, J_{G''}) < p^{2e_G}$, and the number of such pairs is $O(n^{2v_G - t})$. Hence, as in the case of ordinary subgraphs, we have (Exercise!)

$$\mathbb{P}(Y_G = 0) \le \frac{\mathrm{Var}(Y_G)}{(\mathbb{E}\,Y_G)^2} = o(1).$$

There is another way of deducing the above fact. Assume again that $p = o(1)$. By Markov's inequality, $\mathbb{P}\big(X_G - Y_G > \frac{1}{3}\,\mathbb{E}(X_G)\big) = o(1)$ and so *a.a.s.* $Y_G \ge X_G - \frac{1}{3}\,\mathbb{E}(X_G)$. On the other hand, we know that when $\Phi_G \to \infty$, $X_G/\,\mathbb{E}(X_G) \xrightarrow{p} 1$ and, in particular, *a.a.s.* $X_G \ge \frac{2}{3}\,\mathbb{E}(X_G)$. Hence, *a.a.s.* $Y_G \ge \frac{1}{3}\,\mathbb{E}(X_G) > 0$.

The presence of an induced copy of G is not a monotone property (except in the trivial cases in which G is either complete or empty). It is not even convex (Exercise!), however, it has a second (disappearence) threshold toward the end of the evolution of $\mathbb{G}(n,p)$. In terms of $q = 1 - p$, it corresponds to the threshold for "$X_{G^c} > 0$" in the complementary random graph $\mathbb{G}(n,q)$, where G^c is the complement of G. Hence, the second threshold is roughly $1 - \Theta(n^{-1/m(G^c)})$ (Exercise!).

Solitary subgraphs

Let us denote by Z_G the number of solitary copies of G. Clearly, $Z_G \le D_G^v$, where D_G^v has been defined in the previous section. Observe that $\mathbb{E}(Z_G) = \mathbb{E}(X_G)\Pi_G$, where Π_G is the conditional probability that a fixed copy of G is solitary, given that it is present in $\mathbb{G}(n,p)$.

For a nonbalanced G, $\lim_{n\to\infty} \mathbb{P}(Z_G > 0) = 0$, since as soon as copies of G emerge, it can be seen by Theorem 2.18 that Π_G is exponentially small with a power of n in the exponent, and hence $\mathbb{P}(Z_G > 0) \le \mathbb{E}(Z_G) = \mathbb{E}(X_G)\Pi_G = o(1)$ (Exercise!). With some additional effort one can prove that for a balanced but not strictly balanced G, we have $\mathbb{P}(Z_G > 0) = o(1)$ as soon as $\mathbb{P}(X_G > 0) \to 1$ and thus $\limsup \mathbb{P}(Z_G > 0)$ is never equal to 1 (Exercise!).

Let us assume now that G is strictly balanced. If $p = \Theta(n^{-1/d(G)})$, then, as we showed in the proof of Theorem 3.19, *a.a.s.* there are no intersecting pairs of G at all, and so $Z_G = X_G$. In other words, when copies of a strictly balanced graph G first emerge, they are all solitary. This holds true even beyond the threshold, when $np^{d(G)} \to \infty$ sufficiently slowly. But the containment of a solitary copy is not monotone either, and with more and more edges in the random graph, the solitary copies of G become rarer until complete extinction occurs. The second (disappearence) threshold was detected for strictly K_1-balanced graphs by Suen (1990), and for a slightly larger subclass of strictly balanced graphs (including trees) by Kurkowiak and Ruciński (2000). It is determined, roughly, by the equation $\mathbb{E}\, X_G = \Theta(n \log n)$.

The difficulty we are facing here is that the probablity Π_G depends on all pairs in $[n]^2$ and, rather than finding an exact expression, one can only bound it, using results like Theorem 2.18 and Theorem 2.12. We remark that this problem was a motivation for Suen to develop his correlation inequality, some versions of which were discussed in Section 2.3.

Isolated subgraphs

A much simpler situation takes place when one counts the isolated copies of G, that is, assuming G is connected, the connected components of $\mathbb{G}(n,p)$ which are isomorphic to G. Let T_G count the isolated copies of G. This time $\mathbb{E}(T_G) = \mathbb{E}(Y_G)(1-p)^{v_G(n-v_G)} = O(n^{v_G}p^{e_G}e^{-v_G np})$; hence, if $e_G > v_G$, we have $\mathbb{P}(T_G > 0) \le \mathbb{E}(T_G) = o(1)$, and there are *a.a.s.* no isolated copies of G.

The same is true if $e_G = v_G$ and further $p \ll 1/n$ or $p \gg 1/n$. We urge the reader (Exercise!) to show that T_G has a limiting Poisson distribution when G is connected with $e_G = v_G$ and $p \sim c/n$, $0 < c < \infty$ (Erdős and Rényi 1960); see Example 6.29.

It remains to consider connected graphs G such that $e_G < v_G$, that is, trees. These are the only small graphs which *a.a.s.* become components of a random graph. Instead of focusing on a single tree, we will count all trees of a given order at once. Let T_v denote the number of all v-vertex isolated trees in $\mathbb{G}(n,p)$, $v = 1, 2, \ldots$. Then, provided $np^2 \to 0$,

$$\mathbb{E}(T_v) = \binom{n}{v} v^{v-2} p^{v-1} (1-p)^{v(n-v)+\binom{v}{2}-v+1} \sim \frac{v^{v-2}}{v!} n^v p^{v-1} e^{-vnp}.$$

This quantity converges to a constant if either $n^v p^{v-1} \to c \geq 0$ or $vnp - \log n - (v-1)\log\log n \to c \in (-\infty, \infty)$. The following result, proved already by Erdős and Rényi (1960), asserts that these two conditions determine two thresholds for the property $T_v > 0$. (See also Theorem 6.38 and Example 6.29.)

Theorem 3.30. *Let* $c_n = vnp - \log n - (v-1)\log\log n$. *Then*

$$\mathbb{P}(T_v > 0) \to \begin{cases} 0 & \text{if } n^v p^{v-1} \to 0 \text{ or } c_n \to \infty, \\ 1 & \text{if } n^v p^{v-1} \to \infty \text{ and } c_n \to -\infty. \end{cases}$$

Moreover, if $n^v p^{v-1} \to c \in (-\infty, \infty)$ *or* $c_n \to c > 0$, *then* $T_v \xrightarrow{d} \mathrm{Po}(\lambda)$, *where* $\lambda = \lim_{n\to\infty} \mathbb{E}(T_v) \in (0, \infty)$. ∎

The case $v = 1$ is special here. The random variable T_1 is the number of isolated vertices in $\mathbb{G}(n,p)$ and $n^v p^{v-1} = n$. Hence there is only one threshold. Furthermore, the $\log\log n$ term drops out and we arrive at the following corollary.

Corollary 3.31. *Let* $c_n = np - \log n$ *and let* T_1 *be the number of isolated vertices in* $\mathbb{G}(n,p)$. *Then*

$$\mathbb{P}(T_1 > 0) \to \begin{cases} 0 & \text{if } c_n \to \infty, \\ 1 & \text{if } c_n \to -\infty. \end{cases}$$

Moreover, if $c_n \to c \in (-\infty, \infty)$, *then* $T_1 \xrightarrow{d} \mathrm{Po}(e^{-c})$. ∎

The proof of Theorem 3.30 follows the lines of those of Theorems 3.4, 3.19 and 3.22 (Exercise!). Another proof will be given in Example 6.28.

4

Matchings

Perfect matchings play an important role in graph theory. On the one hand, they find a broad spectrum of applications. On the other hand, they are the subject of elegant theorems. The two results which characterize their existence – Hall's and Tutte's theorems – are truly beautiful pearls of the theory. No wonder that Erdős and Rényi, after settling the question of connectivity (Erdős and Rényi 1959, 1961), turned their attention to the problem of finding the thresholds for existence of perfect matchings in random graphs (Erdős and Rényi 1964, 1966, 1968).

The results they obtained reveal a special feature of random graphs one could call "the minimum degree phenomenon." Namely, it is frequently true that if a minimum degree condition is necessary for a property to hold, then *a.a.s.* the property holds in a random graph as soon as the condition is satisfied. This phenomenon is discussed to larger extent in Section 5.1. Here we just mention that as soon as the last isolated vertex disappears, the random graph becomes connected (Erdős and Rényi (1959) and, for the hitting version, Bollobás and Thomason (1985)). Moreover, as we will learn later in this chapter (Corollary 4.5 and Theorem 4.6), provided n is even, from that very moment the random graph also contains a perfect matching (Erdős and Rényi 1966, Bollobás and Thomason 1985).

In the first section of this chapter we present a new proof of the threshold theorem for perfect matchings, which is based on Hall's rather than Tutte's theorem. This approach was originally designed to solve the problem of perfect tree-matchings (Łuczak and Ruciński 1991), a special instance of G-factors, where one looks for a vertex-disjoint union of copies of a given graph G which covers all vertices of $\mathbb{G}(n,p)$. Ordinary matchings correspond to the case

$G = K_2$. Section 4.2 collects results on G-factors and partial G-factors in random graphs. Finally, in Section 4.3 we present recent advances on two long-standing open problems: finding the threshold for triangle-factors in $\mathbb{G}(n,p)$ and for perfect matchings in random 3-uniform hypergraphs, the latter known as the Shamir problem.

4.1 PERFECT MATCHINGS

A matching in a graph can be identified with a set of disjoint edges. A *perfect matching* is one which covers every vertex of the graph. Sometimes a perfect matching is called a 1-*factor*, because it is, in fact, a 1-regular spanning subgraph. A necessary condition for the existence of a perfect matching in a graph is the absence of isolated vertices. It turned out that in a random graph this is *a.a.s.* sufficient. Because of the simplicity of Hall's condition, it is quite straightforward to prove this result for random bipartite graphs.

Random bipartite graphs

Recall that a *random bipartite graph*, denoted here by $\mathbb{G}(n,n,p)$, is the reliability network with the initial graph being the complete bipartite graph $K_{n,n}$ with bipartition (V_1, V_2), $|V_1| = |V_2| = n$. In other words, $\mathbb{G}(n,n,p)$ is obtained from $K_{n,n}$ by independent removal of each edge with probability $1-p$. Assume that $\log n - \log\log n \le np \le 2\log n$ and suppose that the random graph $\mathbb{G}(n,n,p)$ does not have a perfect matching. Then, by the Hall Theorem, there is a set $S \subset V_i$ for some $i = 1,2$, which violates Hall's condition, that is, $|S| > |N(S)|$, where $N(S)$ is the set of all vertices adjacent to at least one vertex from S. Let S be a minimal such set. Then (Exercise!),

(i) $|S| = |N(S)| + 1$,

(ii) $|S| \le \lceil n/2 \rceil$,

(iii) every vertex in $N(S)$ is adjacent to at least two vertices of S.

Set $s = |S|$. If $s = 1$, then S is an isolated vertex. If $s = 2$, then S consists of two vertices of degree 1 adjacent to the same vertex. Let us call such a structure a *cherry* (see Figure 4.1)) and let X count cherries in $\mathbb{G}(n,n,p)$. Then

$$\mathbb{E}(X) = O(n^3p^2e^{-2np}) = O\left(n^3\left(\frac{\log n}{n}\right)^2\frac{\log^2 n}{n^2}\right) = o(1), \tag{4.1}$$

meaning that *a.a.s.* there are no cherries in $\mathbb{G}(n,n,p)$.

Let $\mathcal{A}$ denote the event that there is a minimal set S of size $s \ge 3$ which violates Hall's condition. Then, using (i)–(iii), and bounding by $\binom{s}{2}^{s-1}$ the

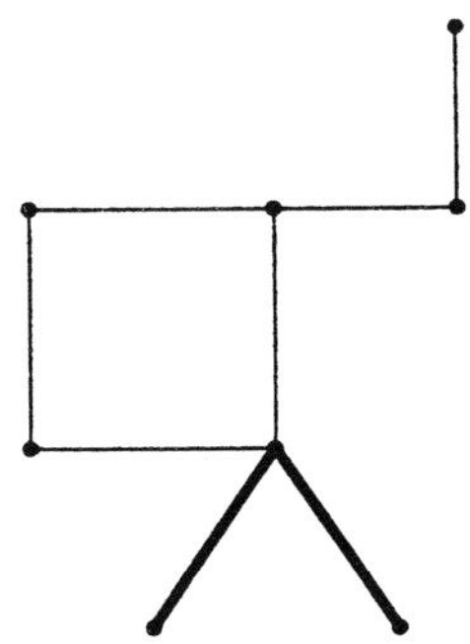

Fig. 4.1 A cherry in a graph marked with bold lines.

number of choices to realize (iii), we obtain

$$\begin{aligned}
\mathbb{P}(\mathcal{A}) &\le \sum_{s=3}^{\lceil \frac{n}{2} \rceil} \binom{n}{s}\binom{n}{s-1}\binom{s}{2}^{s-1} p^{2s-2}(1-p)^{s(n-s+1)} \\
&< \sum_{s\ge 3} \left(\frac{en}{s}\right)^s \left(\frac{en}{s-1}\right)^{s-1} s^{2(s-1)} \\
&\quad \times \left(\frac{2\log n}{n}\right)^{2(s-1)} \exp\left(-\frac{\log n}{2}s + \frac{\log\log n}{2}s\right) \\
&< \frac{n}{(2\log n)^2} \sum_{s\ge 3} \left(\frac{4e^2 \log^{5/2} n}{\sqrt{n}}\right)^s = O\left(\frac{\log^{11/2} n}{\sqrt{n}}\right) = o(1). \qquad (4.2)
\end{aligned}$$

In conclusion, the threshold for having a perfect matching coincides with that for the disappearance of isolated vertices. The latter can be routinely found (Exercise!) by the method of moments (see Corollary 3.31 and Example 6.28). Thus, we obtain the result of Erdős and Rényi (1964).

Theorem 4.1.

$$\mathbb{P}(\mathbb{G}(n,n,p) \text{ has a perfect matching}) \to \begin{cases} 0 & \text{if } np - \log n \to -\infty, \\ e^{-2e^{-c}} & \text{if } np - \log n \to c, \\ 1 & \text{if } np - \log n \to \infty. \end{cases} \qquad \blacksquare$$

Remark 4.2. Consider a random bipartite graph process $\{\mathbb{G}(n,n,M)\}_{M=0}^{n^2}$ defined in analogy with the standard random graph process (Section 1.1) and define the hitting times $\tau_1 = \min\{M : \delta(\mathbb{G}(n,n,M)) \ge 1\}$ and $\tau_{\mathrm{pm}} = \min\{M : \mathbb{G}(n,n,M)$ has a perfect matching$\}$. (Thus, trivially, $\tau_1 \le \tau_{\mathrm{pm}}$.)

The proof of Theorem 4.1 above yields also the stronger result that *a.a.s.* $\tau_{\mathrm{pm}} = \tau_1$. Indeed, if for convenience we instead consider the corresponding continuous time random graph process $\{\mathbb{G}(n,n,t)\}_{0\le t\le 1}$, the same calculations as in (4.1) and (4.2) show that *a.a.s.* there is no minimal subset S of size $s \geq 2$ violating Hall's condition for any $t \in [(\log n - \log\log n)/n, 2\log n/n]$, and the result follows easily (Exercise!).

Remark 4.3. For future reference we explicitly state the error probability estimate

$$\mathbb{P}(\mathbb{G}(n,n,p) \text{ has no perfect matching}) = O\big(ne^{-np}\big), \tag{4.3}$$

which is valid for all n and p. This too follows by the argument above; the probabilities of having an isolated vertex or a cherry are easily seen to be of this order (Exercise!), while a minor modification of (4.2) shows that, assuming, as we may, $np \geq \log n$,

$$\mathbb{P}(\mathcal{A}) \leq \sum_{s\geq 3} e^{2s} n^{2s-1} p^{2s-2} e^{-snp/2} = O\big(n^5 p^4 e^{-3np/2}\big) = O\big(ne^{-np}\big).$$

Ordinary random graphs

Let us return now to the ordinary random graph $\mathbb{G}(n,p)$, n even. Fixing an $n/2$ by $n/2$ bipartition of the vertex set and ignoring the edges within each of the two sets, we immediately see that Theorem 4.1 implies that $\mathbb{G}(n,p)$ has a perfect matching *a.a.s.* as soon as $np - 2\log n \to \infty$. We will show how to reduce the above value of p by half so that, again, the threshold is the same as for the disappearance of isolated vertices (*cf.* Corollary 3.31).

In fact, we will show an even stronger result due to Bollobás and Thomason (1985). But first let us extend the notion of a perfect matching by saying that a graph satisfies property PM if there is a matching covering all but at most one of the nonisolated vertices. It can be routinely checked (Exercise!) that as soon as $2np - \log n - \log\log n \to \infty$, there are only isolated vertices outside the giant component (*cf.* Chapter 5). Note that this holds already when the number of edges in $\mathbb{G}(n,p)$ is only about $\frac{1}{4}n\log n$ – roughly half the threshold for the disappearance of isolated vertices.

However, the main obstacle for PM is now the presence of cherries. Two or more of them make it impossible. If there is exactly one cherry, PM is still possible, provided the number of nonisolated vertices (as well as isolated vertices) is odd. The expected number of cherries is

$$3\binom{n}{3}p^2(1-p)^{2(n-3)} < n^3p^2e^{-2np+6p} = o(1)$$

if $2np - \log n - 2\log\log n \to \infty$, which also holds already when there are about $\frac{1}{4}n\log n$ edges in $\mathbb{G}(n,p)$. Again, as proved by Bollobás and Thomason (1985), this trivial necessary condition becomes *a.a.s.* sufficient.

Theorem 4.4. *Let* $y_n = 2np - \log n - 2\log\log n$. *Then*

$$\mathbb{P}(\mathbb{G}(n,p) \in \mathsf{PM}) \to \begin{cases} 0 & \text{if } y_n \to -\infty, \\ (1 + \frac{1}{16}e^{-c})e^{-\frac{1}{8}e^{-c}} & \text{if } y_n \to c, \\ 1 & \text{if } y_n \to \infty. \end{cases}$$

Consequently we obtain a result of Erdős and Rényi (1966).

Corollary 4.5.

$$\mathbb{P}(\mathbb{G}(n,p) \text{ has a perfect matching}) \to \begin{cases} 0 & \text{if } np - \log n \to -\infty, \\ e^{-e^{-c}} & \text{if } np - \log n \to c, \\ 1 & \text{if } np - \log n \to \infty. \end{cases}$$ ■

The proof below is easily modified to give the corresponding hitting time result too (Exercise!).

Theorem 4.6. *The random graph process* $\{\mathbb{G}(n,M)\}_M$ *is* a.a.s. *such that the hitting times* $\tau_1 = \min\{M : \delta(\mathbb{G}(n,M)) \geq 1\}$ *and* $\tau_{\mathrm{pm}} = \min\{M : \mathbb{G}(n,M)$ *has a perfect matching*$\}$ *coincide.* ■

The original proofs of the 1-statements of Theorem 4.4 and Corollary 4.5 were both based on Tutte's theorem. Here we propose an alternative approach, via Hall's theorem, by Łuczak and Ruciński (1991). This approach relies on the following technical lemma. Given two disjoint sets of vertices in a graph, *the bipartite subgraph induced by them* consists of all edges with one endpoint in each set.

Lemma 4.7. *Let* $np = \Theta(\log n)$. *For every* $c > 0$, a.a.s. *every bipartite subgraph, induced in* $\mathbb{G}(n,p)$ *by two sets of equal size and with minimum degree at least* $c\log n$, *contains a perfect matching.*

Proof. Set $u = \frac{n(\log\log n)^2}{\log n}$. By the first moment method it is easy to check (Exercise!) that *a.a.s.*

(i) for every pair of disjoint subsets of vertices of size bigger than u there is in $\mathbb{G}(n,p)$ an edge between them,

(ii) every set S of at most $2u$ vertices induces in $\mathbb{G}(n,p)$ fewer than $(\log\log n)^3|S|$ edges.

The rest of the proof of Lemma 4.7 is purely deterministic. Suppose G is an n-vertex graph satisfying (i) and (ii), and B is a bipartite subgraph of G induced by the bipartition (W_1, W_2), $|W_1| = |W_2| = w$, $\delta(B) \geq c\log n$, but without a perfect matching. Then, by Hall's theorem there is $S \subseteq W_1$ such that $|N_B(S)| < |S|$.

Case 1: $|S| \leq u$.
Then $|S \cup N_B(S)| < 2u$, but since all the edges with one endpoint in S have the other endpoint in $N_B(S)$, there are at least $c\log n|S|$ such edges – a contradiction with (ii).

Case 2: $|N_B(S)| \geq w - u$.
Then $|W_1 \setminus S| < |W_2 \setminus N_B(S)| \leq u$ and all the edges with one endpoint in $W_2 \setminus N_B(S)$ have the other endpoint in $W_1 \setminus S$ – again a contradiction with (ii).

Case 3: $|S| > u, |N_B(S)| < w - u$.
Now $|W_2 \setminus N_B(S)| > u$, but there is no edge between S and $W_2 \setminus N_B(S)$ – a contradiction with (i). ■

Proof of Theorem 4.4. Throughout the proof we assume that $\frac{1}{2}\log n \le np \le 2\log n$ and n is even.

If $y_n \to -\infty$, the second moment method yields that *a.a.s.* there are many cherries in $\mathbb{G}(n,p)$ (Exercise!). Since already the presence of two cherries makes PM impossible, the 0-statement follows.

If $y_n \to c$, the number of cherries has asymptotically the Poisson distribution with expectation $\frac{1}{8}e^{-c}$. It can also be proved that the probability that there is exactly one cherry and the number of isolated vertices is odd, converges to $\frac{1}{16}e^{-c}\exp(-\frac{1}{8}e^{-c})$. (Advanced Exercise! *Hint*: Apply the two-round exposure – *cf.* Section 1.1 – with p_2 so small that during the second round a fixed cherry cannot be destroyed.)

If $y_n \to \infty$, then, as was shown above, *a.a.s.* there are no cherries at all. Thus, in order to prove the two latter parts of the theorem, we have to show that the only obstruction for PM is the presence of either at least two cherries or one cherry while the number of isolated vertices is even.

The idea of the proof is as follows: Suppose that we have a graph on n vertices which has either no cherries at all or exactly one cherry, but then the number of isolated vertices is odd. Fix an arbitrary bipartition of the vertex set into two halves called *sides*. Call a vertex *bad* if it has either fewer than $\frac{1}{200}\log n$ neighbors within its own side or fewer than $\frac{1}{200}\log n$ neighbors on the other side. Assume that the graph satisfies the hypothesis of Lemma 4.7 (say, with $c = \frac{1}{300}$) and some other properties held *a.a.s.* by $\mathbb{G}(n,p)$ (*viz.* Claim below).

Match the bad vertices first (except for the isolated vertices, of course). If there is an odd number of them, leave one out. If the graph has a cherry, the left-out vertex must be a degree one vertex of that cherry. Remove all bad vertices and their partners (which do not need to be bad) from the graph. Then adjust the bipartition so that it becomes even again, but so that the minimum degree of the induced bipartite subgraph has not dropped much. Finally, apply Lemma 4.7 and obtain a matching covering all but at most one of the nonisolated vertices.

Now we turn to the details. Let X be the number of bad vertices in $\mathbb{G}(n,p)$, and let $Y \in \mathrm{Bi}(\frac{n}{2},p)$. Then $\mathbb{E}(X) = n\,\mathbb{P}(\text{vertex 1 is bad})$ and, by (2.6), for sufficiently large n,

$$\mathbb{P}(\text{vertex 1 is bad}) \le 2\,\mathbb{P}(Y < \tfrac{1}{200}\log n) < n^{-0.21}.$$

Hence, by Markov's inequality (1.3), *a.a.s.* $X < n^{4/5}$.

We still need to distinguish another class of vertices of low degree. Call a vertex *small* if its degree in $\mathbb{G}(n,p)$ is at most three. We claim that neither small nor bad vertices can cling together too much.

Claim. *For every fixed integer k,* a.a.s. *there is no k-vertex tree in $\mathbb{G}(n,p)$ which contains more than two small vertices or more than four bad vertices.*

Proof. Let $Z \in \text{Bi}(n-k+1, p)$. It is straightforward to show (Exercise!) that for every fixed z

$$\mathbb{P}(Z \leq z) = O\big((np)^z e^{-np}\big) = O\left(\frac{\log^z n}{\sqrt{n}}\right) = O(n^{-0.49}).$$

The expected number of k-vertex trees containing three or more small vertices can now be bounded from above by

$$\binom{n}{k} k^{k-2} p^{k-1} \binom{k}{3} [\mathbb{P}(Z \leq 2)]^3 = O(n^k p^{k-1} n^{-1.47}) = o(1).$$

Similarly, the expected number of k-vertex trees with more than four bad vertices is

$$\binom{n}{k} k^{k-2} p^{k-1} \binom{k}{5} [2\,\mathbb{P}(Y' \leq \tfrac{1}{200}\log n)]^5 = O\big(n^k p^{k-1} (n^{-0.21})^5\big) = o(1),$$

where $Y' \in \text{Bi}(\frac{n}{2} - k + 1, p)$. ■

As a final preparatory step, it can be easily checked that *a.a.s.* the maximum degree of our random graph is at most $8 \log n$ (Exercise!).

We are now ready to complete the proof of Theorem 4.4. Consider a graph on n vertices which has either no cherries at all or exactly one cherry, but then the number of isolated vertices is odd. Suppose further that this graph satisfies the hypothesis of Lemma 4.7 (with $c = \frac{1}{300}$) and, for a fixed bipartition, the hypothesis of the above claim for k up to 11. Furthermore, assume that there are fewer than $n^{4/5}$ bad vertices and that the maximum degree Δ is at most $8 \log n$. Note that all these properties hold *a.a.s.* for the random graph $\mathbb{G}(n,p)$. We will show that each such a graph satisfies property PM.

Remove the isolated vertices and, if there is an odd number of them, remove one additional vertex of degree one, destroying the cherry if there is any. Order the remaining bad vertices by degrees, from low to high:

$$\deg(v_1) \leq \deg(v_2) \leq \ldots \deg(v_l).$$

We will match them one by one with some vertices $u_1, u_2, \ldots, u_l$, always choosing as u_i a vertex of a smallest possible degree. We begin with isolated edges as their endpoints are matched naturally.

Suppose $v_1, \ldots, v_{i-1}$ are already matched with $u_1, \ldots, u_{i-1}$ (some v_j may be matched with some v_s, but then, clearly, $u_j = v_s$ and $u_s = v_j$). Let V_{i-1} denote the set $\{v_1, u_1, \ldots, v_{i-1}, u_{i-1}\}$. If $v_i \in V_{i-1}$, it is already matched to some vertex; otherwise we choose u_i as follows.

If $\deg(v_i) = 1$, then take as u_i the neighbor of v_i. It is available, since we follow the degrees from low to high, and since there are no cherries left in the graph.

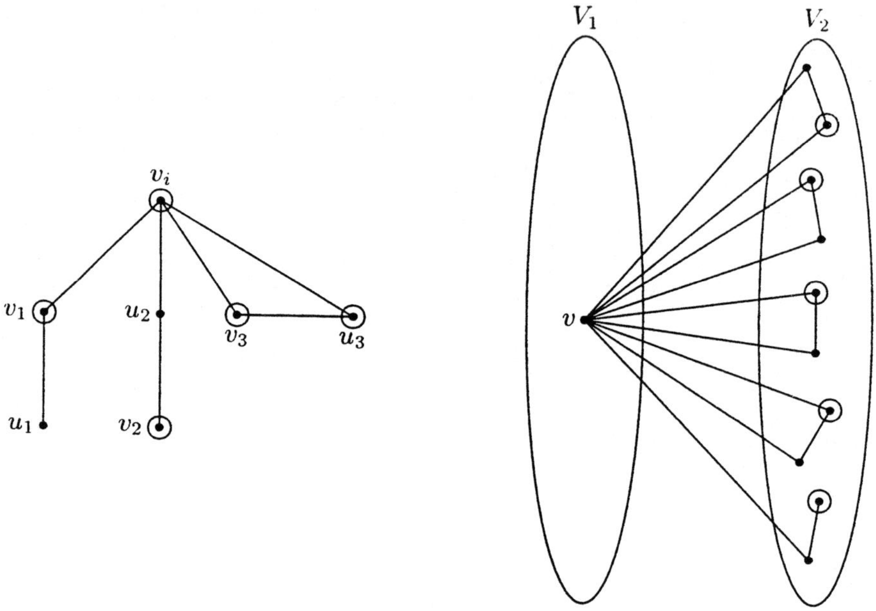

(a) (b)

Fig. 4.2 Scenes from the proof of Theorem 4.4.

If $2 \leq \deg(v_i) \leq 3$, then v_i has at most one neighbor within the set V_{i-1}, since otherwise there would be three small vertices on a small tree (Exercise!). Thus we may choose u_i as one of the neighbors of v_i outside V_{i-1}.

If $\deg(v_i) \geq 4$, then v_i has at most three neighbors within the set V_{i-1}, since otherwise there would be five bad vertices on a small tree (Exercise! – see Figure 4.2(a)). Again, we may choose u_i as one of the neighbors of v_i outside V_{i-1}.

Hence, *a.a.s.* all nonisolated bad vertices can be matched. The removal of bad vertices and their partners does not affect the degrees of other vertices by more than eight (Exercise! – see Figure 4.2(b)). However, the remaining vertices may no longer form an even bipartition. In order to apply Lemma 4.7 we have to balance them back by moving across up to $n^{4/5}$ carefully chosen vertices. To minimize the effect on degrees in the induced bipartite graph, we use for this purpose a 2-*independent set of vertices*, that is, an independent set of vertices no two of which have a common neighbor (thus, the degree of a vertex may drop further by at most one). Trivially, there is always such a set of size at least $n/(\Delta^2 + 1)$ (Exercise!), which is more than is needed. The obtained bipartite subgraph has, therefore, minimum degree at least

$$\frac{1}{200}\log n - 9 > \frac{1}{300}\log n,$$

and, by Lemma 4.7, it contains a perfect matching, which together with the previously constructed matching $\{v_1, u_1\}, \ldots, \{v_l, u_l\}$ forms a matching covering all but at most one of the nonisolated vertices. This completes the proof of Theorem 4.4. ∎

Disjoint 1-factors

Erdős and Rényi, after establishing a threshold for the existence of at least one perfect matching in the bipartite random graph $\mathbb{G}(n, n, p)$, went on and generalized their result to the existence of at least r disjoint 1-factors in $\mathbb{G}(n, n, p)$ (Erdős and Rényi 1966). Trivially, if a graph possesses fewer than r disjoint 1-factors, then the removal of all the edges of a maximal family of disjoint 1-factors results in a graph with no 1-factor at all and with all the degrees decreased by at most $r - 1$. Thus, if the original graph had minimum degree at least r, then, after the removal, there would be no isolated vertices left, and Hall's condition would have to be violated in a nontrivial way. By an argument similar to that used in the proof of Theorem 4.1 this can be shown to be unlikely for $\mathbb{G}(n, n, p)$ (Exercise!), and it follows that the threshold for containing at least r disjoint 1-factors in $\mathbb{G}(n, n, p)$ coincides with that for minimum degree r. The latter can easily be found by the method of moments (Exercise!). The respective hitting time result holds too.

The corresponding problem for an ordinary random graph $\mathbb{G}(n, p)$ was solved much later by Shamir and Upfal (1981) via an algorithmic approach. Since every even Hamilton cycle is a union of two disjoint 1-factors, the solution also follows, together with the hitting time version, from a result of Bollobás and Frieze (1985) (see Section 5.1). However, the proof of Theorem 4.4 presented above can easily be adapted to yield the threshold for r disjoint 1-factors as well. It boils down to showing that *a.a.s.* after removing the edges of an arbitrary subgraph of maximum degree at most $r - 1$, the remainder of $\mathbb{G}(n, p)$ will contain a perfect matching. The definition of a *bad* vertex remains unchanged, while a *small* vertex is now one with degree at most $r + 2$. The details are left to the reader (Exercise!). It follows that *a.a.s.* the hitting time for having r disjoint 1-factors coincides with the hitting time for having minimum degree at least r.

4.2 *G*-FACTORS

In this section we study thresholds for containment of spanning (or almost spanning) subgraphs in $\mathbb{G}(n, p)$ which are unions of vertex disjoint copies of a given graph. For a graph G, every disjoint union of copies of G is called

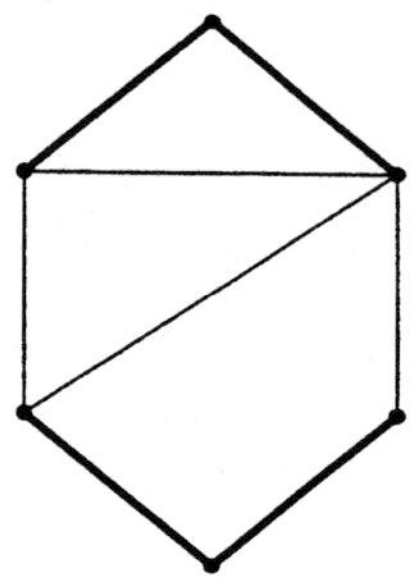

Fig. 4.3 A graph with a P_2-factor marked with bold lines.

a *partial G-factor*. If $G = K_2$, this is the notion of a matching. A partial G-factor which is a spanning subgraph of a graph F is called a *G-factor* in F (see Figure 4.3). Our main objective is to find a threshold for the property of containing a G-factor by the random graph $\mathbb{G}(n,p)$. Observe that when $G = K_2$, this is just the property of containing a perfect matching. Note further that using the notation of Section 3.5, $\mathbb{G}(n,p)$ contains a G-factor if and only if $D_G^v = n/v_G$.

Łuczak and Ruciński (1991) have shown that for every nontrivial tree T, the threshold for possessing a T-factor is the same as that for the disappearance of isolated vertices. (Again, the corresponding hitting time result holds too.)

Theorem 4.8. *For every tree T with $t \geq 2$ vertices, assuming n is divisible by t,*

$$\mathbb{P}(\mathbb{G}(n,p) \textit{ has a } T\textit{-factor}) \to \begin{cases} 0 & \textit{if } np - \log n \to -\infty, \\ e^{-e^{-c}} & \textit{if } np - \log n \to c, \\ 1 & \textit{if } np - \log n \to \infty. \end{cases} \qquad \blacksquare$$

The proof, which we omit (Advanced Exercise!), is very similar to that of Theorem 4.4. Instead of a bipartition, we now take a t-partition, and construct a T-factor from $t-1$ perfect matchings between the appropriate sets of the partition. The existence of these perfect matchings follows by the same argument as in the case $T = K_2$.

Clearly, Theorem 4.8 remains true for forests without isolated vertices. In general, the threshold for the property of having a G-factor is not known and the triangle $G = K_3$ is the smallest unknown case. But already for the whisk graph K_3^+ (see Figure 3.3), the problem becomes relatively easy. It seems that the structural asymmetry of K_3^+ helps here. In fact, there is a broad family of graphs G for which the threshold has been found.

Partial G-factors

Before we prove this result, we will consider the related, weaker property $\mathsf{F}_G(\varepsilon)$ of having a partial G-factor covering all but at most εn vertices of $\mathbb{G}(n,p)$. For this property we can pinpoint the threshold very precisely (Ruciński 1992a). By Theorem 3.29, for $\mathsf{F}_G(\varepsilon)$ to hold, one needs to have $\Phi_G = \Omega(n)$. Recall that the condition $\Phi_G \to \infty$ *a.a.s.* guarantees the existence of at least one copy of G in $\mathbb{G}(n,p)$, and is equivalent to assuming that $n^{-1/m(G)} = o(p)$ (see Theorem 3.4 and Lemma 3.6). Similarly, one can show that $\Phi_G = \Omega(n)$ if and only if $p = \Omega(n^{-1/m^{(1)}(G)})$, where $m^{(1)}(G)$ is defined in (3.17). This is indeed the right threshold.

Theorem 4.9. *For every graph G with at least one edge and for every $\varepsilon > 0$ there are positive constants c and C such that*

$$\lim_{n\to\infty} \mathbb{P}(\mathbb{G}(n,p) \in \mathsf{F}_G(\varepsilon)) = \begin{cases} 0, & \text{if } p \le cn^{-1/m^{(1)}(G)}, \\ 1, & \text{if } p \ge Cn^{-1/m^{(1)}(G)}. \end{cases}$$

Proof. By the monotonicity of $\mathsf{F}_G(\varepsilon)$ we may assume that $\Phi_G \to \infty$. If $p \le cn^{-1/m^{(1)}(G)}$, then $\Phi_G < c'n$, where c' can be made arbitrarily small by picking c small enough. Let H achieve the minimum in Φ_G. Then $\Phi_G = \Phi_H = \mathbb{E}(X_H)$ and, by Chebyshev's inequality and by Lemma 3.5 (see Remark 3.7),

$$\mathbb{P}(|X_H - \mathbb{E}(X_H)| > \tfrac{1}{10}\,\mathbb{E}(X_H)) < \frac{100\,\mathrm{Var}(X_H)}{(\mathbb{E}(X_H))^2} = O\left(\frac{1}{\Phi_H}\right) = o(1).$$

Hence *a.a.s.* $X_H < \frac{11}{10}c'n < \frac{(1-\varepsilon)n}{v_G}$ for c small enough, and it is impossible to cover all but at most εn vertices of $\mathbb{G}(n,p)$ by vertex disjoint copies of G.

Let $p \ge Cn^{-1/m^{(1)}(G)}$ and suppose that $\mathbb{G}(n,p) \notin \mathsf{F}_G(\varepsilon)$. Then there exists a subset of at least εn vertices which does not contain any copy of G. The probability that this happens is, by Theorem 3.9, at most

$$\binom{n}{\lceil \varepsilon n \rceil} \mathbb{P}(\mathbb{G}(\lceil \varepsilon n \rceil, p) \not\supset G) \le 2^n e^{-c''\Phi_G(\lceil \varepsilon n \rceil, p)},$$

where $c'' > 0$ depends on G only. By choosing C large enough, $\Phi_G(\lceil \varepsilon n \rceil, p) \ge (c'')^{-1}n$, and we conclude that the above quantity, and thus also $\mathbb{P}(\mathbb{G}(n,p) \notin \mathsf{F}_G(\varepsilon))$, converges to 0. ■

Thresholds for G-factors

Based on the last result, one can find the threshold for the property of containing a G-factor for a broad class of graphs G. Let $\delta(G)$ stand for the minimum degree of G. The following result was proved independently by Alon and Yuster (1993) and Ruciński (1992a).

Theorem 4.10. *Let G be a graph with v vertices, satisfying $\delta(G) < m^{(1)}(G)$. There are positive constants c and C such that*

$$\lim_{n\to\infty} \mathbb{P}(\mathbb{G}(vn,p) \text{ has a } G\text{-factor}) = \begin{cases} 0, & \text{if } p \leq cn^{-1/m^{(1)}(G)}, \\ 1, & \text{if } p \geq Cn^{-1/m^{(1)}(G)}. \end{cases}$$

Proof. The 0-statement follows immediately from the 0-statement in Theorem 4.9, so we only have to prove the 1-statement. In this proof we utilize the "two-round exposure" technique described in Section 1.1.

For clarity we will demonstrate the proof in the smallest case $G = K_3^+$. Here $m^{(1)}(G) = 3/2$. By Theorem 4.9 (with $\varepsilon = 1/4$), there are *a.a.s.* n disjoint triangles in $\mathbb{G}(4n, p_1)$, where $p_1 = p/2$. This is our property $\mathcal{B}$. Now, fix one graph F satisfying $\mathcal{B}$ and choose one vertex from each triangle of a collection of n vertex disjoint triangles in F.

Let A be the set of chosen vertices, and let B denote the set of vertices not belonging to any of these triangles. If there is a matching M of n edges, each with one endpoint in A and the other one in B, then these edges, together with the triangles, constitute a K_3^+-factor. To prove the existence of M, consider the random bipartite graph $\mathbb{G}(n, n, p_2)$ with vertex classes A and B. As

$$p_2 \geq p/2 = \Omega(n^{-2/3}) \gg \frac{\log n}{n},$$

by Theorem 4.1 there is *a.a.s.* a perfect matching in $\mathbb{G}(n, n, p_2)$.

When $\delta(G) = h > 1$, we treat h-tuples of vertices as single vertices of the bipartite graph (side A) and require that $p^h \gg \frac{\log n}{n}$. This argument remains valid for any G as long as $0 < \delta(G)/m^{(1)}(G) < 1$ (Exercise!). When $\delta(G) = 0$, we are done already after the first round. ■

Remark 4.11. The same technique can be extended to spanning subgraphs other than G-factors. For example, the proof of Theorem 4.10 presented above, gives for free the existence of a "triangle necklace" of length n, that is, a cycle of length n with every vertex adjacent to one vertex of a triangle, the triangles mutually disjoint and also disjoint from the cycle (see Figure 4.4). The reason is that the random graph on the set B has the edge probability p_2 large enough to ensure (*a.a.s.*) the existence of a Hamilton cycle (*cf.* Section 5.1).

Remark 4.12. Alon and Yuster (1993) observed that the technique from the above proof can be used to enlarge the family of graphs G for which $n^{-1/m^{(1)}(G)}$ is a threshold for the property of possessing a G-factor. Indeed, such a family can be recursively constructed by first including all graphs G for which $\delta_G < m^{(1)}(G)$, and then producing new members by splitting an existing member into two unions of its components, and inserting fewer than $m^{(1)}(H)$ edges between them. The proof of this statement is by induction on the number of applications of the above recursive rule (Exercise!). Both sets

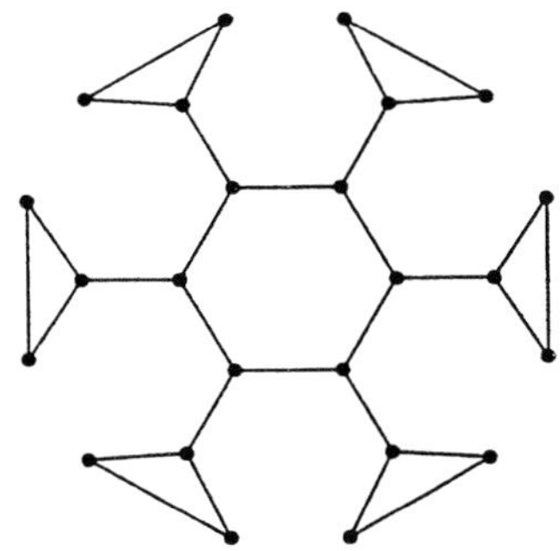

Fig. 4.4 The triangle necklace.

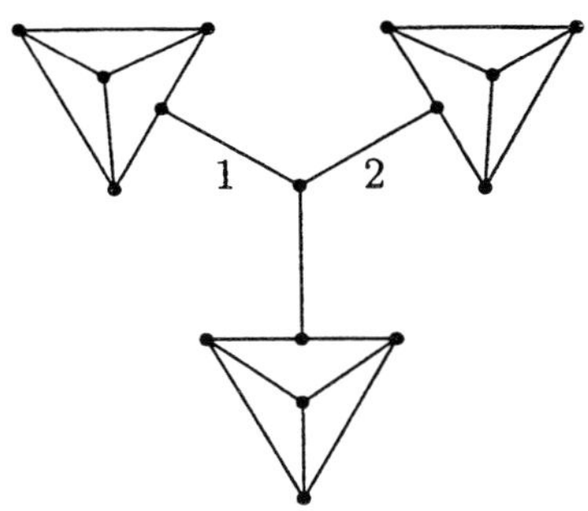

Fig. 4.5 A cubic graph obtained from graphs satisfying $\delta(G) < m^{(1)}(G)$ by the procedure described in Remark 4.12; the added edges are numbered in the order of addition.

of vertices of the auxiliary bipartite graph, A and B, appearing in the proof of Theorem 4.10, can now consist of sets of the original vertices. Surprisingly, some highly regular graphs can be obtained this way (see Figure 4.5).

Despite all these efforts, there are still many graphs for which the threshold for containing a G-factor is not known. Theorem 4.9 provides a lower bound. A better lower bound can sometimes be obtained by looking at the threshold for the property COV_G that every vertex belongs to a copy of G (*cf.* Theorem 3.22), which is a natural necessary condition. Although the conjecture that these two thresholds must always coincide (as is the case of trees) turned out to be false (e.g., the case $G = K_3^+$; see Example 3.24 and Theorem 4.10), it is still plausible to hope that it is so at least for complete graphs. If $G = K_3$, the threshold for the property COV_G is, by Example 3.24, $(\log n)^{1/3} n^{-2/3}$, a logarithmic improvement over the lower bound given by Theorem 4.9.

General spanning subgraphs

A general upper bound on the threshold for the existence of a spanning subgraph in terms of its maximum degree is provided by the following argument from Alon and Füredi (1992). Let H_n be a sequence of n-vertex graphs with maximum degree $\Delta = \Delta_n$, $n = 1, 2, \ldots$. We are interested in the increasing property "$\mathbb{G}(n,p) \supseteq H_n$". The result below says that the threshold does not exceed $(\frac{\log n}{n})^{1/\Delta}$.

Theorem 4.13. *If* $\frac{n}{\Delta^2+1}p^\Delta - \log n \to \infty$, *then*

$$\lim_{n\to\infty} \mathbb{P}(\mathbb{G}(n,p) \supseteq H_n) = 1.$$

In particular, this holds for $p \geq 3(\frac{\log n}{n})^{1/\Delta}$.

Proof. Recall that the Hajnal–Szemerédi Theorem (Hajnal and Szemerédi 1970) (*cf.* Bollobás (1978)) asserts that the vertex set of every graph G can be partitioned into D independent sets of size $\lfloor |V(G)|/D \rfloor$ or $\lceil |V(G)|/D \rceil$, for each integer D satisfying $\Delta(G) < D \leq |V(G)|$. Recall also that in a graph an independent set is *2-independent* if the neighborhoods of its members are mutually disjoint, and that *the square of a graph G* is a graph on the same vertex set with edges between those pairs of vertices which are at distance at most two in G. Now, let $u = \lfloor n/(\Delta^2+1) \rfloor$. By applying the Hajnal–Szemerédi Theorem to the square of H_n one can split $V(H_n)$ into $D = \Delta^2 + 1$ 2-independent sets $U_1, \ldots, U_D$, of size u or $u+1$ (Exercise!).

Let us make a corresponding split $[n] = V_1 \cup \cdots \cup V_D$ with $|V_i| = |U_i|$, $i = 1, \ldots, D$. We will show by induction on i that with probability at least $1 - (i-1)Q$, where $Q = O(u \exp(-up^\Delta))$, the subgraph $\mathbb{G}(n,p)[V_1 \cup \cdots \cup V_i]$ contains a copy of the subgraph $H_n^{(i)} = H_n[U_1 \cup \cdots \cup U_i]$ (property $\mathcal{A}_i$). This is trivial for $i = 1$, and for $i = D$ it implies Theorem 4.13, as $(D-1)Q = O(n \exp(-up^\Delta)) = o(1)$ by assumption.

We expose the edges of $\mathbb{G}(n,p)$ in rounds, first the edges in $[V_1]^2$, then the edges in $[V_1 \cup V_2]^2 \setminus [V_1]^2$, and so forth. Assume that $i \geq 2$ and that $\mathcal{A}_{i-1}$ holds. Set $V = V_1 \cup \cdots \cup V_{i-1}$ and fix a copy of $H_n^{(i-1)}$ on V. (The choice of this copy should depend only on the edges exposed so far.) Let N_x be the set of vertices of V corresponding to the neighbors of $x \in U_i$ in the graph $H_n^{(i)}$.

Consider the auxiliary bipartite random graph with bipartition (U_i, V_i), where an edge is drawn between $x \in U_i$ and $y \in V_i$ if and only if y is adjacent in $\mathbb{G}(n,p)$ to all vertices in N_x (see Figure 4.6). It should be clear now that we are after a perfect matching in this bipartite graph.

Due to the 2-independence of U_i, the appearances of the edges in the auxiliary graph are independent events with probabilities bounded from below by p^Δ. Hence we may look at the random bipartite graph $\mathbb{G}(u_i, u_i, p^\Delta)$ instead, where $u_i = |U_i| \in \{u, u+1\}$. By Remark 4.3, we conclude that with probability at least $1 - O(u \exp(-up^\Delta)) = 1 - Q$ there is a perfect matching in the auxiliary bipartite graph, and thus $\mathcal{A}_i$ holds. Consequently, $\mathbb{P}(\mathcal{A}_i) \geq (1-Q)\,\mathbb{P}(\mathcal{A}_{i-1})$ and, by induction, $\mathbb{P}(\mathcal{A}_i) \geq (1-Q)^{i-1} \geq 1-(i-1)Q$, which completes the proof. ∎

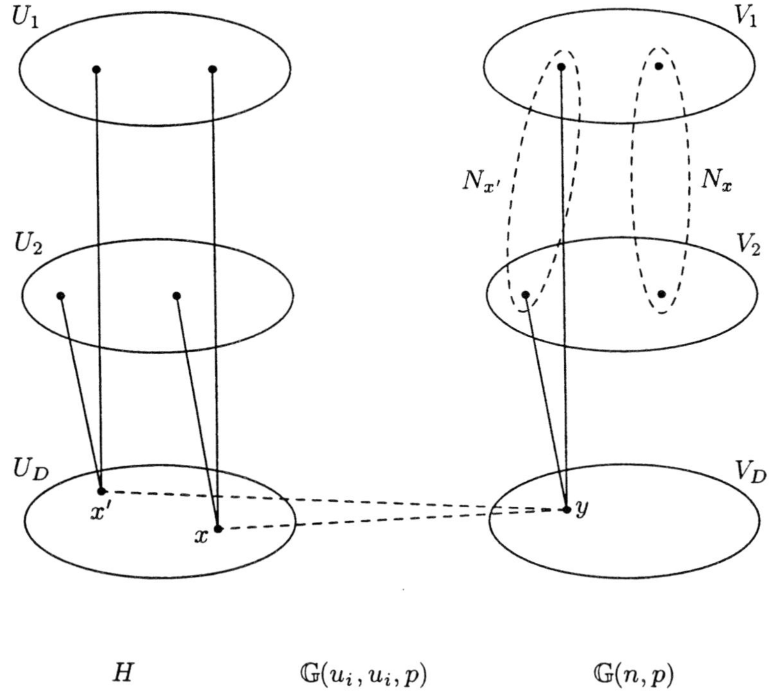

Fig. 4.6 The battleground of the proof of Theorem 4.13.

Example 4.14. The above result is so general that it can be applied to spanning d-cubes. It follows that *a.a.s.* there exists such a d-cube in $\mathbb{G}(n,p)$, $n = 2^d$, if $p > 1/2$ is fixed (Exercise!). Estimates of the expectation have suggested (*cf.* Alon and Füredi (1992)) that the threshold should be around $p = 1/4$ (Exercise!). This has been recently confirmed by Riordan (2000), who applied the second moment method supported by a detailed analysis of variance.

Remark 4.15. In the case in which H_n is a union of disjoint copies of G, that is, when we are after a G-factor, the above bound can be easily improved to $O(\log n/n)^{1/D(G)}$, where $D(G) = \max_{H \subseteq G} \delta_H$ is the degeneracy number of G. Indeed, for any graph G one can order its vertices so that each vertex has at most $D(G)$ neighbors among its predecessors (Exercise!). Now, provided $np^{D(G)} - v(G) \log n \to \infty$, one can repeat the above proof with U_i being the set of the i-th vertices taken from all the copies of G which make up H_n. The sets U_i are clearly 2-independent.

This, however, does not improve the bounds for the threshold for the existence of a K_3-factor. The above results yield only that, ignoring some logarithmic terms, the threshold lies somewhere between $n^{-2/3}$ and $n^{-1/2}$. Real progress on this problem has been made recently by Krivelevich (1996a). In the last section of this chapter we outline his ingenious approach.

4.3 TWO OPEN PROBLEMS

Triangles in graphs and triples in 3-uniform hypergraphs are, undoubtedly, related combinatorial objects. One can build a 3-uniform hypergraph, the edges of which are the triangles of a graph and, conversely, the triples of a 3-uniform hypergraph can be replaced by triangles to form a graph. Two of the most challenging, unsolved problems in the theory of random structures are finding the thresholds for the existence of a K_3-factor in the random graph $\mathbb{G}(n,p)$ and for the existence of a perfect matching (a collection of $n/3$ disjoint triples) in a random 3-uniform hypergraph. The latter, known as the Shamir problem, goes back to at least Schmidt and Shamir (1983). The former, discussed in greater generality in the previous section, was probably first stated in Ruciński (1992b).

Some believe that these two problems are immanently related and a solution of one of them will yield a solution of the other one. Let us point out, however, that in the hypergraph case the triples are (in the binomial model) independent from each other, while the triangles of $\mathbb{G}(n,p)$ are not. What certainly links these two problems is that, after a quiet period, recently significant progress was made with respect to both of them. In this section an account on this progress is given.

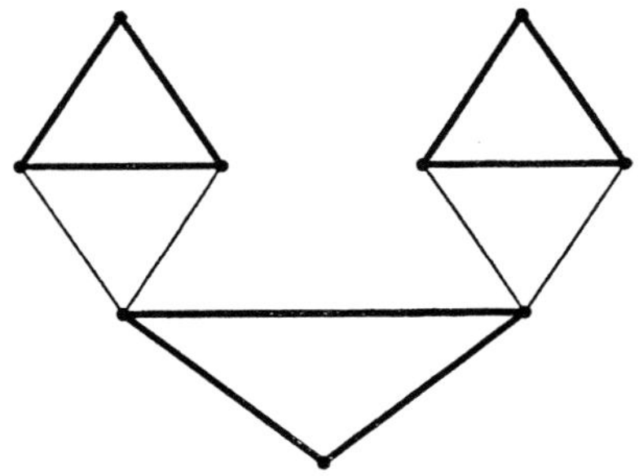

Fig. 4.7 Two diamonds and a vertex, linked by a triangle, contain a K_3-factor (designated by bold lines).

Triangle-factors

We begin with the result of Krivelevich (1996a). Throughout, it is assumed that n is divisible by three.

Theorem 4.16. *There exists a constant $C > 0$ such that if $p = Cn^{-3/5}$, then* a.a.s. *the random graph $\mathbb{G}(n,p)$ contains a K_3-factor.*

Proof (*Outline*). Let us explain the mysterious fraction 3/5 right away. It is simply the reciprocal of the parameter $m^{(1)}(K_4^-) = 5/3$ of the graph K_4^- obtained from the complete graph K_4 by removing one edge. This graph, called here the *diamond*, is the 'building block' of a K_3-factor in the proof. The removal of a vertex of degree two from it leaves a triangle. Therefore, the two vertices of degree two are called *removable*.

The key observation is that two diamonds, D_1 and D_2, and one vertex v, linked together by a triangle with one vertex being v and the other two being removable vertices of, respectively, D_1 and D_2, form a subgraph that contains a K_3-factor (see Figure 4.7).

A naive strategy for proving Theorem 4.16 could thus be as follows (for simplicity we assume that n is divisible by nine). Apply the two-round exposure. In round one, by the 1-statement of Theorem 4.9, with $\varepsilon = 1/9$, *a.a.s.* there is a partial diamond-factor of $\mathbb{G}(n,p)$ consisting of $2n/9$ diamonds. All we need in the second round is to link each of the outstanding $n/9$ vertices, via a triangle, with two diamonds as described above. It seems that we are on the right track, because, for a fixed vertex v, the expected number of such triangles is $O(n^2p^3) = O(n^{0.2})$, and by Theorem 2.18(i), *a.a.s.* there is at least one for each vertex v (Exercise!).

Unfortunately, the $n/9$ pairs of diamonds must be disjoint, so we have to proceed greedily one by one. The problem we immediately face is that toward the end of this procedure the expected number of available triangles drops dramatically. At the very extreme, it is only $4p^3$ for the last vertex. And here comes a second crucial idea. At this late phase we need to use something bigger than diamonds. This bigger structure should have many removable

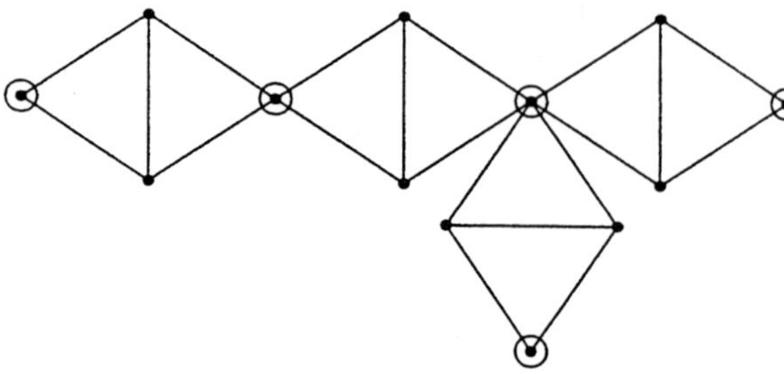

Fig. 4.8 A diamond tree; the removable vertices are designated by open circles.

vertices and be likely to occur frequently in $\mathbb{G}(n,p)$. Both these requirements are provided by *diamond trees*, which we now define recursively.

We call a vertex in a graph *removable* if the removal of this vertex leaves a graph with a K_3-factor. The diamond itself is the smallest diamond tree. Given any diamond tree T and a removable vertex v in it, a new diamond tree is obtained by taking the union of T and a copy of the diamond in which v is a vertex of degree two and the other three vertices do not belong to T. Each time a diamond is added in this way to a diamond tree, the number of vertices increases by three, while the number of removable ones increases by one (the other vertex of degree two in the diamond). Hence, in every diamond tree more than 1/3 of the vertices are removable (Exercise! – see Figure 4.8).

Note that, given two disjoint diamond trees T_1, T_2, and a vertex v, the expected number of triangles with one vertex at v and the other two being removable vertices of, respectively, T_1 and T_2, is at least (Exercise!)

$$\tfrac{1}{9}|V(T_1)||V(T_2)|p^3 = \Theta\big(|V(T_1)||V(T_2)|n^{-9/5}\big).$$

Thus, both diamond trees should be large enough to guarantee the presence of the desired triangle. And we would need many of them.

Since, obviously, one cannot fit too many large, vertex-disjoint subgraphs within the frame of n vertices, we will build the desired K_3-factor in rounds, using originally more, but smaller diamond trees, and gradually turning to the bigger ones. In any case, we need to have many disjoint diamond trees at hand. As the expected number of diamonds is linear in n, on average each vertex belongs to a (large) constant number of them, which allows the diamond trees to grow without any bound. This is specified in the following technical lemma. We refer the reader to Krivelevich (1996a) for the proof.

Lemma 4.17. *If $p = Cn^{-3/5}$, $C > 6$, then for every integer $k = k(n)$ satisfying $4 \le k \le n/6$ and $k \equiv 1 \bmod 3$, the random graph $\mathbb{G}(n,p)$ contains* a.a.s. *$\lfloor n/(6k)\rfloor$ vertex disjoint diamond trees, each of order k.* ∎

Equipped with this lemma, we can now furnish the proof of Theorem 4.16 in just twenty steps. Except for the first, simple step, and for the last one, all these steps are quite similar and, therefore, we organize them into an inductive statement, with the first step serving as a kick-off.

Lemma 4.18. *For every $i = 1, 2, \ldots, 19$, there is a constant C_i such that if $p = C_i n^{-3/5}$ then* a.a.s. *the random graph $\mathbb{G}(n,p)$ contains a partial K_3-factor covering all but at most $n^{1-i/20}$ vertices.*

Before we outline the proof of Lemma 4.18, let us show the last step of the proof of Theorem 4.16. Apply the two-round exposure. After round one with $p = Cn^{-3/5}$, $C > 6$, by Lemma 4.17 there are *a.a.s.* $2n^{0.05}$ vertex disjoint diamond trees, each of order $n^{0.95}/12$ (we ignore floors here). These diamond trees occupy together a vertex set V of size $n/6$.

Round two is generated in two subrounds, as in the proof of Theorem 4.13. First, expose the pairs of $[n] \setminus V$ with $p = C_{19}(5n/6)^{-3/5}$ and conclude by Lemma 4.18 ($i = 19$) applied to $\mathbb{G}([n] \setminus V, p)$, that there is a partial K_3-factor $\mathcal{T}$ covering all but $n^{1/20}$ vertices of $[n] \setminus V$. (We assume for simplicity that $n^{1/20}$ is an integer divisible by three.)

In the second subround expose the pairs with at least one element in V and repeatedly connect each of the uncovered vertices of $[n] \setminus V$ with two diamond trees, by a triangle linking that vertex with one removable vertex in each of the two trees. The probability of failure, by an application of Theorem 2.18(i), is smaller than $n^{1/20}e^{-\Theta(n^{0.1})}$ (Exercise!). Thus, *a.a.s.* all remaining vertices are included to $\mathcal{T}$, creating a K_3-factor of $\mathbb{G}(n,p)$. ■

The proof of Lemma 4.18 is quite similar to the above argument, although a little bit more involved.

Proof of Lemma 4.18. We use induction on i. The case $i = 1$ follows, as in the proof of the 1-statement of Theorem 4.9, by using Theorem 3.9 (Exercise!).

Assume that the lemma is true for some i, $1 \leq i < 19$. Apply the two-round exposure. In round one, with $p = Cn^{-3/5}$, $C > 6$, apply Lemma 4.17 to have *a.a.s.* $2n^{1-i/20} + n^{1-(i+1)/20}$ vertex disjoint diamond trees of order $k \sim n^{i/20}/12$ each. Let us denote the union of the vertex sets of these diamond trees by V. Notice that $|V| \sim n/6$.

In round two take $p = C_i(n - |V|))^{-3/5}$ and first conclude, by the induction assumption, that *a.a.s.* there is a partial K_3-factor $\mathcal{T}$ covering all but $n^{1-i/20}$ vertices of $[n] \setminus V$. Then, reduce the number of uncovered vertices by incorporating them, together with some diamond trees, into $\mathcal{T}$ at a rate of two diamond trees per vertex. Since the diamond trees are smaller now, we seek the linking triangles among all triangles with one vertex being a fixed uncovered vertex of $[n] \setminus V$ and the other two vertices belonging to *any* two diamond trees which are available at this stage.

During this procedure the number of available diamond trees decreases steadily by two, but, owing to the excess we have, even at the end there are still at least $n^{1-(i+1)/20}$ disjoint diamond trees around. Thus, at any given time, the expected number of such triangles is at least of the order

$$\Theta(n^{2-(i+1)/10}n^{i/10}p^3) = \Theta(n^{1/10}) ,$$

and the probability of failure is as small as before (Exercise!). The procedure ends when all vertices of $[n] \setminus V$ are covered by $\mathcal{T}$, leaving at that point $n^{1-(i+1)/20}$ vertex disjoint diamond trees. Each of the diamond trees can be broken into a subgraph with a K_3-factor and a single vertex. Summarizing, *a.a.s.* there is a partial K_3-factor covering all but at most $n^{1-(i+1)/20}$ vertices of $\mathbb{G}(n,p)$. ∎

It is worthwhile to notice that the above proof can be applied to K_r-factors as well as to the Shamir problem; however, in the latter case, it yields a result only slightly stronger than that of Schmidt and Shamir (1983) (*cf.* Krivelevich (1996a)).

Perfect matchings in random hypergraphs

Let $\mathbb{H}_3(n,M)$ be the random hypergraph $([n]^3)_M$, defined as the uniform model of a random subset, where the initial set is the set of all triples of $[n]$. Schmidt and Shamir (1983) showed that if $M/n^{3/2} \to \infty$ then *a.a.s.* there is a perfect matching in $\mathbb{H}_3(n,M)$. This was improved by Frieze and Janson (1995).

Theorem 4.19. *If $M/n^{4/3} \to \infty$ then* a.a.s. *there is a perfect matching in $\mathbb{H}_3(n,M)$.*

Proof (*Outline*). Let us begin by introducing an interim model of a random hypergraph, simpler to analyze than $\mathbb{H}_3(n,M)$, but sufficient for our task. It is based on a random sequence $\mathbf{x} = (x_1,\ldots,x_{3M})$, chosen uniformly at random from the set $\Omega(n,M)$ of all $3M$-element sequences of integers from $[n]$. Then we define a random hypergraph $\mathbb{H}(\mathbf{x})$ as the hypergraph on vertex set $[n]$ with the edges being consecutive triples of elements of $\mathbf{x}$, that is, $\{x_1,x_2,x_3\}$, $\{x_4,x_5,x_6\}$, ... , $\{x_{3M-2},x_{3M-1},x_{3M}\}$. Observe that the hypergraph $\mathbb{H}(\mathbf{x})$ may have repeated edges as well as deficient edges with less than three vertices. Therefore, let $\bar{\mathbb{H}}(x)$ be the hypergraph obtained from $\mathbb{H}(\mathbf{x})$ by deleting both repeated and deficient edges.

For $M' \leq M$, conditioning on the event that $|\bar{\mathbb{H}}(x)| = M'$, $\bar{\mathbb{H}}(x)$ is distributed exactly as $\mathbb{H}_3(n,M')$, because each hypergraph with M' triples and vertex set $[n]$ arises from the same number of sequences $\mathbf{x}$ in this way (Exercise!). Moreover, if $\mathbb{H}(\mathbf{x})$ has a perfect matching then $\bar{\mathbb{H}}(x)$ does (Exercise!). The above two facts, together with the monotonicity of the property of containing a perfect matching, imply that

$$\begin{aligned}
&\mathbb{P}\big(\mathbb{H}(\mathbf{x}) \text{ has a perfect matching}\big) = \mathbb{P}\big(\bar{\mathbb{H}}(x) \text{ has a perfect matching}\big)\\
&\quad = \sum_{M' \le M} \mathbb{P}\big(\bar{\mathbb{H}}(x) \text{ has a perfect matching} \mid |\bar{\mathbb{H}}(x)| = M'\big)\, \mathbb{P}(|\bar{\mathbb{H}}(x)| = M')\\
&\quad = \sum_{M' \le M} \mathbb{P}\big(\mathbb{H}_3(n, M') \text{ has a perfect matching}\big)\, \mathbb{P}(|\bar{\mathbb{H}}(x)| = M')\\
&\quad \le \mathbb{P}\big(\mathbb{H}_3(n, M) \text{ has a perfect matching}\big).
\end{aligned}$$

Thus, our goal is to show that

$$\mathbb{P}\big(\mathbb{H}(\mathbf{x}) \text{ has a perfect matching}\big) \to 1$$

as $M/n^{4/3} \to \infty$. This will be achieved by breaking the space $\Omega(n, M)$ according to the degree sequence.

The degree of an element $v \in [n]$ in a sequence $\mathbf{x}$ is defined as $d_v = \deg_{\mathbf{x}}(v) = |\{i : x_i = v\}|$. For $\varepsilon > 0$, a degree sequence $\mathbf{d} = (d_1, \ldots, d_n)$ is called *ε-smooth* if the degrees $d_1, \ldots, d_n$ do not fluctuate too much, in a precise technical sense for which we refer the reader to Frieze and Janson (1995). It can be routinely proved that *a.a.s.* the sequence $\deg_{\mathbf{x}}(v)$ is $n^{4/3}/M$-smooth.

Hence, to complete the proof it suffices to show that given for each n an ε-smooth sequence $\mathbf{d}$, where $\varepsilon = \varepsilon(n) \to 0$, a random sequence $\mathbf{x}$ chosen uniformly from the family $\mathcal{X}(\mathbf{d}) = \{\mathbf{x} \in \Omega(n, M) : \deg_{\mathbf{x}} = \mathbf{d}\}$ yields *a.a.s.* a random hypergraph $\mathbb{H}(\mathbf{x})$ which contains a perfect matching. (Here the probability of failure should be $o(1)$ uniformly for all ε-smooth sequences $\mathbf{d}$.)

This can be shown using a configuration model similar to that discussed in detail in Chapter 9. In this model, surprisingly, the second moment method works! Indeed, after some tedious calculations, it was shown in Frieze and Janson (1995) that if Y denotes the number of perfect matchings in the random hypergraph $\mathbb{H}(\mathbf{x})$ with $\mathbf{x}$ chosen uniformly from $\mathcal{X}(\mathbf{d})$, then $\mathbb{E}(Y)^2/\mathbb{E}(Y^2) \to 1$ as $n \to \infty$ and thus, by (3.2), $\mathbb{P}(Y > 0) \to 1$. ∎

Remark 4.20. The idea of the proof of Theorem 4.19 relies on the observation that although the second moment method does not apply directly to the unconditional number of perfect matchings (the right-hand side of (3.2) does not tend to zero unless $M/n^{3/2} \to \infty$), it can be used if we first condition on a suitable variable (in this case the degree sequence) which is responsible for most of the variance. A previous instance of combining the second moment method with conditioning was given by Robinson and Wormald (1992), see Chapter 9 of this book.

It is believed that the actual threshold for the existence of a perfect matching in $\mathbb{H}_3(n, M)$ coincides with that for the disappearance of isolated vertices, that is, it occurs around $\frac{1}{3}n \log n$.

Remark 4.21. The fractional version of Shamir's problem asks for the existence of a nonnegative function, defined on the triples of a hypergraph, which totals $n/3$, and which for every vertex totals 1 on all the triples containing that vertex. The existence of a perfect matching is easily seen to be equivalent to the existence of such a function taking the values 0 and 1 only. It was proved by Krivelevich (1996b) that the threshold for the presence of a perfect fractional matching in $\mathbb{H}_3(n, M)$ is roughly $\frac{1}{3}n \log n$. Moreover, Krivelevich also provided the expected hitting time version: in the naturally defined random hypergraph process (see Remark 1.22), *a.a.s.* a perfect fractional matching exists as soon as the last isolated vertex disappears.

5

The Phase Transition

Undoubtedly, the most important and by far the most influential paper about random graphs which has ever appeared was the article of Erdős and Rényi (1960), where the authors studied the changes in the structure of $\mathbb{G}(n, M)$ as M grows from 0 to $\binom{n}{2}$, identifying main features of the evolution of the random graph. A large part of their impressive work was devoted to the phase transition, the spectacular period of the random graph evolution when the size of the largest component of $\mathbb{G}(n, M)$ rapidly grows from $\Theta_C(\log n)$ to $\Theta_C(n)$. In this chapter we try to describe and understand this intriguing phenomenon. We begin with some highlights of the evolution of the random graph and make a "historic" journey reproving, at least partly, Erdős and Rényi's result on the sudden "jump" of the size of the largest component. Then we ourselves jump over twenty years ahead to Bollobás's paper (1984a), which opened a new era of study of the phase transition in $\mathbb{G}(n, M)$. Finally, in the last four sections of the chapter, we present more recent developments concerning various features of the random graph process in this fascinating period.

5.1 THE EVOLUTION OF THE RANDOM GRAPH

The tale of $\mathbb{G}(n, M)$

Let us consider how the properties of $\mathbb{G}(n, M)$ vary when n is fixed but large, and M grows from 0 to $\binom{n}{2}$. Clearly, when the random graph becomes denser its properties change; the moment when a new property appears (or disappears) can often be characterized by the threshold function (see Sections 1.5 and 1.6 for a more elaborate treatment of this subject and the proof that for every monotone property the threshold function exists). Since the publication of Erdős and Rényi (1960), identifying the threshold functions for different properties has been a major task in the theory of random graphs. Nowadays, the threshold functions for most (but by no means all) important graph properties have been found and the picture of the evolution of the random graph is fairly complete.

The beginning stages of the random graph process are easy to study and describe. It is an immediate consequence of Theorem 3.4 that for a fixed $k \geq 2$ and every sequence $M = M(n)$ such that $n^{(k-2)/(k-1)} \ll M \ll n^{(k-1)/k}$, *a.a.s.* $\mathbb{G}(n, M)$ is a forest, which contains copies of all trees of size at most k and no trees with more than k vertices. If $M \ll n$ but $M = n^{1-o(1)}$ then *a.a.s.* $\mathbb{G}(n, M)$ has no cycles (Exercise! – Note that this fact does *not* follow from Theorem 3.4), and the size of the largest component, although clearly unbounded, is $o_p(\log n)$.

The evolution of $\mathbb{G}(n, M)$ for $M = \Theta(n)$ is far more interesting. Let $M = cn/2$ where c is a positive constant. If c is small, *a.a.s.* all components of $\mathbb{G}(n, M)$ are trees or unicyclic, the largest of them having $\Theta_C(\log n)$ vertices. As the process evolves the components increase their size, however, as long as $c < 1$, the largest components of $\mathbb{G}(n, M)$ merge mainly with small trees of size $O_C(1)$; thus they grow slowly and quite smoothly. Nonetheless, at some point of the process, the largest components become so large that it is likely for a new edge to connect two of them. Note that the addition of such an edge can increase the size of the largest component significantly; furthermore, a new component resulting from such a fusion has greater chances to be joined to another component of a similar size. Thus, fairly quickly, all the largest components of $\mathbb{G}(n, M)$ merge into one giant component, much larger than any of the remaining ones. This spectacular phenomenon, now called "the phase transition", is the main theme of the following sections of this chapter. In particular, we will learn that the giant component is formed from smaller ones during the so called *critical period*, or the *critical phase*, where $M = n/2 + O(n^{2/3})$. The critical period separates the *subcritical phase*, where $M - n/2 \ll -n^{2/3}$, from the *supercritical phase*, where $M - n/2 \gg n^{2/3}$. We will also soon see (Theorem 5.4) that the random graphs $\mathbb{G}(n, 0.49n)$ and $\mathbb{G}(n, 0.51n)$ are dramatically different: $\mathbb{G}(n, 0.49n)$ has no components larger than $O_C(\log n)$, while the giant component of $\mathbb{G}(n, 0.51n)$ already has $\Theta_C(n)$ vertices.

As M increases, the giant component of $\mathbb{G}(n, M)$ grows, catching other components of the graph. Because larger components are easier game and they are less frequent than smaller ones, they disappear from the graph earlier, merging with the giant. In particular, if M is about $n \log n/4$, then *a.a.s.* $\mathbb{G}(n, M)$ consists only of the giant component and some number of isolated vertices. Finally, when the last isolated vertex joins the giant, which occurs when $M = n \log n/2 + O_p(n)$, the graph becomes connected. At the very same moment a perfect matching *a.a.s.* can be found in $\mathbb{G}(n, M)$ (provided, of course, that the number of vertices n is even). All these results were already shown by Erdős and Rényi (1959, 1966), but their strongest "hitting time" versions (see Theorem 4.6) were proved much later by Bollobás and Thomason (1985) (see also Bollobás (1985, Chapter VII)).

Soon after the random graph becomes connected, for $M = \frac{n}{2}(\log n + (k-1)\log\log n + O_p(1))$, where $k \geq 2$ is a fixed natural number, the last vertex of degree $k-1$ vanishes and *a.a.s.* at the very same time $\mathbb{G}(n, M)$ becomes k-connected (see Erdős and Rényi (1961) and Bollobás and Thomason (1985)). The question of whether the thresholds for 2-connectivity and Hamiltonicity coincide remained for a long time the major open problem of the theory of random graphs and, finally, was settled in the affirmative by Komlós and Szeméredi (1983) and Bollobás (1984b). (Later, Luczak (1991e) proved that at this threshold $\mathbb{G}(n, M)$ becomes pancyclic, i.e. contains cycles of all lengths $\ell = 3, 4, \ldots, n$.) More generally, let M_k denote the property that a graph G on n vertices contains $\lfloor k/2 \rfloor$ edge-disjoint Hamilton cycles and, if k is odd, a matching of size $\lfloor n/2 \rfloor$, disjoint from the cycles. Then, *a.a.s.* the random graph $\mathbb{G}(2n, M)$ has M_k at the very moment when the last vertex of degree smaller than k disappears (Bollobás and Frieze 1985).

As the process evolves, $\mathbb{G}(n, M)$ becomes denser and denser. Its minimum degree and the connectivity grow, and dense subgraphs gradually appear (see Chapter 3). When $M^d n^{-d-1} = 2^{1-d}\log n + O_p(1)$ and $d \geq 2$, its diameter drops from $d+1$ to d (Burtin (1973), Bollobás (1981a); see also Bollobás (1985, Chapter X)). For $M \sim a\binom{n}{2}$, where $0 < a < 1$ is a constant, any two vertices share $\Theta_C(n)$ neighbors and the largest complete subgraph of $\mathbb{G}(n, M)$ has $\Theta_C(\log n)$ vertices (Theorem 7.1). Finally, for $M = \binom{n}{2}$, $\mathbb{G}(n, M)$ becomes a complete graph.

The k-core of $\mathbb{G}(n, M)$

Note that for the connectivity, the existence of a perfect matching, and the existence of a Hamilton cycle, an obvious necessary condition that the minimum degree is large enough turns out to be *a.a.s.* sufficient for the random graph $\mathbb{G}(n, M)$. How deep is this "probabilistic equivalence" of, say, the property M_k and the property that a graph has minimum degree at least k? This problem can be addressed in two ways. Bollobás, Fenner and Frieze (1990) studied the structure of $\mathbb{G}(n, M)|_{\delta \geq k}$ – a graph chosen at random from the family of all graphs with vertex set $\{1, 2, \ldots, n\}$ and M edges which have minimum degree at least k. In particular, they proved that for a fixed $k \geq 1$, the threshold for the property that $\mathbb{G}(n, M)|_{\delta \geq k}$ has M_k is $M = \Theta_C(n \log n)$ and is related to the fact that at this moment some "local" obstruction for M_k disappears from $\mathbb{G}(n, M)|_{\delta \geq k}$. However, $\mathbb{G}(n, M)|_{\delta \geq k}$ is typically very different from $\mathbb{G}(n, M)$; moreover, there is no obvious way to obtain $\mathbb{G}(n, M+1)|_{\delta \geq k}$ from $\mathbb{G}(n, M)|_{\delta \geq k}$. An alternative, more natural approach was proposed by Bollobás (1984b). It is based on the elementary observation that if a graph G

contains a subgraph H of minimum degree k, then the maximal subgraph of G with this property is unique (Exercise!). We call such a maximal subgraph the *k-core* of G and denote it by $\mathrm{cr}_k(G)$. (If G contains no subgraph with minimum degree k we say that the k-core of G is empty.) Thus, in particular, $\mathrm{cr}_1(G)$ is obtained from G by removing all its isolated vertices, while $\mathrm{cr}_2(G)$ consists of all cycles of G and the paths joining them. Now, instead of $\mathbb{G}(n, M)$, consider the behavior of $\mathrm{cr}_k(\mathbb{G}(n, M))$ as M grows from 0 to $\binom{n}{2}$. Is it *a.a.s.* k-connected? When does the 1-core of $\mathbb{G}(n, M)$ *a.a.s.* contain a matching saturating all but at most one of its vertices? For which M is the 2-core of $\mathbb{G}(n, M)$ *a.a.s.* Hamiltonian?

The study of the evolution of the k-core $\mathrm{cr}_k(\mathbb{G}(n, M))$ was initiated by Bollobás (1984b), who noticed that if $k \geq 3$ is fixed then *a.a.s.* $\mathrm{cr}_k(\mathbb{G}(n, M))$ is k-connected even at very early stages of the evolution, when $M = O(n)$. This result was strengthened by Łuczak (1991d), who proved that for some constant $a > 0$, which depends neither on n nor on k, *a.a.s.* the random graph process is such that for every M, $0 \leq M \leq \binom{n}{2}$, and every k, $3 \leq k \leq n-1$, the k-core of $\mathbb{G}(n, M)$ is either empty, or larger than an and k-connected. Note that, in particular, this result implies that the k-core emerges very rapidly: for almost every graph process one can find a critical moment M_k^{cr} such that the k-core of the graph at the M_k^{cr}-th stage of the process is empty, while at the very next stage the size of the k-core "jumps" to $\Theta_C(n)$. The proof that M_k^{cr} exists is not very hard, especially for large k; the reader is invited to show that for every $\varepsilon > 0$ there exists k_ε such that for $k \geq k_\varepsilon$ we have *a.a.s.* $|M_k^{\mathrm{cr}} - kn/2| \leq \varepsilon n$ (Exercise!). On the other hand, determining M_k^{cr} for a given $k \geq 3$ up to a factor of $1 + o_p(1)$ is a challenging task accomplished only recently by Pittel, Spencer and Wormald (1996), who also provided precise bounds for the size of the k-core at the moment it emerges.

Unlike in the case of k-connectivity, replacing $\mathbb{G}(n, M)$ by its k-core does not help very much with respect to the property M_k. The threshold for $\mathrm{cr}_1(\mathbb{G}(n, M))$ to have a matching which covers all except at most one of its vertices occurs only when $M \sim n \log n/4$ and is related to the existence of pairs of vertices of degree one adjacent to the same vertex (see Theorem 4.4). More generally, for arbitrary k, the k-core of $\mathbb{G}(n, M)$ has property M_k when M is of the order $n \log n$ (Łuczak 1987) – as in the case of $\mathbb{G}(n, M)|_{\delta \geq k}$, at this point a certain local obstruction disappears from $\mathrm{cr}_k(\mathbb{G}(n, M))$. The property that the chromatic number of $\mathbb{G}(n, M)$ is at least $k + 1$, for some $k \geq 3$, does not seem to be related to the existence of any "local" substructure of $\mathbb{G}(n, M)$ at all. Clearly, the k-core of a graph with such a chromatic number must be non-empty (see Lemma 7.6), but the fact that $\mathrm{cr}_k(\mathbb{G}(n, M)) \neq \emptyset$ is not *a.a.s.* sufficient for $\chi(\mathbb{G}(n, M)) \geq k + 1$. For large k this follows immediately from the fact that M_k^{cr} is close to $kn/2$, and thus $\chi(\mathbb{G}(n, M_k^{\mathrm{cr}}))$ is about $k/2 \log k$ (see Theorem 7.16). However, for small k, especially for $k = 3$, the problem whether *a.a.s.* the thresholds for the properties $\mathrm{cr}_k(\mathbb{G}(n, M)) \neq \emptyset$ and $\chi(\mathbb{G}(n, M)) \geq k + 1$ coincide is more involved and was settled (in the negative) only recently by Molloy (1996) and Achlioptas and Molloy (1997) for details see Section 7.5.

5.2 THE EMERGENCE OF THE GIANT COMPONENT

In this section we reprove a part of Erdős and Rényi's theorem on the (un)expected "jump" of the size of the largest component which occurs in the random graph when it has about $n/2$ edges. Although soon we will give a much more precise description of this phenomenon, we feel that the branching process argument used here can provide a better understanding of this feature and, most importantly, it explains why the abrupt change of the structure of the random graph takes place when the average degree of its vertices approaches one.

Branching processes

Since our approach is based on branching processes let us first recall some elementary definitions and facts concerning them. (For proofs and a more elaborate treatment of this topic see Athreya and Ney (1972), or any textbook on probability theory.) Let X be a random variable which takes values in the non-negative integers. The *Galton–Watson branching process* defined by X starts with a single particle, which produces Z_1 other particles, where the number Z_1 of first-generation particles has the same distribution as X. Each of the offspring particles produces, in turn, its own children, whose number has distribution X, independently for each particle, and so on. If by Z_i we denote the number of offspring in the i-th generation, then $Z_0 \equiv 1$, while for $i \geq 1$ the variable Z_i is the sum of Z_{i-1} independent copies of X; clearly, this observation can also be used as an equivalent definition of the random variables Z_i. Note that if $Z_n = 0$ for some n, then $Z_m = 0$ for all $m \geq n$.

The most basic fact about branching processes states that if the expectation of X is larger than one, then with positive probability the process will continue forever, while otherwise, except for the degenerate case, with probability one the process will die out, that is, for some n we have $Z_n = 0$. More precisely, let $f : [0,1] \to \mathbb{R}$ denote the probability generating function of X, defined as

$$f_X(x) = f(x) = \sum_{i\geq 0} x^i \, \mathbb{P}(X = i) .$$

Moreover, let $Z = \sum_{i\geq 0} Z_i$ be the total number of offspring in the branching process. The probability $\rho = \rho_X$ of extinction of the branching process is defined as

$$\rho = \mathbb{P}(Z < \infty) = \lim_{n\to\infty} \mathbb{P}(Z_n = 0) .$$

Then the following holds.

Theorem 5.1. *For* $\mathbb{E}X \leq 1$ *we have* $\rho_X = 1$, *unless* $\mathbb{P}(X = 1) = 1$. *If* $\mathbb{E}X > 1$ *and* $\mathbb{P}(X = 0) > 0$, *then* $\rho_X = x_0$, *where* x_0 *is the unique solution of the equation* $f(x) = x$ *which belongs to the interval* $(0, 1)$. ■

Example 5.2. Let $X \in \mathrm{Po}(c)$. Then

$$f_X(x) = \sum_{i=0}^{\infty} \frac{c^i x^i}{i!} e^{-c} = \exp(c(x-1)).$$

Thus, if $c > 1$, the probability ρ_X that the branching process defined by X dies out is equal to $1 - \beta(c)$, where $\beta = \beta(c) \in (0, 1)$ is uniquely determined by the equation

$$\beta + e^{-\beta c} = 1. \tag{5.1}$$

Example 5.3. Let $Y_n \in \mathrm{Bi}(n, p)$, where $np \to c > 1$ as $n \to \infty$. Since

$$f_{Y_n}(x) = \sum_{i=0}^{n} \binom{n}{i} x^i p^i (1-p)^{n-i} = (1 - p + xp)^n,$$

for every real number x we have

$$\lim_{n\to\infty} f_{Y_n}(x) = \exp(c(x-1)) = f_X(x),$$

that is, the probability generating function of Y_n tends pointwise to the probability generating function of $X \in \mathrm{Po}(c)$, precisely as one might expect. Thus, as $n \to \infty$, the probability of extinction $\rho(n, c)$ of the branching process defined by Y_n converges to $1 - \beta(c)$, where $\beta(c)$ is defined as in (5.1).

The giant component

We will use branching processes to study the rapid growth of the size of the largest component in $\mathbb{G}(n,p)$ – the analysis of the behavior of $\mathbb{G}(n, M)$ is similar, but the fact that in $\mathbb{G}(n,p)$ edges appear independently from each other makes the argument simpler. Thus, let $p = p(n) = c/n$, where c is a positive constant. We reveal the component structure of $\mathbb{G}(n,p)$ step by step, using the following procedure. Choose a vertex v in $\mathbb{G}(n,p)$, find all neighbors $v_1, \ldots, v_r$ of v, and mark v as *saturated.* Then, generate all vertices $\{v_{11}, \ldots, v_{1s}\}$ from $[n] \setminus \{v, v_1, \ldots, v_r\}$ which are adjacent to v_1 in $\mathbb{G}(n,p)$, so v_1 becomes saturated, and continue this process until all vertices in the component of $\mathbb{G}(n,p)$ containing v are saturated.

If during the above procedure we saturate first the vertices which lie closer to v, the process resembles very much the branching process. However, in our case, the number $X_i = X_i(n, m, p)$ of new vertices we add to the component in the i-th step, provided m of its elements have already been found, has binomial distribution $\mathrm{Bi}(n - m, p)$, whereas in the branching process the distribution of the immediate offspring of a particle does not depend on the previous history of the process. Nonetheless, while m is not very large, the process of generating the component containing a given vertex can be closely approximated by the branching process defined by a variable with binomial distribution $\mathrm{Bi}(n,p)$. Thus, one may expect that the probability that a vertex is contained in a "small" component is roughly given by the probability that the process dies out, which happens with probability 1 for $c<1$. On the other hand, if $c>1$, then with some positive probability $1-\rho_c$ the process continues for a long time and thus we may expect that $(1-\rho_c+o(1))n$ vertices of $\mathbb{G}(n,p)$ belong to one giant component.

Theorem 5.4. *Let $np=c$, where $c>0$ is a constant.*

(i) *If $c<1$, then* a.a.s. *the largest component of $\mathbb{G}(n,p)$ has at most $\frac{3}{(1-c)^2}\log n$ vertices.*

(ii) *Let $c>1$ and let $\beta=\beta(c)\in(0,1)$ be defined as in (5.1). Then $\mathbb{G}(n,p)$ contains a giant component of $(1+o_p(1))\beta n$ vertices. Furthermore,* a.a.s. *the size of the second largest component of $\mathbb{G}(n,p)$ is at most $\frac{16c}{(c-1)^2}\log n$.*

Proof. Let us assume first that $pn=c$ and $c<1$. Note that the probability that a given vertex v belongs to a component of size at least $k=k(n)$ is bounded from above by the probability that the sum of $k=k(n)$ random variables X_i is at least $k-1$. Furthermore, X_i can be bounded from above by X_i^+, where all X_i^+ have the same binomial distribution $\mathrm{Bi}(n,p)$ and the random variables $X_1^+,\dots,X_k^+$ are independent; note that for such random variables we have $\sum_{i=1}^k X_i^+\in\mathrm{Bi}(kn,p)$. Thus, from (2.5) we infer that for n large enough the probability that $\mathbb{G}(n,p)$ contains a component of size at least $k\geq 3\log n/(1-c)^2$ is bounded from above by

$$
\begin{aligned}
n\,\mathbb{P}\Big(\sum_{i=1}^k X_i^+\geq k-1\Big) &= n\,\mathbb{P}\Big(\sum_{i=1}^k X_i^+\geq ck+(1-c)k-1\Big)\\
&\leq n\exp\Big(-\frac{((1-c)k-1)^2}{2(ck+(1-c)k/3)}\Big)\leq n\exp\Big(-\frac{(1-c)^2}{2}k\Big)=o(1)\,.
\end{aligned}
$$

Now let $c>1$. Set $k_-=\frac{16c}{(c-1)^2}\log n$ and $k_+=n^{2/3}$. First we will show that *a.a.s.* for every k, $k_-\leq k\leq k_+$, and all vertices v of $\mathbb{G}(n,p)$, either the process described above which starts at v terminates after fewer than k_- steps, or at the k-th step there are at least $(c-1)k/2$ vertices in the component containing v that have been generated in the process but which are not yet saturated. In particular, no component of $\mathbb{G}(n,p)$ has k vertices, with $k_-\leq k\leq k_+$. Note first that in order to check if the process which starts at v produces after each k step at least $(c-1)k/2$ unsaturated vertices in the component containing v, we need only to identify at most $k+(c-1)k/2=(c+1)k/2$ vertices of this component. Hence, as in the previous case, for each

i, where $1 \le i \le k$, we can bound X_i from below by $X_i^- \in \mathrm{Bi}(n - \frac{c+1}{2}k^+, p)$, where all variables X_i^- are independent. Furthermore, the probability that either after the k first steps we produce fewer than $(c-1)k/2$ saturated vertices, or that the process dies out after the first k steps, is smaller than the probability that

$$\sum_{i=1}^{k} X_i^- \le \sum_{i=1}^{k} X_i \le k - 1 + \frac{(c-1)k}{2}.$$

Thus, from the large deviation inequality (2.6) the probability that it happens for some vertex v of $\mathbb{G}(n, M)$ and for some k, $k_- \le k \le k_+$, is, for n large enough, bounded from above by

$$n \sum_{k=k_-}^{k_+} \mathbb{P}\Big(\sum_{i=1}^{k} X_i^- \le k - 1 + \frac{(c-1)k}{2}\Big) \le n \sum_{k=k_-}^{k_+} \exp\Big(-\frac{(c-1)^2k^2}{9ck}\Big)$$
$$\le nk_+ \exp\Big(-\frac{(c-1)^2}{9c}k_-\Big) = o(1).$$

Now let us consider a pair of vertices v' and v'' which belong to components of size at least k_+. What is the probability that they belong to different components? Let us run the process of identifying vertices of the component of $\mathbb{G}(n, M)$ containing v' for the first k_+ steps. According to the fact we have just proved, at the end of this procedure we are left with some set V' of vertices of the component containing v', such that at least $(c-1)k_+/2$ vertices from V' are unsaturated. Let us now run a similar process starting at the vertex v''. Then, either we join v'' to some of the vertices which belong to V', or end up with some set of vertices V'' of the component containing v'', among which at least $(c-1)k_+/2$ have yet to be saturated. Now the probability that there are no edges between as yet unsaturated vertices of V' and V'' is bounded from above by

$$(1-p)^{[(c-1)k_+/2]^2} \le \exp(-(c-1)^2cn^{1/3}/4) = o(1/n^2).$$

Consequently, the probability that $\mathbb{G}(n, M)$ contains two vertices v' and v'' which belong to two different components both of size at least k_+ tends to 0 as $n \to \infty$.

Thus, we have shown that *a.a.s.* the vertices of $\mathbb{G}(n,p)$ can be divided into two classes: "small" ones, which belong to components of size at most k_-, and "large" ones, contained in one large component of size at least k_+. Now to complete the proof we need to estimate the number of small vertices. Observe that the probability $\rho(n,p)$ that a vertex is small is bounded from above by the extinction probability $\rho_+ = \rho_+(n,p)$ of the branching process, in which the distribution of the immediate offspring of a particle is given by the binomial distribution $\mathrm{Bi}(n - k_-, p)$. On the other hand, $\rho(n,p)$ is bounded from below by $\rho_- + o(1)$, where $\rho_- = \rho_-(n,p)$ is the probability of extinction for the branching process with distribution $\mathrm{Bi}(n, p)$ (the term $o(1)$ bounds the probability that the branching process dies after more than k_- steps.) We know (Example 5.3) that if $np = c$ and $n \to \infty$, then both ρ_- and ρ_+ converge to $1 - \beta$, with $\beta = \beta(c) < 1$ defined as in (5.1). Hence the expectation of the number Y of small vertices is $(1 - \beta + o(1))n$. Furthermore,

$$\mathbb{E}(Y(Y-1)) \leq n\rho(n,p)(k_- + n\rho(n - O(k_-), p)) = (1 + o(1))(\mathbb{E}Y)^2 .$$

Hence from Chebyshev's inequality (1.2), $\mathbb{G}(n,p)$ contains $(1 - \beta + o_p(1))n$ small vertices and the assertion follows. ∎

The double jump

Although Theorem 5.4 tells about the behavior of $\mathbb{G}(n,p)$, from the equivalence of $\mathbb{G}(n,p)$ and $\mathbb{G}(n,M)$ (Proposition 1.13) we infer that a similar sudden change of the size of the largest component occurs also for $\mathbb{G}(n,M)$, the model of the random graph used by Erdős and Rényi (1960). Hence, it remains to study the structure of $\mathbb{G}(n,M)$ when the expected degree of each of its vertices is close to 1, that is, for $M \sim n/2$. Erdős and Rényi suggested that in this case a "double jump" occurs: the largest component of $\mathbb{G}(n,M)$ has $\Theta_p(n^{2/3})$ vertices and near the point $M \sim n/2$ the largest component changes its size twice - first from $O_C(\log n)$ to $\Theta_p(n^{2/3})$, and then from $\Theta_p(n^{2/3})$ to $\Theta_C(n)$. (Note that, as was mentioned in the previous section, in the evolution of the k-core of $\mathbb{G}(n,M)$ for $k \geq 3$, a somewhat similar "single jump" can be observed). Thus, in particular, Erdős and Rényi expected that whatever function $M = M(n)$ we choose, the number of vertices $L_1(n,M)$ in the largest component of $\mathbb{G}(n,p)$ can only be either $O_C(\log n)$, or $\Theta_p(n^{2/3})$, or maybe $\Theta_C(n)$. However, it seems that the proof of Theorem 5.4 works also for $c = 1 + \varepsilon(n)$, provided that $\varepsilon(n) > 0$ tends to 0 with n slowly enough. As can be easily checked expanding $\beta(c) = \beta(1+\varepsilon)$ in a Taylor's series, in this case the largest component should *a.a.s.* have about $2\varepsilon n$ vertices (Exercise!).

Let us mention yet another piece of evidence against such an abrupt change in the size of the largest component. Choose any function $r = r(n)$, say $r = n^{4/5}$. Now, for every n and each of the $\binom{n}{2}!$ increasing sequences of graphs $\tilde{G} = (G_0, G_1, \dots, G_N)$, where $N = \binom{n}{2}$ and the graph G_i has i edges for $0 \leq i \leq N$, choose the maximum value $M(\tilde{G})$ such that the largest component in G has at most r vertices. Finally, let $M_r = M_r(n)$ be a median of all $N!$ values of $M(\tilde{G})$. Then, by the choice of M_r,

$$\mathbb{P}(L_1(n, M_r) \leq r) \geq 1/2 ,$$

but at the same time we have also

$$\mathbb{P}(L_1(n, M_r) \geq r/2) \geq \mathbb{P}(L_1(n, M_r + 1) \geq r) \geq 1/2 ,$$

because by adding one edge to a graph we can at most double the number of vertices in the largest component. Hence, $L_1(n,M)$ must grow more or less "smoothly" with n. (Note, by the way, that this argument cannot be applied to the evolution of the k-core, when $k \geq 2$; in this case the addition of one edge can immensely increase the size of the k-core of a graph.) It is somewhat surprising that the fact that near the point $M \sim n/2$ the size of the largest component must grow gradually with M was not noticed, or at least was not studied, for over twenty years. It was addressed only by Bollobás (1984a), who first described in detail the behavior of $\mathbb{G}(n, M)$ for $M \sim n/2$.

Let us also mention that the size of the largest component of $\mathbb{G}(n, M)$ is indeed $\Theta_p(n^{2/3})$ when $M = n/2$ (see Theorem 5.20 below); thus, if $M = cn$ and c is required to be a constant independent of n, as the value of c grows from zero to infinity, at $c = 1$ a "double jump" does occur, precisely as described by Erdős and Rényi.

5.3 THE EMERGENCE OF THE GIANT: A CLOSER LOOK

Theorem 5.4 stated that if $M = cn/2$ and $c < 1$ then *a.a.s.* $\mathbb{G}(n, M)$ consists of small components, while for $c > 1$ *a.a.s.* the structure of $\mathbb{G}(n, M)$ is dominated by one giant component which contains a positive fraction of all vertices. The aim of this section is to study to what extent this description remains true when $2M/n \to 1$. Thus, we show that in the subcritical phase, when $M = n/2 - s$ and $s \gg n^{2/3}$, no component of $\mathbb{G}(n, M)$ is significantly larger than the remaining ones. More precisely, if by $L_r(n, M)$ we denote the number of vertices in the r-th largest component of $\mathbb{G}(n, M)$, then in the subcritical phase we have $L_2(n, M) \geq (1 + o_p(1))L_1(n, M) - 1$ (Theorem 5.6). On the other hand, in the supercritical phase when $M = n/2 + s$ and $s \gg n^{2/3}$, the largest component exceeds by far all its competitors, that is, $L_2(n, M) = o_p(L_1(n, M))$ (Theorems 5.7 and 5.12). Finally, it should be mentioned that a systematic study of the phase transition was started by a remarkable paper of Bollobás (1984a) which described the most characteristic features of this phenomenon.

The subcritical phase

Let us first introduce some notation. A component H of a graph is an ℓ-*component* if it has k vertices and $k + \ell$ edges for some $k \geq 1$; in such a case we call ℓ the *excess* of H. Note that we always have $\ell \geq -1$. Furthermore, $\ell = -1$ only for tree components while each 0-component is unicyclic. We call an ℓ-component *complex* if its excess ℓ is positive, that is, if it contains at least two cycles. Our first result states that if M is much smaller than $n/2$, then *a.a.s.* $\mathbb{G}(n, M)$ contains no complex components.

Theorem 5.5. *Let $M = n/2 - s$, where $s = s(n) \geq 0$. Then, the probability that $\mathbb{G}(n, M)$ contains a complex component is smaller than $n^2/4s^3$. In particular, if $s \gg n^{2/3}$, then* a.a.s. *$\mathbb{G}(n, M)$ contains no complex components.*

Proof. If a graph contains a component with at least two cycles, it must contain a subgraph which either consists of two cycles joined by a path (or sharing a vertex), or is a cycle with a "diagonal path". Let $\bar{X}$ be the number of such subgraphs in $\mathbb{G}(n, M)$. Since on k given vertices one can build no more than $k^2 k!$ of them (Exercise!), we have

$$\begin{aligned}\mathbb{P}(\bar{X} > 0) &\le \sum_{k=4}^{n} \binom{n}{k} k^2 k! \binom{\binom{n}{2} - k - 1}{M - k - 1} \binom{\binom{n}{2}}{M}^{-1} \\ &\le \sum_{k=4}^{n} \frac{k^2}{n} \left(\frac{2M}{n}\right)^{k+1} \le \int_0^\infty \frac{x^2}{n} \exp\left(-\frac{2sx}{n}\right) dx = \frac{n^2}{4s^3}\,. \qquad \blacksquare\end{aligned}$$

In the proof of the above result we estimated the number of complex components very crudely; in order to study the phase transition phenomenon, we will need more precise information on the number of ℓ-components at different stages of the random graph process. Thus, let $Y(k,\ell) = Y_{n,M}(k,\ell)$ denote the random variable which counts ℓ-components of $\mathbb{G}(n, M)$ on k vertices. Then, for the expectation of $Y_{n,M}(k,\ell)$, we have

$$\mathbb{E}\, Y_{n,M}(k,\ell) = \binom{n}{k} C(k,\ell) \binom{\binom{n-k}{2}}{M - k - \ell} \binom{\binom{n}{2}}{M}^{-1}, \tag{5.2}$$

where $C(k,\ell)$ is the number of connected labelled graphs with $k+\ell$ edges on a given set of k vertices. We estimate the value of $\mathbb{E}\, Y(k,\ell)$ using *Stirling's formula*

$$n! = (1 + O(1/n))\sqrt{2\pi n}(n/e)^n\,, \tag{5.3}$$

the expansion of the logarithm which for $0 \le x < 1/2$ gives

$$1 - x = \exp(-x - x^2/2 - x^3/3 - O(x^4))\,, \tag{5.4}$$

and the asymptotic formula for the falling factorial which follows from them

$$(n)_k = n^k \exp\left(-\frac{k^2}{2n} - \frac{k^3}{6n^2} - O\left(\frac{k}{n} + \frac{k^4}{n^3}\right)\right). \tag{5.5}$$

In order to simplify our further calculations, let us assume that $\ell \ge -1$ does not depend on n, $k = O(n^{2/3})$ and $M = n/2 + s$, where $k \ll |s| = o(n)$. Then, using (5.5), from (5.2) we get

$$\begin{aligned}\mathbb{E}\, Y_{n,M}(k,\ell) \sim \frac{C(k,\ell)}{k!} &\left(\frac{n-k}{n}\right)^{2(M-k-\ell)} \frac{(M)_{k+\ell}}{M^{k+\ell}} \\ &\times \left(\frac{2M}{n}\right)^k \left(\frac{2M}{n^2}\right)^\ell \exp\left(-\frac{k^2}{2n} - \frac{k^3}{6n^2}\right),\end{aligned} \tag{5.6}$$

while (5.4) and (5.5) give

$$\begin{aligned}\left(\frac{n-k}{n}\right)^{n+2s-2k-2\ell} &\sim \exp\left(-k+\frac{3k^2}{2n}+\frac{2k^3}{3n^2}-\frac{2sk}{n}-\frac{sk^2}{n^2}\right)\\ \frac{(M)_{k+\ell}}{M^{k+\ell}} &\sim \exp\left(-\frac{k^2}{n}-\frac{2k^3}{3n^2}+\frac{2sk^2}{n^2}+O\left(\frac{s^2k^2}{n^3}\right)\right) \qquad (5.7)\\ \left(\frac{2M}{n}\right)^k &= \exp\left(\frac{2sk}{n}-\frac{2s^2k}{n^2}+O\left(\frac{s^3k}{n^3}\right)\right).\end{aligned}$$

Hence, (5.6) becomes

$$\mathbb{E}\,Y_{n,M}(k,\ell) \sim \frac{C(k,\ell)}{n^\ell\, k!}\exp\left(-k-\frac{2s^2k}{n^2}+\frac{sk^2}{n^2}-\frac{k^3}{6n^2}+O\left(\frac{s^3k}{n^3}\right)\right). \qquad (5.8)$$

Finally, let us recall that the number of forests with vertex set $\{1,2,\ldots,k\}$ which consist of i trees such that vertices $1,2,\ldots,i$ belong to different trees is given by ik^{k-i-1}. Thus, in particular, $C(k,-1)=k^{k-2}$, while for $C(k,0)$ we have (Exercise!)

$$C(k,0)=\sum_{i=3}^{k}\binom{k}{i}\frac{(i-1)!}{2}ik^{k-i-1}=(1+O(1/\sqrt{k}))\sqrt{\pi/8}\,k^{k-1/2}. \qquad (5.9)$$

Although (5.8) applies only for $k=O(n^{2/3})$, it is not hard to show that if $|s|\gg n^{2/3}$ the expected number of isolated trees and unicyclic components larger than $n^{2/3}$ quickly tends to 0 as $n\to\infty$.

Note also that for any given ℓ

$$\mathbb{E}\,Y_{n,M}(k,\ell)\big(Y_{n,M}(k,\ell)-1\big)=\mathbb{E}\,Y_{n,M}(k,\ell)\,\mathbb{E}\,Y_{n-k,M-k-\ell}(k,\ell),$$

while for $k_1\neq k_2$

$$\mathbb{E}\,Y_{n,M}(k_1,\ell)Y_{n,M}(k_2,\ell)=\mathbb{E}\,Y_{n,M}(k_1,\ell)\,\mathbb{E}\,Y_{n-k_1,M-k_1-\ell}(k_2,\ell).$$

Hence, for any $k_-\le k_+=O(n^{2/3})$, the variance of the number of trees or unicyclic components of size k, where $k_-\le k\le k_+$, can be estimated using calculations similar to the ones above. Using a somewhat more precise version of (5.8), the first moment method and Chebyshev's inequality, one can, after some work, arrive at the following result.

Theorem 5.6. *Let $M=n/2-s$, where $s=s(n)$ is such that $n^{2/3}\ll s\ll n$. Moreover, let r be a fixed natural number, which does not depend on n, and finally, let $\alpha=\alpha(n)\le 1/3$ but $\alpha\gg\max\{s/n,\log^{-1/2}(s^3/n^2)\}$. Then, for n large enough, with probability at least $1-(n^2/s^3)^\alpha$ the r-th largest component of $\mathbb{G}(n,M)$ is a tree and*

$$(1-2\alpha)\frac{n^2}{2s^2}\log\frac{s^3}{n^2}\le L_r(n,M)\le(1+2\alpha)\frac{n^2}{2s^2}\log\frac{s^3}{n^2}. \qquad \blacksquare$$

Thus, for $M=n/2-s$, $n^{2/3}\ll s\ll n$, the r largest components of $\mathbb{G}(n,M)$ are all trees with $(1/2+o_p(1))(n^2s^{-2})\log(s^3n^{-2})$ vertices. As a matter of fact, one can use (5.8) to show a much more precise result on the limit distribution of $L_r(n,M)$ (see Łuczak (1990c, 1996)).

The supercritical phase

As we have just proved, in the subcritical phase, when $M = n/2 - s$ and $s \gg n^{2/3}$, *a.a.s.* $\mathbb{G}(n, M)$ consists of small trees and unicyclic components (Theorems 5.5 and 5.6), and thus its structure is rather easy to study. The properties of $\mathbb{G}(n, M)$ in the *supercritical phase*, when $M = n/2 + s$ and $s \gg n^{2/3}$, are much harder to investigate.

Let us start with a simple but profound observation on the formula (5.8): if $k = O(n^{2/3})$, then the leading factor containing s, $\exp(-2s^2k/n^2)$, does not depend on the sign of s. Thus, all estimates of the moments of $Y(k, \ell)$ for $k = O(n^{2/3})$ which are true for the subcritical phase are expected to hold also in the supercritical phase and the behavior of the components of size $k = O(n^{2/3})$ in both $\mathbb{G}(n, n/2 + s)$ and $\mathbb{G}(n, n/2 - s)$ should be similar. We will soon see that this is indeed the case and this vague remark can be stated in a rigorous way as a symmetry rule (Theorem 5.24). Here we only mention that estimates of the moments of $Y_{n,M}(k, \ell)$, similar to those which led us to Theorem 5.6, give the following result.

Theorem 5.7. *Let $M = n/2 + s$, where $s = s(n)$ is such that $n^{2/3} \ll s \ll n$. Furthermore, let $r \geq 1$ be fixed and let $\alpha = \alpha(n) \leq 1/3$ be such that $\alpha \gg \max\{s/n, \log^{-1/2}(s^3/n^2)\}$. Then, for n large enough, with probability at least $1 - (n^2/s^3)^\alpha$ among all trees and unicyclic components of $\mathbb{G}(n, M)$ the r-th largest is a tree, of size contained between $(1/2 - \alpha)n^2 s^{-2} \log(s^3 n^{-2})$ and $(1/2 + \alpha)n^2 s^{-2} \log(s^3 n^{-2})$.* ∎

Now it is time to look at the behavior of complex components of $\mathbb{G}(n, M)$ in the supercritical case. Note that our sketchy proof of Theorem 5.6 is implicitly based on the fact that in the subcritical phase there are many components of sizes close to $L_1(n, M)$. Thus, we can count them, show that their expected number tends to infinity, and then use Chebyshev's inequality to show that their number is close to its expected value. Theorem 5.4 suggests that in the supercritical phase, if $M = n/2 + o(n)$ and the term $o(n)$ is positive and tends to zero slowly enough, then *a.a.s.* $\mathbb{G}(n, M)$ contains precisely one large complex component and so computing moments of $Y_{n,M}(k, \ell)$ does not seem to be of much use. Hence, instead of counting complex components of $\mathbb{G}(n, M)$ at one value of M, we look at the stages of the random graph process $\{\mathbb{G}(n, M)\}_M$ when such components have been created.

Let G be a graph and $\{v, w\}$ be a pair of its vertices which is *not* an edge of G. We call $\{v, w\}$ a *k-internal juncture* if both v and w belong to the same unicyclic component of G of k vertices. Similarly, we call a pair $\{v, w\}$ a *(k_1, k_2)-proper juncture* if v belongs to a unicyclic component on k_1 vertices and w is a vertex of a different unicyclic component on k_2 vertices. Let $Z'_n(M_1, M_2; k)$ $[Z''_n(M_1, M_2; k_1, k_2)]$ denote the number of M's, $M_1 \leq M \leq M_2$, such that the edge added to the graph at the M-th step of the random graph process $\{\mathbb{G}(n, M)\}_M$ is a k-internal juncture $[(k_1, k_2)$-

proper juncture] of $\mathbb{G}(n, M-1)$. Furthermore, let

$$Z_n(M_1, M_2; k) = Z'_n(M_1, M_2; k) + \sum_{k_1=3}^{k/2} Z''_n(M_1, M_2; k_1, k-k_1)$$

and

$$Z_n(M_1, M_2) = \sum_{k=4}^{n} Z_n(M_1, M_2; k)\,.$$

Note that adding a juncture to a graph is the only way of creating a "new" complex component. Thus, at each moment of the random graph process, the number of complex components is bounded from above by the number of junctures added to the graph so far, and for every M, $1 \le M \le \binom{n}{2}$, we have

$$Y(n, M) = \sum_k \sum_{\ell \ge 1} Y_{n,M}(k, \ell) \le Z_n(1, M)\,. \tag{5.10}$$

We use this fact to investigate the behavior of the complex components in $\mathbb{G}(n, M)$. Observe first that

$$\mathbb{E}\, Z'(M+1, M+1; k) = \binom{n}{k} C(k,0) \binom{\binom{n-k}{2}}{M-k} \frac{\binom{k}{2} - k}{\binom{n}{2} - M} \Big/ \binom{\binom{n}{2}}{M}\,,$$

which looks very much like the formula for $\mathbb{E}\, Y(k, 0)$ given in (5.2). Thus, after calculations similar to those given at the beginning of this section, for $k = O(n^{2/3})$ and $M = n/2 + s$, where $s = o(n)$, we arrive at the following analog of (5.8):

$$\begin{aligned}
\mathbb{E} Z'(M+1, M+1; k) &\sim \frac{k^2\, C(k,0)}{n^2\, k!} \\
&\quad \times \exp\Big(-k - \frac{2s^2k}{n^2} + \frac{sk^2}{n^2} - \frac{k^3}{6n^2} + O\Big(\frac{sk^3}{n^3} + \frac{s^3k}{n^3} + \frac{1}{k}\Big)\Big) \\
&\sim \frac{1}{4}\frac{k}{n^2} \exp\Big(-\frac{2s^2k}{n^2} + \frac{sk^2}{n^2} - \frac{k^3}{6n^2} + O\Big(\frac{1}{\sqrt{k}} + \frac{sk^3}{n^3} + \frac{s^3k}{n^3}\Big)\Big)\,.
\end{aligned}$$

Similarly, for k_1 and k_2 such that $k_1 + k_2 = k$, and $M = n/2 + s$, we get

$$\begin{aligned}
\mathbb{E}\, Z''(M+1, M+1; k_1, k_2) &= \frac{1}{1+\delta_{k_1 k_2}} \binom{n}{k_1} \binom{n-k_1}{k_2} \\
&\quad \times C(k_1, 0) C(k_2, 0) \binom{\binom{n-k_1-k_2}{2}}{M - k_1 - k_2} \frac{k_1 k_2}{\binom{n}{2} - M} \Big/ \binom{\binom{n}{2}}{M} \\
&\sim \frac{1}{1+\delta_{k_1 k_2}} \frac{1}{8n^2} \exp\Big(-\frac{2s^2k}{n^2} + \frac{sk^2}{n^2} - \frac{k^3}{6n^2} \\
&\qquad + O\Big(\frac{1}{\sqrt{k_1}} + \frac{1}{\sqrt{k_2}} + \frac{sk^3}{n^3} + \frac{s^3k}{n^3}\Big)\Big)\,.
\end{aligned}$$

Hence, for such k and M,

$$\mathbb{E}\,Z(M+1, M+1; k) \sim \frac{5k}{16n^2}\exp\Big(-\frac{2s^2k}{n^2} + \frac{sk^2}{n^2} - \frac{k^3}{6n^2} + O\Big(\frac{1}{\sqrt{k}} + \frac{sk^3}{n^3} + \frac{s^3k}{n^3}\Big)\Big). \quad (5.11)$$

We will use (5.11) to prove the following result on complex components (Janson 1993).

Theorem 5.8. *Let $\omega = \omega(n) > 1$ be a function of n.*

(i)
$$\lim_{n\to\infty} \mathbb{E}\,Z\Big(1, \binom{n}{2}\Big) = \frac{5\pi}{8\sqrt{3}} = 1.134\ldots. \quad (5.12)$$

In particular,

$$\liminf_{n\to\infty} \mathbb{P}\Big(Z\Big(1, \binom{n}{2}\Big) = 1\Big) \geq 2 - \frac{5\pi}{8\sqrt{3}} = 0.865\ldots, \quad (5.13)$$

and, for n large enough,

$$\mathbb{P}\Big(Z\Big(1, \binom{n}{2}\Big) \geq \omega\Big) \leq \frac{2}{\omega}. \quad (5.14)$$

(ii) *The probability that during the random graph process $\{\mathbb{G}(n, M)\}_M$ for some M, $0 \leq M \leq \binom{n}{2}$, the graph $\mathbb{G}(n, M)$ contains a complex component with fewer than $n^{2/3}/\omega$ vertices is, for large n, smaller than $1/\omega$.*

(iii) *If $M_\pm = n/2 \pm \omega n^{2/3}$, then, for large n,*

$$\mathbb{P}\Big(Z\Big(1, M_-\Big) + Z\Big(M_+, \binom{n}{2}\Big) > 0\Big) \leq 1/\omega. \quad (5.15)$$

Remark 5.9. Janson (1993) used the method of moments (Section 6.1) to show that the random variable $Z(1, \binom{n}{2})$ converges in distribution and so the limit in (5.13) exists (for its value see Theorem 5.29, below).

Proof. Elementary but somewhat tedious calculations, which we omit here, show that the main contribution to

$$\mathbb{E}\,Z\Big(1, \binom{n}{2}\Big) = \sum_k \sum_M \mathbb{E}\,Z(M+1, M+1; k)$$

comes from the terms for which $k = O(n^{2/3})$ and $M = n/2 + O(n^{2/3})$. Thus, using (5.11) and replacing the sums over k and s with integrals over $x = sn^{-2/3}$ and $y = kn^{-2/3}$ one arrives at

$$\mathbb{E}\, Z\Big(1, \binom{n}{2}\Big) = (1+o(1))\frac{5}{16}\int_0^\infty \int_{-\infty}^\infty y \exp(-2x^2y + xy^2 - y^3/6)\, dx\, dy$$
$$= (1+o(1))\frac{5}{16}\int_0^\infty \int_{-\infty}^\infty y \exp\big(-2y(x-y/4)^2 - y^3/24\big)\, dx\, dy\,,$$

which, after elementary calculations, gives (5.12). Note that the random graph process ends with a complete graph, so $Z_n(1, \binom{n}{2}) \geq 1$. This fact, together with (5.12), implies (5.13), while (5.14) is an immediate consequence of (5.12) and Markov's inequality (3.1). Furthermore, (ii) follows using the estimate

$$\frac{5}{16}\int_0^{1/\omega} \int_{-\infty}^\infty y \exp(-2x^2y + xy^2 - y^3/6)\, dx\, dy < 1/\omega\,.$$

Finally, in order to show (iii) we observe that

$$\frac{5}{16}\int_0^{\infty} \int_{|x|\geq\omega} y \exp(-2x^2y + xy^2 - y^3/6)\, dx\, dy < 1/\omega\,. \qquad \blacksquare$$

Let us note an important consequence of the above statement.

Theorem 5.10. *Let $M \geq n/2 + s$, where $s \gg n^{2/3}$. Then with probability at least $1 - 6n^{2/9}/s^{1/3}$ the random graph $\mathbb{G}(n, M)$ contains exactly one complex component.*

Proof. Let $M_- = n/2 + \lfloor s/2 \rfloor$ and set $\omega = s^{1/3}n^{-2/9}$. Theorem 5.8(i,ii) implies that with probability at least $1 - 3/\omega$, $\mathbb{G}(n, M_-)$ contains at most ω complex components, each of at least $n^{2/3}/\omega$ vertices. On the other hand, Theorem 5.8(iii) states that with probability at least $1 - 2/\omega$ no new complex components appears in the random graph process after the moment M_-. Thus, since the final stage of the random graph process, the complete graph on n vertices, has a positive excess (provided $n \geq 4$), there must be at least one complex component in $\mathbb{G}(n, M_-)$.

Now consider the graphs $\mathbb{G}(n, M_-)$ and $\mathbb{G}(n, M)$ as two stages of the same random graph process. Given that $\mathbb{G}(n, M_-)$ is as in the preceding paragraph, the probability that some pair of complex components of $\mathbb{G}(n, M_-)$ is not joined in $\mathbb{G}(n, M)$ by at least one edge is bounded from above by

$$\binom{\omega}{2}\binom{\binom{n}{2} - M_- - n^{2/3}\omega^{-1}n^{2/3}\omega^{-1}}{M - M_-} \Big/ \binom{\binom{n}{2} - M_-}{M - M_-}$$
$$\leq \tfrac{1}{2}\omega^2 \exp\Big(-\frac{s}{2}n^{4/3}\omega^{-2} \Big/ \binom{n}{2}\Big) \leq \tfrac{1}{2}\omega^2 \exp(-\omega) \leq 1/\omega\,.$$

Since, as we have already observed, with probability at least $1 - 2/\omega$, no new complex component is created during the process after M_- step, the component obtained from merging all complex components of $\mathbb{G}(n, M_-)$ is the only complex component of $\mathbb{G}(n, M)$. $\blacksquare$

Theorems 5.7, 5.8, and 5.10 tell us that in the supercritical case $\mathbb{G}(n, M)$ consists of some number of small trees and unicyclic components of size $o_p(n^{2/3})$ each, and one complex component of size $\Omega_p(n^{2/3})$. Thus, it remains to estimate the size of the largest component more precisely. In order to do it we follow the original argument of Bollobás (1984a), who used the formula (5.6) to compute the number of vertices which are contained in small components; this, in turn, will give us a precise estimate of the size of the giant complex component. The proof of Bollobás's result, stated below as Theorem 5.11, is very ingenious and quite complicated; thus, instead of presenting it here in full detail, we just say a few words on the main idea behind it.

One can easily check that in the subcritical phase, where $M_- = n/2 - \bar{s}$, and $\bar{s} \gg n^{2/3}$, only a negligible number of vertices (more precisely, $O_p(n^2/\bar{s}^2)$ of them) are contained in unicyclic components and most of the vertices of $\mathbb{G}(n, M)$ belong to isolated trees. Hence, for such an M_-, the value of

$$f(n, M_-) = \sum_{k=1}^{\lceil n^{2/3} \rceil} k \, \mathbb{E}\, Y_{n,M_-}(k, -1)$$

is very close to n. Now suppose that we would like to estimate the value of $f(n, M_+)$ for $M_+ = n/2 + s$. If follows from (5.6)–(5.8) that the leading terms which contain s are: $\exp(-2sk/n)$, which comes from $(1 - k/n)^{n+2s-2k}$, and $(2M/n)^k = (1 + 2s/n)^k$. Thus, let us choose $\bar{s} > 0$ in such a way that

$$\Big(1 - \frac{2\bar{s}}{n}\Big) \exp\Big(\frac{2\bar{s}}{n}\Big) = \Big(1 + \frac{2s}{n}\Big) \exp\Big(-\frac{2s}{n}\Big). \tag{5.16}$$

Then the terms of $f(n, M_+)$ become very similar to those of $f(n, M_-)$, which, as we know, is roughly equal to n. As a matter of fact, the main difference comes from the factor $(2M/n)^\ell$ in (5.6); now, when $\ell = -1$, we may expect that

$$\frac{n/2 - \bar{s}}{n/2 + s} = \frac{M_-}{M_+} \simeq \frac{f(n, M_+)}{f(n, M_-)} \simeq \frac{f(n, M_+)}{n},$$

where here by $a_n \simeq b_n$ we mean that a_n and b_n agree up to the second-order term, in our case $1 - a_n \sim 1 - b_n$. Bollobás (1984a) (see also Bollobás (1985, Chapter VI)) showed that, indeed, the value of the three fractions above are very close to each other; furthermore, he used Chebyshev's inequality to prove that the expectation $f(n, M_+)$ closely approximates the number of vertices contained in small trees of $\mathbb{G}(n, M_+)$.

Theorem 5.11. *Let $M = n/2 + s$, where $s = s(n) \gg n^{2/3}$, and let $\bar{s}$ be the function defined in (5.16). Then, for any $\omega = \omega(n) \to \infty$ and large enough n, with probability at least $1 - 1/\omega$*

$$\sum_{k=3}^{\lceil n^{2/3} \rceil} k Y_{n,M}(k, 0) \le \omega \frac{n^2}{s^2}$$

and

$$\Big|\sum_{k=1}^{\lceil n^{2/3}\rceil} kY_{n,M}(k,-1) - \frac{n-2\bar{s}}{n+2s}n\Big| \le \omega\frac{n}{\sqrt{s}}\,. \qquad \blacksquare$$

Let us remark that, roughly, the first part of Theorem 5.11 says that the total number of vertices contained in small unicyclic components is $O_p(n^{2/3})$ and thus negligible. Furthermore, since from the Taylor expansion we get

$$\frac{\bar{s}}{n} = \frac{s}{n} - \frac{4}{3}\frac{s^2}{n^2} + O\Big(\frac{s^3}{n^3}\Big), \tag{5.17}$$

Theorem 5.11 implies that if $s \gg n^{2/3}$, then, up to an error of $o_p(n^{2/3})$,

$$n - \frac{n-2\bar{s}}{n+2s}n = \frac{2(s+\bar{s})n}{n+2s} = 4s + O\Big(\frac{s^2}{n}\Big) \tag{5.18}$$

vertices belong to components which are either larger than $n^{2/3}$ or contain more than one cycle.

From Theorems 5.7, 5.10, and 5.11 we immediately get the main result of this section characterizing the structure of $\mathbb{G}(n,M)$ for $M = n/2+s$, where $n^{2/3} \ll s \ll n$. The theorem below was proved by Bollobás (1984a), under the somewhat stronger assumption that $M - n/2 \ge n^{2/3}\sqrt{\log n}/2$, which was later replaced by $M - n/2 \gg n^{2/3}$ by Łuczak (1990c). (Although Theorem 5.7 is valid only for $s \ll n$, we state the result for all $s \gg n^{2/3}$, so it covers the whole supercritical phase of the evolution of the random graph.)

Theorem 5.12. *Let $M = n/2+s$, where $s = s(n) \gg n^{2/3}$ and let $\bar{s}$ be defined as in (5.16). Then, for large enough n, with probability at least $1-7n^{2/9}s^{-1/3}$,*

$$\Big|L_1(n,M) - \frac{2(s+\bar{s})n}{n+2s}\Big| \le \frac{n^{2/3}}{5}$$

and the largest component is complex, while all other components are either trees or unicyclic components, smaller than $n^{2/3}$. $\blacksquare$

Let us mention that the proof of Theorem 5.12 presented in Bollobás (1984a) (and Łuczak (1990c)) was slightly different and relied strongly on estimates for the expected number of complex components in the supercritical phase. In order to evaluate $\mathbb{E}\,Y_{n,M}(k,\ell)$, Bollobás had to find a way to deal with $C(k,\ell)$, which appears in the formula (5.2). In Bollobás (1984a) he obtained a particularly useful upper bound for $C(k,\ell)$, showing that, for some absolute constant A, and $k,\ell \ge 1$,

$$C(k,\ell) \le \Big(\frac{A}{\ell}\Big)^{\ell/2} k^{k+(3\ell-1)/2}\,. \tag{5.19}$$

Łuczak (1990b) observed that the right-hand side of the above inequality with $A = e/12$ approximates the value of $C(k,\ell)$ quite precisely, as long as ℓ is large but $\ell = o(k)$; at the same time, Bender, Canfield and McKay (1990) found a fairly complicated asymptotic formula for $C(k, \ell)$ for every function $\ell = \ell(k)$ as $k \to \infty$. One can use their powerful result to provide a description of the asymptotic distribution of $L_1(n, M)$ better than that given by Theorem 5.12. For instance, one can show that for $M = n/2 + s$, where $n^{2/3} \ll s = O(n)$, the distribution of the random variable $L_1(n, M)$ is asymptotically normal. The idea of the proof is very simple. For a constant r, $0 < r < 1$, choose a function $k_r = k_r(n, M)$ such that for the random variable

$$Y_r = Y_r(n, M) = \sum_{\ell \geq 1} \sum_{k \leq k_r} Y_{n,M}(k, \ell)$$

we have $\mathbb{E}\, Y_r \to r$, and for each $i \geq 2$ the factorial moment $\mathbb{E}(Y_r)_i$ tends to 0 as $n \to \infty$. Then, from a special (and, in fact, obvious) case of the method of moments (see Theorem 6.7) we infer that

$$\mathbb{P}(Y_r = 1) \sim \mathbb{P}(L_1(n, M) \leq k_r) \to r$$

as $n \to \infty$. We should remark, however, that, because the formula for $C(k, \ell)$ is quite involved, finding $k_r(n, M)$ in this way is a long and not very exciting task (see Pittel (1990) for another approach to this problem).

5.4 THE STRUCTURE OF THE GIANT COMPONENT

In this section we consider the behavior of $\mathbb{G}(n, M)$ in the "early supercritical phase", when $M = n/2 + s$ and $n^{2/3} \ll s \ll n$. In particular, we study the structure of the giant component soon after it emerges. Unfortunately, most of the results given here have lengthy and complicated proofs; thus, this part of the chapter consists mainly of heuristic arguments which, hopefully, shed some light on the nature of this intriguing period of the evolution of the random graph.

We already know that in the supercritical phase the largest component is complex, but what we can say about its excess $\kappa(n, M)$? Note that *a.a.s.* no new complex component can emerge in the supercritical phase (Theorem 5.8). Thus, an edge added to $\mathbb{G}(n, M)$ in the supercritical stage can increase the value of $\kappa(n, M)$ only by one, if it either connects two vertices of the largest component, or joins the giant to one of the unicyclic components. We first estimate the number $\kappa'(n, M)$ of edges which, at the moment they are added to the graph, have both ends in the largest component of the graph. It is not hard to check (Exercise!) that for $M_0 = n/2 + O(n^{2/3})$ we have $\mathbb{E}\, \kappa'(n, M_0) = O(1)$ and thus $\kappa'(n, M_0) = O_p(1)$. Hence, it is enough to study the evolution of $\mathbb{G}(n, M)$ for $M - n/2 \gg n^{2/3}$.

Let $M_i = n/2 + i$. Clearly, $\kappa'(n, M)$ is the sum of the random variables X_i, where $X_i = 1$ if the edge added at the M_i-th stage of the random process is contained in the largest component of $\mathbb{G}(n, M_i - 1)$, and $X_i = 0$ otherwise.

Since we restrict ourselves to the supercritical period of the random graph process, when, by Theorem 5.12 and (5.18), the size of the largest component of $\mathbb{G}(n, M_i)$ is close to $4i$, we have

$$\mathbb{E}\, X_i = \mathbb{P}(X_i = 1) \sim \frac{\binom{4i}{2}}{\binom{n}{2} - n/2 - i} \sim \frac{16i^2}{n^2}\,,$$

and thus

$$\mathbb{E}\, \kappa'(n, M) \sim \sum_{i=\lfloor n^{2/3} \rfloor}^{s} \mathbb{E}\, X_i \sim \sum_{i=\lfloor n^{2/3} \rfloor}^{s} \frac{16i^2}{n^2} \sim \frac{16s^3}{3n^2}\,.$$

Furthermore, it is easy to check that if $M_i = n/2 + i$ and $i \gg n^{2/3}$, then $\Theta_p(n^2/i^2)$ vertices of $\mathbb{G}(n, M_i - 1)$ belong to unicyclic components. Thus, arguing as before, the expected number of edges added to the process before the moment $M = n/2 + s$, which joined the largest component of the graph with one of the unicyclic components can be bounded from above by

$$\sum_{i=\lfloor n^{2/3} \rfloor}^{s} \frac{4i\, O(n^2/i^2)}{\binom{n}{2} - n/2 - i} = O(\log(s/n^{2/3})) \ll s^3/n^2\,.$$

Thus, we expect that if $M = n/2 + s$ and $n^{2/3} \ll s \ll n$, then the value of $\kappa(n, M)$ should be about $16s^3/3n^2$. The following theorem by Łuczak (1990c) states that this is indeed the case. (See also Janson, Knuth, Łuczak and Pittel (1993, Lemma 5) for a sharper concentration result when $s \le 0.5n^{3/4}$.)

Theorem 5.13. *Let $\omega = \omega(n) \to \infty$ and $M = n/2 + s$, where $s \ge n^{2/3}\omega$ but $s \le n/\omega$. Then, with probability at least $1 - \omega^{-0.1}$,*

$$\left| \kappa(n, M) - \frac{16s^3}{3n^2} \right| \le \omega^{-0.1} \frac{s^3}{n^2}\,. \qquad \blacksquare$$

Remark 5.14. The random variable $\kappa(n, M)$, appropriately normalized, has asymptotically normal distribution (see Janson, Knuth, Łuczak and Pittel (1993, Theorem 13); one can also show this fact using the "straightforward approach" described at the end of the previous section). Then, for the standardization of $\kappa(n, M)$, instead of $16s^3/3n^2$ one should use the more precise expression $2(s^2 - \bar{s}^2)/(n + 2s)$ with $\bar{s}$ defined by (5.16), which closely approximates the value of $\kappa(n, M)$ for all $M = n/2 + s$, where $n^{2/3} \ll s = O(n)$.

Before we describe the internal structure of the giant component let us recall that the *2-core* or simply the *core* of a graph G, denoted by $\mathrm{cr}(G)$, is the maximal subgraph of G with minimum degree two. By the *kernel* $\ker(G)$ of a graph G whose components are all complex we mean a multigraph, possibly with loops, obtained from the core of G by replacing each path whose internal vertices are all of degree two by a single edge (see Figure 5.1). Note that the

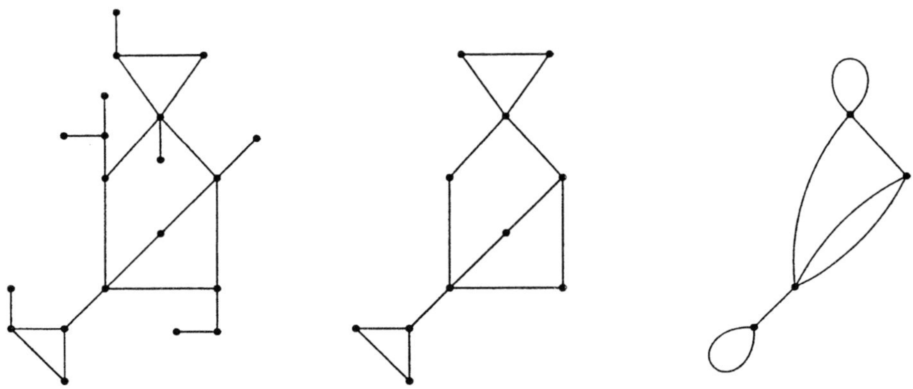

Fig. 5.1 A graph, its core and its kernel.

minimum degree of a kernel is at least three and both the core and the kernel of G have the same excess as the graph itself.

Let us consider first the behavior of $|\operatorname{cr}^L(n,M)|$, the random variable which counts vertices in the core of the largest component of $\mathbb{G}(n,M)$ for $M = n/2+s$, where $n^{2/3} \ll s \ll n$. Theorems 5.12 and 5.13 state that for such an M the largest component of $\mathbb{G}(n,M)$ has $k = (4+o_p(1))s$ vertices and $k+\ell$ edges, where $\ell = (16+o_p(1))s^3/3n^2$. Furthermore, once we condition on the event that the largest component $\mathcal{L}(n,M)$ of $\mathbb{G}(n,M)$ has vertex set V' and $k+\ell$ edges, all connected graphs with vertex set V' and $k+\ell$ edges are equally likely to appear as $\mathcal{L}(n,M)$. Thus, the size and structure of $|\operatorname{cr}^L(n,M)|$ should be similar to that of the core of $\mathbb{C}(k,\ell)$, the graph chosen uniformly at random from the family of all *connected* graphs with vertex set $[k]$ and $k+\ell$ edges, provided $k \sim 4s$ and $\ell \sim 16s^3/3n^2$. One can show that if $\ell = \ell(k)$ tends to infinity as $k \to \infty$, then with probability tending to 1 as $k \to \infty$, all except a negligible fraction of edges in the core of $\mathbb{C}(k,\ell)$ lie on cycles. Hence, since there are no cycles outside the core, the number of edges in the core of $\mathbb{C}(k,\ell)$ should be roughly the same as the number of edges of $\mathbb{C}(k,\ell)$ whose removal does not disconnect the graph. Note that deleting such an edge from a connected graph on k vertices and $k+\ell$ edges results in a connected graph on the same number of vertices and $k+\ell-1$ edges. Furthermore, each such graph can be obtained as the result of this deletion procedure from precisely $\binom{k}{2}-k-\ell+1$ connected graphs with $k+\ell$ edges. Hence, the average size of the core of $\mathbb{C}(k,\ell)$ should be close to

$$\binom{k}{2}\frac{C(k,\ell-1)}{C(k,\ell)},$$

which, because of (5.19) and the comment following it, is roughly equal to $\sqrt{3k\ell}$. Consequently, as long as $\ell(k) = o(k)$, we may expect that the core of $\mathbb{C}(k,\ell)$ contains about $\sqrt{3k\ell}-\ell \sim \sqrt{3k\ell}$ vertices, which, in turn, gives $8s^2/n$

as an estimate for $|\operatorname{cr}^L(n,M)|$, for $M = n/2 + s$ and $n^{2/3} \ll s \ll n$. As we will soon see this is indeed the case.

Since so far our intuitive argument works quite well, let us proceed further and estimate the number of vertices of degree three and four in $\operatorname{cr}^L(n,M)$. Recall that $\kappa(n,M) = (1+o_p(1))16s^3/3n^2$, so the number of vertices of degree at least three in the core (and thus in the kernel) is at most $2\kappa(n,M) \sim 32s^3/3n^2$. Furthermore, if the core of the largest component has roughly $8s^2/n$ vertices, the average size of a tree rooted at any vertex of the core is about $n/2s$. A moment of reflection reveals that the most probable way to obtain a vertex of degree four in the core of the largest component is to connect a vertex from a tree rooted at a vertex of degree three in the core with some other vertex of the giant. The probability that it happens at the M-th stage of the process, where $M = n/2 + s$, should be close to

$$4s\,\frac{32s^3}{3n^2}\,\frac{n}{2s}\Big/\binom{n}{2} \sim \frac{128s^3}{3n^3}.$$

Thus, the expected number of vertices of degree at least four in the core created in the process in the first $M = n/2 + s$ stages can be approximated by

$$\sum_{t=\lfloor n^{2/3}\rfloor}^{s} \frac{128t^3}{3n^3} \sim \frac{32s^4}{3n^3}.$$

Note that the number of vertices of degree at least four in the core is, for $s = o(n)$, much smaller than the number of vertices of degree at least three. Thus, we should expect that the core contains roughly $32s^3/3n^2$ vertices of degree three in it. Observe also that the expectation of the number of vertices of degree at least four tends to 0 for $s \ll n^{3/4}$; thus, they do not appear in the process until the moment when $M = n/2 + \Omega(n^{3/4})$. The same heuristic can be used to guess the number of vertices of degree i in the core. Nonetheless, we should mention that the following result from Łuczak (1991a), which confirms our speculations, has been proved by somewhat different techniques.

Theorem 5.15. *Let $M = n/2 + s$, where $n^{2/3} \ll s \ll n$, and, for $i \geq 2$, let $D_i = D_i(n,M)$ denote the random variable which counts vertices of degree i in the core of $\mathbb{G}(n,M)$. Then*

$$D_2 = (1 + o_p(1))\frac{8s^2}{n},$$

$$D_3 = (1 + o_p(1))\frac{32s^3}{3n^2}.$$

Moreover, for a given $i \geq 4$, $D_i = O_p\big(s^i/n^{i-1}\big)$. If $s \ll n^{1-1/i}$, a.a.s. *no vertex of $\mathbb{G}(n,M)$ has i neighbors (or more) in $\operatorname{cr}(\mathbb{G}(n,M))$, while for $n^{1-1/i} \ll s \ll n$, we have*

$$D_i = (1 + o_p(1))\frac{2^{2i}}{i!}\frac{s^i}{n^{i-1}}. \qquad \blacksquare$$

Remark 5.16. Since in the supercritical phase *a.a.s.* $\mathbb{G}(n,M)$ contains no complex components except the giant one, *a.a.s.* all vertices which have more than three neighbors in the core belong to the largest component of the random graph. Furthermore, the expected number of vertices that belong to the cycles in the unicyclic components is of the order n/s (Exercise!). Thus, in the supercritical phase $|\operatorname{cr}(\mathbb{G}(n,M))| = (1+o_p(1))|\operatorname{cr}^L(n,M)|$, and the vast majority of the vertices of $\mathbb{G}(n,M)$ which have two neighbors in the core of $\mathbb{G}(n,M)$ is contained in its largest component.

The above result implies that in the supercritical phase, as long as $n^{2/3} \ll s \ll n$, most of the vertices of the kernel of the largest component of $\mathbb{G}(n, n/2+s)$ are of degree three. A stronger *3-regularity principle* was shown by Łuczak (1991a). It states, roughly speaking, that if $M = n/2+s$ and $n^{2/3} \ll s \ll n$, then $\mathbb{G}(n,M)$ contains an induced topological copy of a random cubic graph on $(32+o_p(1))s^3/3n^2$ vertices, where, clearly, all vertices of this copy of degree three must belong to $\ker(\mathbb{G}(n,M))$. On the other hand, as was proved by Robinson and Wormald (1992) (see Theorem 9.20) *a.a.s.* the random cubic graph contains a Hamilton cycle. Thus, *a.a.s.* there exists a cycle that goes through nearly all vertices of the kernel, and thus contains roughly two-thirds of its edges. However, vertices of degree two in the core are placed on the edges of the kernel in a random manner. Thus, roughly two-thirds of all vertices of $\operatorname{cr}^L(n,M)$, that is, about $16s^2/3n$ of them, should lie on the cycle which corresponds to a Hamilton cycle of the kernel. On the other hand, each cycle must be contained in the core, so it cannot have more than $|\operatorname{cr}^L(n,M)| = (8+o_p(1))s^2/n$ vertices. In fact, we can slightly improve this bound, since at least one-third of the paths which connect vertices of degree at least three in the core do not belong to such a cycle. Thus, after some computations, one can arrive at the following result (Łuczak 1991a).

Theorem 5.17. *Let $M = n/2+s$, where $n^{2/3} \ll s \ll n$. Then the length of the longest cycle in $\mathbb{G}(n,M)$ lies between $(16+o_p(1))s^2/3n$ and $(7.496+o_p(1))s^2/n$.* ∎

The above theorem estimates the length of the longest cycle contained in the giant component; we conclude this section with a few words about other cycles that emerge in $\mathbb{G}(n,M)$. As one can expect, the subcritical case is easy to study. If $M = n/2 - s$ and $n^{2/3} \ll s \ll n$ the largest unicyclic component of $\mathbb{G}(n,M)$ is of size $\Theta_p(n^2/s^2)$, and the longest cycle has length $\Theta_p(n/s)$. (Exercise! *Hint*: use (5.9). See also Remark 5.16.) It is not much harder to check that in the supercritical phase, when $M = n/2+s$ and $s \gg n^{2/3}$, the size of the largest cycle of $\mathbb{G}(n,M)$ which is not contained in its largest component is $\Theta_p(n/s)$ as well (Exercise!). Hence, in the supercritical phase, typically the size of the largest cycle outside the largest component decreases when the process evolves; thus we infer that the unicyclic components with long cycles must merge with the largest components quite quickly and, consequently, the shortest cycle inside the giant must be of length $O_p(n/s)$. It is somewhat surprising at first sight that at the same time the largest component of $\mathbb{G}(n, M)$ contains no cycles that are much shorter than n/s. Hence, in the supercritical phase, basically all long cycles are contained in the giant component, while all short ones lie outside it (Łuczak 1991a).

Theorem 5.18. *Let $s = s(n)$ be a function such that $n^{2/3} \ll s \ll n$.*

(i) *If $M = n/2 - s$, then the length of the longest cycle in $\mathbb{G}(n, M)$ is $\Theta_p(n/s)$.*

(ii) *If $M = n/2 + s$, then the length of the longest cycle not contained in the largest component of $\mathbb{G}(n, M)$ is $\Theta_p(n/s)$. Furthermore, the same is true for the length of the shortest cycle contained in the giant component of $\mathbb{G}(n, M)$.* ∎

5.5 NEAR THE CRITICAL PERIOD

Now (at last!) we have a quick look at the critical phase, crucial for the phase transition phenomenon, when the giant component is just about to emerge. From Theorems 5.8 and 5.12 we know that this is the only time of the evolution when more than one complex component may simultaneously appear in $\mathbb{G}(n, M)$. Theorem 5.8 tells us also that all complex components created in the critical period must be quite large and they are not very numerous. Let us first show that all of them are of size $O_p(n^{2/3})$ (in fact, $\Theta_p(n^{2/3})$ – see Theorem 5.8) and their total excess is bounded in probability.

Theorem 5.19. *Let $M = n/2 + O(n^{2/3})$ and and let r_ℓ denote the number of ℓ-components in $\mathbb{G}(n, M)$. Then $\sum_{\ell \geq 3} \ell r_\ell = O_p(1)$ and all complex components of $\mathbb{G}(n, M)$ have $O_p(n^{2/3})$ vertices combined.*

Proof. Let $\omega = \omega(n)$ be any function such that $\omega \to \infty$ as $n \to \infty$ but, say, $\omega = o(\log n)$, and let $M_+ = n/2 + \omega^{1/4} n^{2/3}$. Consider $\mathbb{G}(n, M_+)$ as the M_+-th stage of the random graph process $\{\mathbb{G}(n, M)\}_M$. Theorem 5.12 states that *a.a.s.* $\mathbb{G}(n, M_+)$ contains only one complex component of at most $\omega n^{2/3}$ vertices and Theorem 5.13 says that *a.a.s.* the excess of this component $\kappa(n, M)$ is less than ω. Since all complex components that appear at the earlier stages of the process $\{\mathbb{G}(n, M)\}_M$ are vertex-disjoint subgraphs of the giant component of $\mathbb{G}(n, M_+)$, the assertion follows. ∎

Somewhat surprisingly, not much beyond what follows from Theorems 5.8 and 5.19 can be said with probability tending to 1 as $n \to \infty$ about the component structure of $\mathbb{G}(n, M)$ in the critical period. It turns out that with probability bounded away from both 0 and 1, every possible configuration not excluded by the above results can appear in $\mathbb{G}(n, M)$, provided we do not restrict the sizes of largest components up to a factor of $1 + o(1)$. In order to make this vague statement precise, let us consider a family of triples

$T = (\ell; a, b)$, where $\ell \geq -1$ is an integer and $0 \leq a < b \leq \infty$. We call such a triple *regular* if $\ell \geq 1$, or $\ell = -1, 0$, but $a > 0$. Two triples $T = (\ell; a, b)$ and $T' = (\ell'; a', b')$ are *non-intersecting* if either $\ell \neq \ell'$, or the intervals $[a, b]$ and $[a', b']$ are disjoint. For a positive constant d, a family of regular pairwise non-intersecting triples $\underline{T} = ((\ell_1; a_1, b_1), (\ell_2; a_2, b_2) \ldots, (\ell_m; a_m, b_m))$ and a sequence of natural numbers $\underline{t} = (t_1, t_2, \ldots, t_m)$, let $\mathcal{A}(d, \underline{T}, \underline{t})$ denote the event that $\mathbb{G}(n, M)$ contains precisely t_i ℓ_i-components whose sizes lie between $a_i n^{2/3}$ and $b_i n^{2/3}$, for $i = 1, 2, \ldots, m$, in $\mathbb{G}(n, M)$ there are no other complex components and, finally, no other isolated trees and unicyclic components of size at least $dn^{2/3}$ appear in $\mathbb{G}(n, M)$. The following result has been proved by Łuczak, Pittel and Wierman (1994) and Łuczak (1996).

Theorem 5.20. *Let c, d be constants such that $-\infty < c < \infty$ and $d > 0$, and $M = n/2 + cn^{2/3}$. Then for every family of regular pairwise non-intersecting triples $\underline{T} = (T_1, T_2, \ldots, T_m)$ and every sequence of natural numbers $\underline{t} = (t_1, t_2, \ldots, t_m)$, the limit*

$$p(c; d, \underline{T}, \underline{t}) = \lim_{n \to \infty} \mathbb{P}\big(\mathcal{A}(d, \underline{T}, \underline{t})\big)$$

exists and $0 < p(c; d, \underline{T}, \underline{t}) < 1$. ∎

The proof of Theorem 5.20 relies on precise estimates of the moments of an appropriate multidimensional random variable, and since it is long and quite technical we omit it here.

A consequence of Theorem 5.20 (almost equivalent to it) is that if we denote the components of $\mathbb{G}(n, M)$, arranged in decreasing order, by $C_1, C_2, \ldots$, and $\mathbf{X}(n)$ is the sequence of pairs $\big(n^{-2/3}|C_1|, e(C_1) - v(C_1)\big), \big(n^{-2/3}|C_2|, e(C_2) - v(C_2)\big), \ldots$, then the random sequence $\mathbf{X}(n)$ converges in distribution, as $n \to \infty$, to some random sequence $\mathbf{X} = \big((X_1', X_1''), (X_2', X_2''), \ldots\big)$. Aldous (1997) gave a proof of this form of the result (for $\mathbb{G}(n, p)$) by a quite different method, which furthermore identifies the limit in terms of the excursions of a certain modified Brownian motion. Aldous's argument is based on exposing the component structure vertex by vertex (as in the branching process argument in Section 5.2) together with martingale convergence techniques.

It follows immediately from Theorems 5.17 and 5.18 that for $M = n/2 + O(n^{2/3})$ each cycle contained in the largest component of $\mathbb{G}(n, M)$ has length $\Theta_p(n^{1/3})$. As a matter of fact, since the core of a complex component is obtained from its kernel by randomly placing vertices of degree two on its edges, it is not hard to see that the following slightly stronger statement holds (Łuczak, Pittel and Wierman 1994).

Theorem 5.21. *Let $M = n/2 + O(n^{2/3})$. Then any two vertices of degree three in the core of $\mathbb{G}(n, M)$ lie at distance $\Theta_p(n^{1/3})$ from each other.* ∎

In particular, in the critical period *a.a.s.* $\mathbb{G}(n, M)$ contains no cycles with diagonals. We challenge the reader to find a heuristic argument that shows that such cycles appear in the process when $M = n/2 + \Theta(n^{3/4})$ (Exercise!), well before $M = n/2+\Theta(n^{4/5})$ as the very first look at Theorems 5.15 and 5.17 may suggest (for the limit probability of this property see Łuczak (1991a)).

Finally, let us say few words about the threshold for the property that a graph is planar, which was addressed already by Erdős and Rényi (1960) (see the comments on their result by Łuczak and Wierman (1989)). Since for $M = n/2 + O(n^{2/3})$, *a.a.s.* all vertices of the core have degree two or three (Theorem 5.15), *a.a.s.* $\mathbb{G}(n, M)$ contains no topological copy of K_5. Does it contain a copy of $K_{3,3}$? Certainly not with probability $1 - o(1)$, since from Theorem 5.20 it follows that at any moment of the critical period, with probability bounded away from 0, $\mathbb{G}(n, M)$ consists only of trees and unicyclic components. However, for a given ℓ, once an ℓ-component appears in $\mathbb{G}(n, M)$, with a non-vanishing probability its kernel can be any 3-regular multigraph with excess ℓ. (Although it is not true that all such multigraphs are equally likely to appear as the kernel of such a component, for a small ℓ the corresponding limit probabilities are easy to work out.) Thus, since Theorem 5.20 implies that the probability that $\mathbb{G}(n, M)$ contains an ℓ-component with $\ell \geq 3$ is larger than some positive constant, the probability that $\mathbb{G}(n, M)$ is planar is also bounded away from zero. With some more work, one can prove the following, somewhat stronger result (Łuczak, Pittel and Wierman 1994).

Theorem 5.22. *Let $M = n/2 + cn^{2/3}$ for some constant c. Then the probability that $\mathbb{G}(n, M)$ is planar tends to a limit $\rho_{\mathrm{pl}}(c)$, where $0 < \rho_{\mathrm{pl}}(c) < 1$. Furthermore,*

$$\lim_{c\to-\infty} \rho_{\mathrm{pl}}(c) = 1 \quad \text{while} \quad \lim_{c\to\infty} \rho_{\mathrm{pl}}(c) = 0 \,. \qquad \blacksquare$$

Unfortunately, we do not know how to find the value of $\rho_{\mathrm{pl}}(c)$, although, in principle, we can approximate it with arbitrary precision. For instance, the best known bounds for $\rho_{\mathrm{pl}}(0)$ are given by Janson, Knuth, Łuczak and Pittel (1993, Theorem 8), who showed that

$$0.987074 \leq \rho_{\mathrm{pl}}(0) \leq 0.999771 \,.$$

5.6 GLOBAL PROPERTIES AND THE SYMMETRY RULE

So far we have dealt mainly with properties that hold *a.a.s.* for the random graph $\mathbb{G}(n, M)$; in this section we will be interested in properties of the random graph process $\{\mathbb{G}(n, M)\}_M$. Thus, for instance, instead of approximating some random variable $X_{n,M}$, defined for $\mathbb{G}(n, M)$, for a given function $M = M(n)$, we will try to obtain uniform bounds for $X_{n,M}$ which *a.a.s.* remain valid at many or all stages of the process $\{\mathbb{G}(n, M)\}_M$.

Let us start with a simple example of one such *global* statement.

Theorem 5.23. *Let $\omega = \omega(n) \to \infty$ as $n \to \infty$, with $\omega \le n^{1/6}$. Moreover, for $M = n/2 + s$, let*

$$k^{\pm}(n, M) = \left(1 \pm \frac{1}{\log\log\omega}\right)\frac{n^2}{2s^2}\log\frac{|s|^3}{n^2}. \tag{5.20}$$

Then a.a.s. *for the random graph process $\{\mathbb{G}(n, M)\}_M$ the following hold:*

(i) *if $-n/\omega \le s \le -\omega n^{2/3}$, then*

$$k^-(n, M) \le L_1(n, M) \le k^+(n, M);$$

(ii) *if $-\omega n^{2/3} \le s \le \omega n^{2/3}$, then*

$$n^{2/3}/\omega^2 \le L_1(n, M) \le 5\omega n^{2/3};$$

(iii) *if $\omega n^{2/3} \le s \le n/\omega$, then*

$$|L_1(n, M) - 4s| \le \omega^{-1/4}s.$$

Proof. Let $s_i = \omega n^{2/3}(1 + 1/\log\omega)^i$, where $i = 0, 1, \ldots, t$ and t is the smallest natural number for which $s_t \ge n/\omega$. Furthermore, let $M_i^- = n/2 - s_i$ and

$$\hat{k}^{\pm}(n, M) = \left(1 \pm \frac{1}{2\log\log\omega}\right)\frac{n^2}{2s^2}\log\frac{|s|^3}{n^2}. \tag{5.21}$$

Then, Theorem 5.6 implies that with probability at least

$$1 - \sum_{i=0}^{t}\left(\frac{n^2}{s_i^3}\right)^{1/4\log\log\omega} = 1 - o(1),$$

for all $i = 0, 1, \ldots, t$, we have $\hat{k}^-(n, M_i^-) \le L_1(n, M_i^-) \le \hat{k}^+(n, M_i^-)$. Now it is enough to observe that for every $M = n/2 - s$, for which $s_{i-1} \le s \le s_i$, we have $k^-(n, M^-) \le \hat{k}^-(n, M_i^-)$ and $k^+(n, M) \ge \hat{k}^+(n, M_{i-1}^-)$. Hence, *a.a.s.* for each such M

$$k^-(n, M) \le L_1(n, M_i^-) \le L_1(n, M) \le L_1(n, M_{i-1}^-) \le k^+(n, M).$$

In the very same way one can deduce (iii) from Theorem 5.12 and (5.18). Finally, the statement (ii) follows immediately from (i) and (iii), applied with $s = \pm\omega n^{2/3}$. ∎

The above argument strongly relies on the fact that the value of the random variable $L_1(n, M)$ cannot decrease during the random process. Suppose, however, that we would like to show a similar uniform bound for the size of the second or, more generally, the r-th largest component of $\mathbb{G}(n, M)$ for a fixed $r \ge 2$. The subcritical phase is rather easy to deal with. Clearly, the uniform upper bound for $L_1(n,M)$ given in Theorem 5.23 is at the same time the upper bound for $L_r(n,M)$. On the other hand, Theorem 5.6 and Theorem 5.23 imply that *a.a.s.* the random graph process is such that for all $M = n/2 - s$ for which $s \gg n^{2/3}$ we have $L_r(n,M) \le L_r(n,M+1)$. It is in the supercritical phase that the problem of studying $L_r(n,M)$ becomes interesting: now the size of $L_r(n,M)$ can increase when the r-th largest component merges with some small component, but the value of $L_r(n,M)$ can also drop significantly, when one of the largest components is joined to the giant one. In this section we learn one method of dealing with such non-monotone random variables: the symmetry rule.

Suppose that $M = n/2 + s$, where $n^{2/3} \ll s \ll n$, and let $\mathbb{G}^L(n,M)$ denote the graph obtained from $\mathbb{G}(n,M)$ by deleting all vertices of the largest component. (In the unlikely case in which there are several components of maximum size we pick one of them uniformly at random.) Then, from Theorem 5.12 and (5.18), $\mathbb{G}^L(n,M)$ has $n' = n - (4 + o_p(1))s$ vertices. Furthermore, due to Theorem 5.13, it has

$$M' = n/2 + s - (4 + o_p(1))s - o_p(s) = n/2 - (3 + o_p(1))s = n'/2 - (1 + o_p(1))s$$

edges. Thus, properties of $\mathbb{G}^L(n,M)$ should be roughly the same as those of $\mathbb{G}(n',M')$, where $n' = (1+o(1))n$ and $M' = n'/2 - s$. The following symmetry rule states this fact in a rigorous way.

Theorem 5.24. *Let $\mathcal{A}$ be any graph property. If $M = n/2 + s$, where $\omega n^{2/3} \le s \le n/\omega$ for some function $\omega = \omega(n) \to \infty$, then for n large enough*

$$\begin{aligned}\mathbb{P}(\mathbb{G}^L(n,M) \text{ has } \mathcal{A}) \le \max\{&\mathbb{P}(\mathbb{G}(n',M') \text{ has } \mathcal{A}) :\\ &|n' - (n-4s)| \le \omega^{-0.9}s,\ |M' - (n'/2 - s)| \le \omega^{-0.9}s\} + 8n^{2/9}s^{-1/3}.\end{aligned}$$

Proof. Let n' and M' denote the number of vertices and edges in $\mathbb{G}^L(n,M)$, respectively. Theorem 5.12 implies that, except for an event with probability at most $7n^{2/9}/s^{1/3}$,

$$\begin{aligned} n' &= \sum_{k=1}^{\lceil n^{2/3}\rceil} kY_{n,M}(k,-1) + \sum_{k=1}^{\lceil n^{2/3}\rceil} kY_{n,M}(k,0),\\ M' &= \sum_{k=1}^{\lceil n^{2/3}\rceil} (k-1)Y_{n,M}(k,-1) + \sum_{k=1}^{\lceil n^{2/3}\rceil} kY_{n,M}(k,0). \end{aligned} \tag{5.22}$$

Assuming this, Theorem 5.11, with ω replaced by $s^{3/2}/n\omega$, and (5.18) show that, except with probability at most $n\omega/s^{3/2} \le n^{1/3}/s^{1/2}$,

$$|n' - (n-4s)| \le \frac{s}{\omega} + \frac{n}{\omega s^{1/2}} + O_C\Big(\frac{s^2}{n}\Big) = O_C\Big(\frac{s}{\omega}\Big).$$

Moreover, by estimates as in Section 5.3, it can be shown that if $\bar{M} = n/2 - \bar{s}$, then

$$\mathbb{E}\Big|\sum_{k=1}^{\lceil n^{2/3}\rceil} Y_{n,M}(k,-1) - \frac{\bar{M}(n-\bar{M})}{M}\Big|^2 = O(n),$$

where

$$\frac{\bar{M}(n-\bar{M})}{M} = \frac{n}{2} - s + \frac{16s^3}{3n^2} + O\Big(\frac{s^4}{n^3}\Big) = \frac{n}{2} - s + O\Big(\frac{s}{\omega^2}\Big),$$

which, in fact, implies Theorem 5.13 when s is large, $s \geq \omega n^{5/6}$ (Exercise!). Thus by Chebyshev's inequality, again assuming (5.22),

$$\mathbb{P}\Big(\Big|n' - M' - \Big(\frac{n}{2} - s\Big)\Big| > \frac{s}{\omega}\Big) = O\Big(\frac{n\omega^2}{s^2}\Big) = O\big(n^{-1/3}\big) = O\Big(\frac{n^{2/3}}{s}\Big).$$

Observe that the graph $\mathbb{G}^L(n, M)$, once we condition on the number n' of its vertices and the number M' of its edges, can be viewed as a graph chosen uniformly at random among all graphs with n' vertices and M' edges in which the largest component is not larger than $n - n'$ (a negligible additional weighting factor emerges in the case when the largest component of $\mathbb{G}^L(n, M)$ has precisely $n - n'$ vertices). Moreover, the values of n' and M' we are dealing with ensure that $\mathbb{G}(n', M')$ is in the subcritical phase of the evolution of the random graph. Therefore, we can use Theorem 5.6 with $\alpha = 1/3$ to infer that, with probability at least

$$1 - \big[(n')^2/(n'/2 - M')^3\big]^{1/3} \geq 1 - 2n^{2/3}/s \geq 1 - 0.5n^{2/9}s^{-1/3},$$

$\mathbb{G}(n', M')$ contains no component larger than $n^{2/3}$, and the result follows. ■

Thus, as we have already anticipated by looking at the formula (5.8), for $n^{2/3} \ll s \ll n$ the graph $\mathbb{G}^L(n, n/2 + s)$ behaves roughly like $\mathbb{G}(n, n/2 - s)$ with respect to any property $\mathcal{A}$ not vulnerable to small changes in size of the random graph (see also Theorem 5.18). It can be shown that an analogous relation between the structures of $\mathbb{G}^L(n, n/2 + s)$ and $\mathbb{G}(n', n'/2 - s')$ remains true also for $s = \Omega(n)$. (However, in this case the dependence of n' and s' on n and s is more involved and the "mirror symmetry rule" is not valid any more.) Thus, the evolution of $\mathbb{G}^L(n, M)$ in the supercritical phase is similar to the evolution of $\mathbb{G}(n, M)$ in the subcritical phase running backwards: for $M/n \to \infty$ all cycles in $\mathbb{G}^L(n, M)$ disappear, the size of the largest component of $\mathbb{G}^L(n, M)$ decreases and, just before $\mathbb{G}(n, M)$ becomes connected, *a.a.s.* $\mathbb{G}^L(n, M)$ consists of isolated vertices. However, the importance of the symmetry rule goes beyond the fact that it could give us better insight into the nature of the evolution of the random graph; it can be also a useful tool in studying global properties of the random graph process.

Theorem 5.25. *Let $r \geq 2$ be a natural number which does not depend on n and let $\omega = \omega(n) \to \infty$ as $n \to \infty$. Moreover, for $M = n/2 + s$, let*

$$k^{\pm}(n, M) = \left(1 \pm \frac{1}{\log\log\omega}\right)\frac{n^2}{2s^2}\log\frac{|s|^3}{n^2}. \tag{5.23}$$

Then a.a.s. *the random graph process $\{\mathbb{G}(n, M)\}_M$ is such that for all $M = n/2 + s$, where $\omega n^{2/3} \leq |s| \leq n/\omega$, we have*

$$k^-(n, M) \leq L_r(n, M) \leq k^+(n, M).$$

Proof. As in the proof of Theorem 5.23, set $s_i = \omega n^{2/3}(1 + 1/\log\omega)^i$, where $i = 0, 1, \ldots, t$ and t is the smallest natural number for which $s_t \geq n/\omega$. Furthermore, let $M_i^- = n/2 - s_i$ and $\hat{k}^{\pm}(n, M)$ be defined as in (5.21). Using Theorem 5.6 as before, *a.a.s.* for all $i = 0, 1, \ldots, t$ we have $\hat{k}^-(n, M_i^-) \leq L_r(n, M_i^-) \leq \hat{k}^+(n, M_i^-)$. Then, using the upper bound for $L_1(n, M_i^-)$ given by Theorem 5.23, and the "*a.a.s.* monotonicity" of $L_r(n, M)$ in the subcritical phase discussed above, we infer that *a.a.s.* for every $i = 1, 2, \ldots, t$, and every $M = n/2 - s$ for which $s_{i-1} \leq s \leq s_i$,

$$\begin{aligned} k^-(n, M) &\leq \hat{k}^-(n, M_i^-) \leq L_r(n, M_i^-) \\ &\leq L_r(n, M) \leq L_r(n, M_{i-1}^-) \leq \hat{k}^+(n, M_{i-1}^-) \leq k^+(n, M). \end{aligned}$$

Now set $M_i^+ = n/2 + s_i$ for $i = 0, 1, \ldots, t$. Then, arguing as before but now using Theorem 5.7, one can show that *a.a.s.* for all $i = 0, 1, \ldots, t$ we have

$$\hat{k}^-(n, M_i^+) \leq L_r(n, M_i^+) \leq \hat{k}^+(n, M_i^+).$$

Nonetheless, as we have already observed, this fact does not imply that $k^-(n, M) \leq L_r(n, M) \leq k^+(n, M)$ for all $M_0^+ \leq M \leq M_t^+$, since during the random graph process the largest components merge with the giant component and so the value of $L_r(n, M)$ can go up and down. Thus, for $i = 0, 1, \ldots, t - 1$, let $\mathcal{A}_i'$ be the following property of a graph G: if we add to G $s_{i+1} - s_i$ edges, chosen uniformly at random from all pairs of vertices of G which have not yet become edges of G, then the probability that the largest component of such a graph is larger than $\hat{k}^+(n, M_i^+) = \hat{k}^+(n, M_i^-)$ is smaller than $(n^2 s_i^{-3})^{1/20\log\log\omega}$. Similarly, for $i = 1, 2, \ldots, t$, let $\mathcal{A}_i''$ denote the property that a graph G is such that if we delete from it $s_i - s_{i-1}$ randomly chosen edges then the probability that the r-th largest component of the graph obtained in this way is smaller than $\hat{k}^-(n, M_i^+) = \hat{k}^-(n, M_i^-)$ is at most $(n^2 s_i^{-3})^{1/20\log\log\omega}$. (We define $\mathcal{A}_i'$ and $\mathcal{A}_i''$ false if there is not room in G to add or delete this number of edges.) Let us emphasize that, although probability is involved in the definition of $\mathcal{A}_i'$ and $\mathcal{A}_i''$, they are purely "deterministic" graph properties; for instance, $\mathcal{A}_i''$ says, roughly, that G contains a lot of large trees whose sizes do not drop rapidly when we delete from the graph a moderate number of randomly chosen edges.

Now suppose that $(1-5/\omega)n \le n' \le n$ and $|s'-s_i| \le \omega^{-0.9}s_i$. We will show first that the probability that the property $\mathcal{A}_i'$ does not hold for $\mathbb{G}(n', n'/2-s')$ is bounded from above by

$$\left(\frac{n^2 s_i^3}{(s'+s_i-s_{i+1})^6}\right)^{1/20 \log\log \omega} \le 2\left(\frac{n^2}{s_i^3}\right)^{1/20 \log\log \omega}.$$

Indeed, otherwise the probability that the largest component of $\mathbb{G}(n', n'/2 - s' - s_i + s_{i+1})$ is larger than $\hat{k}^+(n, M_i^-)$ would be bounded from below by

$$\left(\frac{n^2 s_i^3}{(s'+s_i-s_{i+1})^6}\right)^{1/20 \log\log \omega}\left(\frac{n^2}{s_i^3}\right)^{1/20 \log\log \omega}$$
$$= \left(\frac{n^2}{(s'+s_i-s_{i+1})^3}\right)^{1/10 \log\log \omega},$$

contradicting Theorem 5.6. Similarly, the probability that $\mathbb{G}(n', n'/2-s')$ has $\mathcal{A}_i''$ is at least

$$1-2\left(\frac{n^2}{s_i^3}\right)^{1/20 \log\log \omega},$$

since otherwise the graph $\mathbb{G}(n, n/2-s'-s_i+s_{i-1})$ obtained from $\mathbb{G}(n', n'/2-s')$ by deleting s_i-s_{i-1} randomly chosen edges, would contradict Theorem 5.6. Now apply the symmetry rule (Theorem 5.24) to infer that for each $i = 0, 1, \ldots, t-1$, with probability at least

$$1-2\left(\frac{n^2}{s_i^3}\right)^{1/20 \log\log \omega} - \frac{8n^{2/9}}{s_i^{1/3}} \ge 1-3\left(\frac{n^2}{s_i^3}\right)^{1/20 \log\log \omega},$$

$\mathbb{G}^L(n, M_i^+)$ has property $\mathcal{A}_i'$, and for each $i = 1, 2, \ldots, t$, with probability at least

$$1-3\left(\frac{n^2}{s_i^3}\right)^{1/20 \log\log \omega},$$

the property $\mathcal{A}_i''$ holds for $\mathbb{G}^L(n, M_i)$.

Now, in the evolution of $\mathbb{G}(n, M)$ for $M = M_i^+$ to M_{i+1}^+, we add $s_{i+1} - s_i$ edges. Some of them may have one or two endpoints in the largest component of $G(n, M_i)$, which only may decrease the size of the largest component of $\mathbb{G}^L(n, M)$, and since the property that a graph contains a component larger than $\hat{k}^+(n, M_i^+)$ is monotone, it follows that the probability that the largest component of $\mathbb{G}^L(n, M)$ is larger than $k^+(n, M) > \hat{k}^+(n, M_i^+)$ for some $M = n/2 + s$ with $s_i \le s \le s_{i+1}$ is bounded from above by the probability that we create a component larger than $\hat{k}^+(n, M_i^+)$ if we add $s_{i+1} - s_i$ randomly chosen edges to $\mathbb{G}^L(n, M_i^+)$. Thus, from the definition of $\mathcal{A}_i'$ it follows that this probability is at most

$$\mathbb{P}(\mathbb{G}(n, M_i^+) \notin \mathcal{A}_i') + \left(\frac{n^2}{s_i^3}\right)^{1/20 \log\log \omega} \le 4\left(\frac{n^2}{s_i^3}\right)^{1/20 \log\log \omega}.$$

Consequently, the probability that for some $M = n/2 + s$, where $s_0 \le s \le s_t$, the largest component of $\mathbb{G}^L(n, M)$ is larger than $k^+(n, M)$ is bounded from above by

$$\sum_{i=0}^{t-1} 4\left(\frac{n^2}{s_i^3}\right)^{1/20 \log\log \omega} = o(1)\,.$$

For the lower bound, we instead for $M_{i+1}^+ \ge M \ge M_i^+$ regard $\mathbb{G}(n, M)$ as obtained from $\mathbb{G}(n, M_{i+1}^+)$ by randomly deleting edges one by one. Some of the removed edges may have belonged to the largest component of $\mathbb{G}(n, M_{i+1}^+)$, but we continue and delete edges until $s_{i+1} - s_i$ of them have been removed from $\mathbb{G}^L(n, M_{i+1}^+)$, and denote the resulting subgraph of $\mathbb{G}^L(n, M_{i+1}^+)$ by H_i. Let $V = V(\mathbb{G}(n, M_{i+1}^+)) = V(H_i)$.

If for some such M the r-th largest component of $\mathbb{G}(n, M)$ is smaller than $k^-(n, M) < \hat{k}^-(n, M_{i+1})$, so that $\mathbb{G}(n, M)$ has less than r components of order at least $\hat{k}^-(n, M_{i+1})$, then the same holds for the subgraph $\mathbb{G}(n, M)[V]$, and if we continue to delete edges, and assume that all components of $\mathbb{G}^L(n, M_{i+1}^+)$ have orders less than $2\hat{k}^-(n, M_{i+1})$, we see that also H_i has less than r components of order at least $\hat{k}^-(n, M_{i+1})$. By the definition of $\mathcal{A}_{i+1}''$, and observing that $2\hat{k}^-(n, M_{i+1}) > \hat{k}^+(n, M_{i+1})$, it follows that the probability that $L_r(n, M) < k^-(n, M)$ for some $M = n/2 + s$, $s_i \le s \le s_{i+1}$, is at most

$$\mathbb{P}(\mathbb{G}^L(n, M_{i+1}^+) \notin \mathcal{A}_{i+1}') + \mathbb{P}(\mathbb{G}^L(n, M_{i+1}^+) \notin \mathcal{A}_{i+1}'') + \left(\frac{n^2}{s_i^3}\right)^{1/20 \log\log \omega}$$

$$\le 7\left(\frac{n^2}{s_i^3}\right)^{1/20 \log\log \omega}.$$

Summing over all $i < t$, we see that the the probability that $L_r(n, M) < k^-(n, M)$ for some $M = n/2 + s$, $s_0 \le s \le s_t$ is $o(1)$ as $n \to \infty$. ■

5.7 DYNAMIC PROPERTIES

In the previous section we showed how to use the symmetry rule to extend results about $\mathbb{G}(n, M)$ to theorems about the behavior of the random graph process $\{\mathbb{G}(n, M)\}_M$. Now we mention a few "genuine" properties of the random graph process which do not correspond to any property of $\mathbb{G}(n, M)$.

The first problem we will consider is Erdős's question about the length ξ_n of the first cycle which appears during the random graph process. The limit distribution of ξ_n is given by the following result of Janson (1987) (see also Bollobás (1988b)).

Theorem 5.26. *For every $j \ge 3$*

$$\lim_{n\to\infty} \mathbb{P}(\xi_n = j) = \frac{1}{2}\int_0^1 t^{j-1} e^{t/2+t^2/4}\sqrt{1-t}\,dt\,.$$

Proof. We just give the idea of the proof, omitting all technical details and computations which can be easily filled in by the reader (Exercise!). Thus, let $\eta > 0$. Choose $j \geq 3$ and a large constant A. For $i = 0, 1, \ldots, \lceil (1-\eta)A \rceil$ set $M_i = t_i n/2$, where $t_i = i/A$. One can apply the method of moments (Section 6.1) to show that for every such M_i the probability that $\mathbb{G}(n, M_i)$ contains no cycle tends to

$$\exp\left(-\sum_{k=3}^{\infty} t_i^k/2k\right) = e^{t_i/2+t_i^2/4}\sqrt{1-t_i}\,.$$

Furthermore, one can use Chebyshev's inequality to verify that for every i the number of pairs of vertices of $\mathbb{G}(n, M_i)$ connected by a path of length $j-1$ is equal to $(1/2 + o_p(1))t_i^{j-1}n$. Note that the property that a graph contains no cycle is decreasing and that the number of pairs of vertices joined by a path of length $j-1$ can only increase when a new edge is added to the graph. Thus, since A can be chosen arbitrarily large, we infer that for every $\varepsilon > 0$ *a.a.s.* for every $M = tn/2$, where $0 \leq t \leq 1-\eta$, the number of pairs of vertices of $\mathbb{G}(n, M)$ connected by paths of length $j-1$ is contained between $(1/2-\varepsilon)t^{j-1}n$ and $(1/2+\varepsilon)t^{j-1}n$. Moreover, if $M = tn/2$, where $0 \leq t \leq 1-\eta$, then the probability that $\mathbb{G}(n, M)$ contains no cycle is $(1+o(1))e^{t/2+t^2/4}\sqrt{1-t}$. Hence the probability that in the first $(1-\eta)n/2$ stages of the process a cycle of length j appears as the first one is given by

$$(1+o(1)) \sum_{M=0}^{(1-\eta)n/2} \frac{(2M/n)^{j-1}n/2}{\binom{n}{2}} e^{M/n+(M/n)^2}\sqrt{1-2M/n}$$
$$= \left(\frac{1}{2} + o(1)\right) \int_0^{1-\eta} t^{j-1}e^{t/2+t^2/4}\sqrt{1-t}\,dt\,.$$

Thus, letting $\eta \to 0$, for each $j \geq 3$ we get

$$\liminf_{n\to\infty} \mathbb{P}(\xi_n = j) \geq \rho_j = \frac{1}{2}\int_0^1 t^{j-1}e^{t/2+t^2/4}\sqrt{1-t}\,dt. \tag{5.24}$$

However, it is easy to check that $\sum_{j\geq 3}\rho_j = 1$, thus (5.24) implies that

$$\lim_{n\to\infty} \mathbb{P}(\xi_n = j) = \rho_j$$

for every $j \geq 3$. ■

Let us remark that the limit distribution of ξ_n has infinite expectation; thus $\mathbb{E}\,\xi_n \to \infty$ and, since $\mathbb{E}\,\xi_n \leq n$, one can ask about the order of $\mathbb{E}\,\xi_n$. Flajolet, Knuth and Pittel (1989) computed all the moments of the length of the first cycle and of the size of the component containing it; in particular, they showed that

$$\mathbb{E}\,\xi_n \sim \frac{\pi^{1/2}\Gamma(1/3)}{2^{1/6}3^{2/3}}n^{1/6}\,.$$

The proof, based on the study of the behavior of a certain generating function, is much harder than that of Theorem 5.26, thus we do not give it here, but refer the interested reader to the original paper (see also Janson, Knuth, Łuczak and Pittel (1993, Section 26)).

Two other problems we would like to mention in this chapter concern the way the giant component emerges. Erdős proposed that we view the evolving graph as a "race of components". In order to express his idea in a rigorous way, we define recursively the *leader* of this race. For a graph with only one edge, the leader is the only component with two vertices. Now suppose that we add a new edge e to a graph G, in which one of the largest components, say, H, has been chosen as the leader. Then, e may miss the leader and connect two other components whose combined size is larger than $|H|$; in this case we say that a *change of leader* occurred and nominate the newly created largest component to be the leader of the graph. In all other cases the leader of $G+e$ is the component that contains all the vertices of the "old" leader H. Erdős asked when in the random graph process the last change of leader takes place. Note that Theorems 5.23 and 5.25 imply that *a.a.s.* no changes of the leader occur when $n^{2/3} \ll M - n/2 \ll n$, and the reader is invited to check that the same remains true for all $M = n/2+\Omega(n)$ (Exercise!). However, it is, in principle, possible that already in the subcritical phase one component starts to dominate, although its superiority becomes evident only later. The following result by Łuczak (1990c), which we give without proof, shows that this is not the case, and that the last change of leader occurs at $M = n/2 + O_p(n^{2/3})$.

Theorem 5.27. *Let* $\mathrm{Lead}(n, M)$ *denote the largest number* r *such that the leader of* $\mathbb{G}(n, M)$ *does not change in the next* r *stages of the random graph process* $\{\mathbb{G}(n, M)\}_M$.

(i) *If* $n^{2/3} \ll s \ll n$, *then*

$$\mathrm{Lead}\Big(n, \frac{n}{2} - s\Big) = \Theta_p\Big(\frac{s}{\log(s^3/n^2)}\Big).$$

(ii) *If* $s \gg n^{2/3}$, *then* a.a.s. $\mathrm{Lead}(n, n/2+s) = \binom{n}{2} - n/2 - s$, *in other words, the largest component of* $\mathbb{G}(n, n/2 + s)$ *will remain the leader until the very end of the random graph process* $\{\mathbb{G}(n, M)\}_M$. ■

Finally, let us look once again at the number of complex components created during the process $\{\mathbb{G}(n, M)\}_M$. Theorem 5.8 says that *a.a.s.* all complex components are created in the critical phase when $M = n/2+O(n^{2/3})$. What is the probability that only one complex component appears in the process $\{\mathbb{G}(n, M)\}_M$ and no other complex component ever emerges? Theorem 5.8, the remark following it, and Theorem 5.20 imply that this probability tends to a limit ν, where $0 < \nu < 1$. In order to find the value of ν, we describe the changes of the structure of $\mathbb{G}(n, M)$ during the critical period in yet another way, as a Markov chain whose stages are "graph configurations".

Thus, we say that $[r_1, r_2, \ldots, r_q]$ is the *configuration* of a graph G, if G contains precisely r_ℓ ℓ-components for $\ell = 1, 2, \ldots, q$ and no ℓ-components with $\ell > q$ appear in G. It is not hard to see that if we add a new edge to a graph the value of $\sum_\ell \ell r_\ell$ increases by at most one; for instance, adding a juncture as an edge to a graph changes its configuration $[r_1, r_2, \ldots, r_q]$ into $[r_1 + 1, r_2, \ldots, r_q]$. The following "switching theorem", proved in Janson, Knuth, Łuczak and Pittel (1993) by a careful analysis of an appropriate generating function, describes quite precisely the way the configuration of $\mathbb{G}(n, M)$ evolves during the random graph process.

Theorem 5.28. *Let $r_1 + 2r_2 + \cdots + qr_q = r$ and $\delta_1 + 2\delta_2 + \cdots = 1$. Then the limit probability that in the random graph process $\{\mathbb{G}(n, M)\}_M$ the configuration $[r_1, r_2, \ldots, r_q]$ is followed by $[r_1 + \delta_1, r_2 + \delta_2, \ldots, r_q + \delta_q, \delta_{q+1}, \ldots]$ is equal to*

$$\begin{array}{rll} \frac{5}{4}/(3r + \frac{1}{2})(3r + \frac{5}{2}) & \textit{if} & \delta_1 = 1 \\ 9j(j+1)r_j/(3r + \frac{1}{2})(3r + \frac{5}{2}) & \textit{if} & \delta_j = -1,\ \delta_{j+1} = 1 \\ 9j^2 r_j(r_j - 1)/(3r + \frac{1}{2})(3r + \frac{5}{2}) & \textit{if} & \delta_j = -2,\ \delta_{2j+1} = 1 \\ 18jkr_jr_k/(3r + \frac{1}{2})(3r + \frac{5}{2}) & \textit{if} & \delta_j = -1,\ \delta_k = -1,\ \delta_{j+k+1} = 1,\ j < k \end{array}$$

and there are no other possibilities. Moreover, asymptotically these probabilities are independent of the history of previous configurations. ■

Note that, in a way, the above theorem nicely supplements Theorem 5.20. Theorem 5.20 implies that, in particular, for any sequence $r_1, \ldots, r_q$ (e.g., $0, 2, 1$) of natural numbers, with probability bounded away from zero, for some M the graph $\mathbb{G}(n, M)$ has configuration $[r_1, \ldots, r_q]$ (e.g., $[0, 2, 1]$). Theorem 5.28 states that we may choose not only such a configuration but also the way it was created (e.g., $[1] \to [0, 1] \to [1, 1] \to [0, 2] \to [0, 1, 1] \to [1, 1, 1] \to [0, 2, 1]$); still the probability that the configurations of $\mathbb{G}(n, M)$ during the random graph process followed such an "evolutionary path" tends to $\rho_0 > 0$ as $n \to \infty$, where $\rho_0 = \rho_0(r_1, \ldots, r_q)$ can be explicitly computed.

Now the question about the limit probability that during the random graph process no two complex components appear at the same time, finds its surprisingly simple answer.

Theorem 5.29. *The probability $\nu(n)$ that in the process $\{\mathbb{G}(n, M)\}_M$ for every M, $0 \le M \le \binom{n}{2}$, $\mathbb{G}(n, M)$ contains at most one complex component tends to*

$$\nu = \prod_{\ell=1}^{\infty} \frac{9\ell(\ell+1)}{(3\ell + 1/2)(3\ell + 5/2)} = \frac{5\pi}{18} = 0.872\ldots$$

as $n \to \infty$.

Proof. Theorems 5.8, 5.10 and 5.13 imply that for every $\varepsilon > 0$ there exists a constant $C = C(\varepsilon)$ such that the following holds: with probability at least

$1-\varepsilon$, the random graph process is such that if in $\mathbb{G}(n, M)$ an ℓ-component with $\ell \geq C$ appears, it will remain the only complex component of the graph until the very end of the random graph process. Thus, the probability $\nu(n)$ can be approximated up to ε by the probability that the first C steps of the evolutionary path are $[1] \to [0,1] \to [0,0,1] \to \cdots \to [0,\ldots,0,1]$. From Theorem 5.28 the probability of this event tends to $\prod_{\ell=1}^{C} 9\ell(\ell+1)/(3\ell+1/2)(3\ell+5/2)$. Now, to complete the proof, it is enough to let $C \to \infty$. ■

6

Asymptotic Distributions

Many questions for random combinatorial structures are qualitative; we ask whether some property is satisfied, for example, the existence of a certain substructure. Other questions are quantitative; we study some numerical characteristic of the random structure, for example, the number of copies of a certain substructure. Since the structure is random, this becomes a random variable and we may ask about its distribution.

Exact formulas for the distributions of interesting combinatorial variables are rare, and even when they exist, they are often too complicated to be of much use. The main interest, therefore, centers on asymptotics and limit theorems. We will in this chapter describe several methods that have been used to prove such results, and illustrate their use with applications to random graphs. We concentrate on presenting the methods rather than presenting new results; thus we prove some results several times by different methods.

The appropriate probabilistic notion is convergence in distribution, as defined in Section 1.2. We will mainly state results on convergence in distribution of single random variables, but most of the results extend easily to joint convergence of several variables. For example, the Cramér–Wold device (see Section 1.2) applies with ease to extend all results in this chapter on normal convergence (Exercise!).

Typically, random variables converge only after rescaling. We use special notation for the most natural and common choice: For a random variable X with finite non-zero variance, we define

$$\widetilde{X} = (X - \mathbb{E}\,X)/(\operatorname{Var} X)^{1/2};$$

thus $\widetilde{X}$ is standardized to have $\mathbb{E}\,\widetilde{X} = 0$ and $\operatorname{Var} \widetilde{X} = 1$.

6.1 THE METHOD OF MOMENTS

The method of moments is one of the oldest methods to prove convergence in distribution, but it is still widely used, both because it is conceptually simple and because it is powerful and well adapted to combinatorial problems. The drawback is that it usually requires long and messy estimates. As a result, many theorems have first been proved by the method of moments but later reproved in more elegant ways by other methods.

The *moments* of a random variable X are the numbers $\mathbb{E}\, X^k$, $k \geq 1$. We consider here only variables such that all moments exist, that is, $\mathbb{E}\,|X|^k < \infty$ for every $k > 0$; we say that such random variables have *finite moments.*

It is clear that the moments of a random variable are determined by the distribution. The converse does not always hold (see, e.g., Chung (1974)), but it holds in many important cases. We thus say that the distribution of X is *determined by its moments* if X has finite moments and every random variable with the same moments as X has the same distribution. A sufficient condition for the distribution of X to be determined by its moments is that the moment generating function $\mathbb{E}\, e^{tX}$ is finite for t in some interval around 0; in particular, this holds if X has a normal or Poisson distribution.

The standard version of the method of moments can be stated as follows; see Chung (1974, Theorem 4.5.5) for a proof.

Theorem 6.1. *Let Z be a random variable with a distribution that is determined by its moments. If $X_1, X_2, \ldots$ are random variables with finite moments such that $\mathbb{E}\, X_n^k \to \mathbb{E}\, Z^k$ as $n \to \infty$ for every integer $k \geq 1$, then $X_n \overset{d}{\to} Z$.* ∎

The method of moments thus requires estimation of all moments of X_n, which often leads to long calculations. Two versions of the method, which are formally equivalent but often more convenient for applications, are given in separate subsections below.

The method of moments also applies to vector-valued variables, or, in other words, to joint convergence in distribution of several random variables; we now have to consider all mixed moments. We write $Z^\alpha = Z_1^{\alpha_1} \cdots Z_d^{\alpha_d}$ for vectors $Z = (Z_1, \ldots, Z_d)$ and $\alpha = (\alpha_1, \ldots, \alpha_d)$.

Theorem 6.2. *Let $Z = (Z_1, \ldots, Z_d)$ be a random vector with a distribution that is determined by its moments. If $X_n = (X_{n1}, \ldots, X_{nd})$ are random vectors with finite moments such that $\mathbb{E}\, X_n^\alpha \to \mathbb{E}\, Z^\alpha$ as $n \to \infty$ for every multi-index $\alpha = (\alpha_1, \ldots, \alpha_d)$, then $X_n \overset{d}{\to} Z$.* ∎

A particularly important case is convergence to a normal distribution. We recall that the semi-factorial $n!!$ is defined to be $n(n-2)\cdots 3\cdot 1 = (2m)!/2^m m!$, when $n = 2m - 1$ is odd.

Corollary 6.3. *If $X_1, X_2, \dots$ are random variables with finite moments and a_n are positive numbers such that, as $n \to \infty$,*

$$\mathbb{E}(X_n - \mathbb{E}\,X_n)^k = \begin{cases} (k-1)!!\,a_n^k + o(a_n^k), & \text{when } k \geq 2 \text{ is even,} \\ o(a_n^k), & \text{when } k \geq 3 \text{ is odd,} \end{cases}$$

then $a_n^{-1}(X_n - \mathbb{E}\,X_n) \xrightarrow{d} \mathrm{N}(0,1)$ and $\widetilde{X}_n \xrightarrow{d} \mathrm{N}(0,1)$.

Proof. Let $X \in \mathrm{N}(0,1)$ be a standard normal variable and let $m_k = \mathbb{E}\,X^k$, $k = 1, 2, \dots$, be its moments. Then, as is well known, $m_k = (k-1)!!$ when k is even and $m_k = 0$ when k is odd.

Let $Y_n = a_n^{-1}(X_n - \mathbb{E}\,X_n)$. Then $\mathbb{E}\,Y_n = 0 = m_1$ and, by the assumption, $\mathbb{E}\,Y_n^k \to m_k$ as $n \to \infty$ for every $k \geq 2$. Consequently, $Y_n \xrightarrow{d} \mathrm{N}(0,1)$ by Theorem 6.1.

Finally, by the assumption with $k = 2$, $\mathrm{Var}(X_n)/a_n^2 \to 1$ and, thus, using Cramér's theorem (Section 1.2), $\widetilde{X}_n = (a_n^2/\operatorname{Var} X_n)^{1/2} Y_n \xrightarrow{d} \mathrm{N}(0,1)$ too. ∎

Example 6.4. Consider the subgraph count X_G in $\mathbb{G}(n,p)$, where $n \to \infty$ and $p = p(n)$ is a function of n. Asymptotic normality under various conditions has been shown by several authors. The complete result (Ruciński 1988) is as follows; as is shown after the proof, the conditions on p are necessary too.

Theorem 6.5. *Let G be a fixed graph with $e_G > 0$. If $n \to \infty$ and $p = p(n)$ is such that $np^{m(G)} \to \infty$ and $n^2(1-p) \to \infty$, then $\widetilde{X}_G \xrightarrow{d} \mathrm{N}(0,1)$.*

Proof. There are $\binom{n}{v} v!/\operatorname{aut}(G)$ copies of G in the complete graph K_n; for each such copy G' we define, as in Chapter 3, the random indicator variable $I_{G'} = \mathbf{1}[G' \subset \mathbb{G}(n,p)]$; note that $\mathbb{E}\,I_{G'} = p^{e_G}$. Then $X_G = \sum_{G'} I_{G'}$ and thus, for every $m \geq 1$,

$$\mathbb{E}(X_G - \mathbb{E}\,X_G)^m = \sum_{G_1,\dots,G_m} \mathbb{E}\big((I_{G_1} - \mathbb{E}\,I_{G_1})(I_{G_2} - \mathbb{E}\,I_{G_2}) \cdots (I_{G_m} - \mathbb{E}\,I_{G_m})\big), \tag{6.1}$$

summing over all m-tuples $G_1, \dots, G_m$ of copies of G in K_n. Let us write

$$T(G_1, \dots, G_m) = \mathbb{E}\big((I_{G_1} - \mathbb{E}\,I_{G_1}) \cdots (I_{G_m} - \mathbb{E}\,I_{G_m})\big),$$

and for each such term in the sum, define a graph $L = L(G_1, \dots, G_m)$ with vertex set $\{1, \dots, m\}$ and an edge ij whenever G_i and G_j have at least one edge in common. Thus L is a dependency graph for the variables $I_{G_1}, \dots, I_{G_m}$, see Example 1.6.

We now group the terms in the sum in (6.1) according to the structure of the graph L. Consider first the case when m is even and L consists of $m/2$ disjoint edges. There are $(m-1)!!$ such graphs L, and each gives the same contribution to the sum in (6.1); moreover it is easy to see that each contribution is $(\operatorname{Var} X_G)^{m/2}(1+O(1/n))$. We claim that every other graph L with m (even or odd) vertices contributes $o(\operatorname{Var} X_G)^{m/2}$; since there are finitely many possible L for each m, the result then follows by summing over L and applying Corollary 6.3, with $a_n^2 = \operatorname{Var} X_G$.

In order to verify the claim, observe first that if L has an isolated vertex i, then every term $T(G_1,\dots,G_m)$ yielding the graph L vanishes, because then $I_{G_i} - \mathbb{E}\, I_{G_i}$ is independent of the product of the other factors. Consequently, such L give no contribution at all.

In the remaining cases, every component of L has at least two vertices, and some component has at least three. We let $c = c(L)$ be the number of components; it follows that $c(L) < m/2$. Moreover, we may, for convenience, assume that the indices are reordered in such a way that the components of L have vertex sets $\{1,\dots,r_1\}$, $\{r_1+1,\dots,r_2\}$, ... , $\{r_{c-1}+1,\dots,r_c=m\}$, and that if $j \notin \{1, r_1+1, r_2+1, \dots, r_{c-1}+1\}$, then L contains an edge ij with $i<j$.

Consider a term $T(G_1,\dots,G_m)$ in (6.1) with such an L, let $G^{(j)} = \bigcup_1^j G_i$, and let F_j be the (possibly empty) subgraph of G which corresponds to $G^{(j-1)} \cap G_j$ under an isomorphism $G_j \cong G$. Note that by our assumption on L, $e(F_j) = 0$ exactly when $j \in \{1, r_1+1, r_2+1, \dots, r_{c-1}+1\}$.

If $p \le 1/2$, we estimate the term $T(G_1,\dots,G_m)$ by taking absolute values:

$$|T(G_1,\dots,G_m)| \le \mathbb{E}\big((I_{G_1} + \mathbb{E}\, I_{G_1})\cdots(I_{G_m} + \mathbb{E}\, I_{G_m})\big).$$

The product can be expanded as a sum of 2^m terms, and it is easily seen that among them, $I_{G_1}\cdots I_{G_m}$ has the largest expectation, namely, $p^{e(G^{(m)})}$. Thus

$$|T(G_1,\dots,G_m)| \le 2^m\, \mathbb{E}(I_{G_1}\cdots I_{G_m}) = O\big(p^{e(G^{(m)})}\big).$$

If $1/2 < p \le 1$, we instead use the estimate

$$|T(G_1,\dots,G_m)| \le \mathbb{E} \prod_{k=1}^{c(L)} |I_{G_{r_k}} - \mathbb{E}\, I_{G_{r_k}}|,$$

keeping only one factor for each component of L. These $c(L)$ factors are independent, and each has the expectation

$$\mathbb{E}\, |I_{G_r} - \mathbb{E}\, I_{G_r}| = 2p^{e_G}(1-p^{e_G}) \le 2(1-p^{e_G}) \le 2e_G(1-p).$$

Consequently,

$$T(G_1,\dots,G_m) = O\big((1-p)^{c(L)}\big).$$

We may combine the two cases by introducing redundant factors in each of the estimates; thus, for all $p \in [0,1]$,

$$T(G_1,\dots,G_m) = O\big((1-p)^{c(L)} p^{e(G^{(m)})}\big).$$

Now, as is easily seen, $e(G^{(j)}) = je_G - \sum_1^j e(F_i)$, so, in particular, $e(G^{(m)}) = me_G - \sum_1^m e(F_i)$. Similarly, $v(G^{(m)}) = mv_G - \sum_1^m v(F_i)$ and thus there are at most $O\big(n^{mv_G - \sum_1^m v(F_i)}\big)$ possible choices of $G_1, \dots, G_m$ yielding L and a particular sequence $F_1, \dots, F_m$. Consequently, fixing L and $F_1, \dots, F_m$ gives a contribution to the sum in (6.1) of the order

$$O\big(n^{mv_G - \sum_1^m v(F_i)}(1-p)^{c(L)} p^{me_G - \sum_1^m e(F_i)}\big). \tag{6.2}$$

We next recall that $c(L)$ of the F_i have no edges, and thus $n^{v(F_i)}p^{e(F_i)} = n^{v(F_i)} \geq 1$, while the $m - c(L)$ others have $e(F_i) \geq 1$ and thus $n^{v(F_i)}p^{e(F_i)} \geq \mathbb{E}(X_{F_i}) \geq \Phi_G$, *cf.* (3.7). Hence, using Lemma 3.5,

$$\begin{aligned} &n^{mv_G - \sum_1^m v(F_i)}(1-p)^{c(L)} p^{me_G - \sum_1^m e(F_i)} \\ &\quad = (1-p)^{c(L)}\big(n^{v_G}p^{e_G}\big)^m \prod_{i=1}^m \big(n^{v(F_i)}p^{e(F_i)}\big)^{-1} \\ &\quad \leq (1-p)^{c(L)}\big(n^{v_G}p^{e_G}\big)^m / \Phi_G^{m-c(L)} \asymp (1-p)^{c(L)}(\mathbb{E}\, X_G)^m \Phi_G^{c(L)-m} \\ &\quad \asymp (\operatorname{Var} X_G)^{m/2}\big((1-p)\Phi_G\big)^{c(L)-m/2}. \end{aligned} \tag{6.3}$$

Now the assumptions $np^{m(G)} \to \infty$ and $n^2(1-p) \to \infty$ imply $(1-p)\Phi_G \to \infty$. Indeed, as remarked in Chapter 3, $np^{m(G)} \to \infty$ is equivalent to $\Phi_G \to \infty$, which implies $(1-p)\Phi_G \to \infty$ provided $p \leq 1/2$. On the other hand, when $p > 1/2$, $\Phi_G \asymp n^2$, and thus $(1-p)\Phi_G \asymp n^2(1-p) \to \infty$.

Moreover, as observed above, $c(L) < m/2$. Consequently, by (6.2) and (6.3), the contribution to the sum in (6.1) by terms corresponding to L and $F_1, \dots, F_m$ is $o\big((\operatorname{Var} X_G)^{m/2}\big)$. Summing over the finitely many possible sequences $F_1, \dots, F_m$, this verifies our claim that the contribution by L is $o\big((\operatorname{Var} X_G)^{m/2}\big)$, which completes the proof as shown above. ∎

To see that the conditions $np^{m(G)} \to \infty$ and $n^2(1-p) \to \infty$ are necessary for the conclusion, we observe that if they are violated, then for some subsequence either $np^{m(G)} \to a < \infty$ or $n^2(1-p) \to b < \infty$. In the first case (along the subsequence) $\sup \Phi_G < \infty$ and thus, by Theorem 3.9, $\inf \mathbb{P}(X_G = 0) > 0$; in the second case $\mathbb{P}(\mathbb{G}(n,p) = K_n) \to e^{-b/2} > 0$, and thus $\inf \mathbb{P}(X_G = f(n,G)) > 0$, where $f(n,G)$ is the number of copies of G in the complete graph K_n. In both cases X_G thus assumes a single value with probability not tending to zero, which obviously rules out asymptotic normality (for any normalization).

Remark 6.6. It is easily seen by the argument above that the conditions $np^{m(G)} \to \infty$ and $n^2(1-p) \to \infty$ are together equivalent to $(1-p)\Phi_G \to \infty$.

Factorial moments

The *factorial moments* of a random variable X (with finite moments) are the numbers

$$\mathbb{E}(X)_k = \mathbb{E}[X(X-1)\cdots(X-k+1)], \qquad k \geq 0$$

(with $\mathbb{E}(X)_0 = 1$). Since the monomials $\{x^j\}_{j=0}^k$ and the (descending) factorials $\{(x)_j\}_{j=0}^k$ form two bases of the vector space of polynomials of degree at most k, there exist numbers a_{kj} and b_{kj} (independent of X), such that

$$\mathbb{E}(X)_k = \sum_{j=0}^{k} a_{kj}\, \mathbb{E}\, X^j$$

and

$$\mathbb{E}\, X^k = \sum_{j=0}^{k} b_{kj}\, \mathbb{E}(X)_j$$

for every random variable X with finite moments. (The coefficients a_{kj} and b_{kj} are the Stirling numbers, see, e.g., Graham, Knuth and Patashnik (1989), but their values do not matter here.) It follows that if $X_1, X_2, \ldots$ and X are random variables, then $\mathbb{E}\, X_n^k \to \mathbb{E}\, X^k$ for every k if and only if $\mathbb{E}(X_n)_k \to \mathbb{E}(X)_k$ for every k. Consequently, Theorem 6.1 can be reformulated as follows.

Theorem 6.7. *Let X be a random variable with a distribution that is determined by its moments. If $X_1, X_2, \ldots$ are random variables with finite moments such that $\mathbb{E}(X_n)_k \to \mathbb{E}(X)_k$ as $n \to \infty$ for every integer $k \geq 1$, then $X_n \xrightarrow{d} X$.* ∎

This form of the theorem is particularly convenient for proving convergence to a Poisson distribution, since the factorial moments of $X \in \mathrm{Po}(\lambda)$ have the simple form $\mathbb{E}(X)_k = \lambda^k$, $k \geq 0$. (In contrast, the moments of a Poisson variable have a more complicated form.)

The method of moments is often applied to counting variables of the form $S = \sum_{\alpha \in A} I_\alpha$, where I_α are indicator variables; in this case $(S)_k$ counts the number of (ordered) k-tuples of objects with $I_\alpha = 1$, that is,

$$(S)_k = \sum_{\alpha_1, \ldots, \alpha_k}^{*} I_{\alpha_1} \cdots I_{\alpha_k},$$

where $\sum^*$ denotes summation over all sequences of distinct indices $\alpha_1, \ldots, \alpha_k$, and thus the factorial moments have the useful expression

$$\mathbb{E}(S)_k = \sum_{\alpha_1, \ldots, \alpha_k}^{*} \mathbb{E}(I_{\alpha_1} \cdots I_{\alpha_k}) = \sum_{\alpha_1, \ldots, \alpha_k}^{*} \mathbb{P}(I_{\alpha_1} = \cdots = I_{\alpha_k} = 1). \tag{6.4}$$

Corollary 6.8. *Let $S_n = \sum_{\alpha \in A_n} I_{n,\alpha}$ be sums of indicator variables $I_{n,\alpha}$. If $\lambda \geq 0$ is such that, as $n \to \infty$,*

$$\mathbb{E}(S_n)_k = \sum_{\alpha_1,\dots,\alpha_k}^{*} \mathbb{P}(I_{n,\alpha_1} = \cdots = I_{n,\alpha_k} = 1) \to \lambda^k, \tag{6.5}$$

for every $k \geq 1$, then $S_n \xrightarrow{d} \mathrm{Po}(\lambda)$. ■

Example 6.9. In Theorem 3.19, we studied the subgraph count X_G of a strictly balanced graph G at the threshold ($np^{m(G)} \to c > 0$), and showed that then (6.5) is satisfied and thus $X_G \xrightarrow{d} \mathrm{Po}(\lambda)$, with $\lambda = c^{v_G} / \mathrm{aut}(G)$.

With only minor modifications in the computation of the expectations, the same argument applies to the random graph $\mathbb{G}(n, M)$, with $n(M/\binom{n}{2})^{m(G)} \to c$.

Theorem 6.7 extends easily to multivariate limits. In particular, for Poisson limits we have the following result.

Theorem 6.10. *Let $(X_n^{(1)}, \dots, X_n^{(m)})$ be vectors of random variables, where $m \geq 1$ is fixed. If $\lambda_1, \dots, \lambda_m \geq 0$ are such that, as $n \to \infty$,*

$$\mathbb{E}\big[(X_n^{(1)})_{k_1} \cdots (X_n^{(m)})_{k_m}\big] \to \lambda_1^{k_1} \cdots \lambda_m^{k_m}$$

for every $k_1, \dots, k_m \geq 0$, then $(X_n^{(1)}, \dots, X_n^{(m)}) \xrightarrow{d} (Z_1, \dots, Z_m)$, where $Z_i \in \mathrm{Po}(\lambda_i)$ are independent. ■

We leave the corresponding extension of Corollary 6.8 to the reader (Exercise!).

Cumulants

Suppose that X is a random variable with finite moments. Then the *characteristic function* $\varphi_X(t) = \mathbb{E}\, e^{itX}$ is infinitely differentiable, and

$$\mathbb{E}\, X^k = i^{-k} \frac{d^k}{dt^k} \varphi_X(0).$$

Similarly, $\log \varphi_X(t)$ is infinitely differentiable in an interval around 0, and we define the *cumulants* (also known as *semi-invariants*) of X by

$$\varkappa_k(X) = i^{-k} \frac{d^k}{dt^k} \log \varphi_X(0).$$

In other words, $\varphi_X(t)$ and $\log \varphi_X(t)$ have the Taylor series $\sum_0^\infty \mathbb{E}\, X^k \frac{(it)^k}{k!}$ and $\sum_0^\infty \varkappa_k(X) \frac{(it)^k}{k!}$, respectively.

Remark 6.11. In general, these Taylor series do not have to converge for any $t \neq 0$. However, we are mainly interested in random variables such that $\mathbb{E}\, e^{t|X|}$ exists for some $t > 0$, and then

$$\mathbb{E}\, e^{zX} = \sum_{0}^{\infty} \mathbb{E}\, X^k \frac{z^k}{k!} = \exp\Bigl(\sum_{1}^{\infty} \varkappa_k(X) \frac{z^k}{k!}\Bigr)$$

with sums converging at least for all complex z with $|z|$ sufficiently small.

Example 6.12. If X has the normal distribution $\mathrm{N}(\mu, \sigma^2)$, then $\varphi_X(t) = \exp(i\mu t - \sigma^2 t^2/2)$, and thus $\varkappa_1 = \mu$, $\varkappa_2 = \sigma^2$ and $\varkappa_k = 0$, $k \geq 3$.

Example 6.13. If X has the Poisson distribution $\mathrm{Po}(\lambda)$, we have $\varphi_X(t) = \exp(\lambda(e^{it} - 1))$, so $\log \varphi_X(t) = \lambda(e^{it} - 1) = \sum_1^\infty \lambda \frac{(it)^k}{k!}$ and $\varkappa_k = \lambda$, $k \geq 1$.

It is obvious by successive differentiations of $\log \varphi_X$ and $\varphi_X = \exp(\log \varphi_X)$ that there are simple algebraic relations between the moments and cumulants: $\varkappa_k = q_k(\mathbb{E}\, X, \dots, \mathbb{E}\, X^k)$ and $\mathbb{E}\, X^k = q_k'(\varkappa_1, \dots, \varkappa_k)$, where q_k and q_k' are some polynomials not depending on X. Explicit expressions are easily given (*cf.* Proposition 6.16(vi, vii) below), but are not important for us, with the exception of

$$\varkappa_1 = \mathbb{E}\, X, \tag{6.6}$$

$$\varkappa_2 = \mathbb{E}\, X^2 - (\mathbb{E}\, X)^2 = \operatorname{Var} X. \tag{6.7}$$

It follows that if $X_1, X_2, \dots$ and X are random variables, then $\mathbb{E}\, X_n^k \to \mathbb{E}\, X^k$ for every k if and only if $\varkappa_k(X_n) \to \varkappa_k(X)$ for every k. Consequently, Theorem 6.1 can be reformulated as follows.

Theorem 6.14. *Let X be a random variable with a distribution that is determined by its moments. If $X_1, X_2, \dots$ are random variables with finite moments such that $\varkappa_k(X_n) \to \varkappa_k(X)$ as $n \to \infty$ for every integer $k \geq 1$, then $X_n \xrightarrow{d} X$.* ∎

Cumulants are particularly convenient for proving convergence to a normal distribution. It follows easily from the definition that if X is a random variable and a and b are real numbers, then

$$\varkappa_k(aX + b) = \begin{cases} a\varkappa_1(X) + b, & k = 1, \\ a^k \varkappa_k(X), & k \geq 2. \end{cases}$$

Hence, Theorem 6.14 and Example 6.12 yield the following result for normal convergence.

Corollary 6.15. *If $X_1, X_2, \dots$ are random variables with finite moments and a_n are positive numbers such that, as $n \to \infty$, $\varkappa_2(X_n)/a_n^2 \to \sigma^2 \geq 0$ and $\varkappa_k(X_n) = o(a_n^k)$ when $k \geq 3$, then $a_n^{-1}(X_n - \mathbb{E}\, X_n) \xrightarrow{d} \mathrm{N}(0, \sigma^2)$ and, provided $\sigma^2 > 0$, $\widetilde{X}_n \xrightarrow{d} \mathrm{N}(0,1)$.* ∎

Mixed cumulants

For the results in the last subsection to be useful, we have to be able to compute, or at least estimate, the cumulants $\varkappa_k(X_n)$. This can often be done using mixed cumulants as follows.

If $X_1, \dots, X_k$, $k \geq 1$, are random variables defined on the same probability space, then their joint characteristic function is

$$\varphi_{X_1,\dots,X_k}(t_1, \dots, t_k) = \mathbb{E}\Big(\exp\Big(\sum_j it_jX_j\Big)\Big).$$

If further $X_1, \dots, X_k$ have finite moments, this function is infinitely differentiable on $\mathbb{R}^k$ and we define the *mixed cumulants* by

$$\varkappa(X_1, \dots, X_k) = i^{-k}\frac{\partial^k}{\partial t_1 \cdots \partial t_k} \log \varphi_{X_1,\dots,X_k}(0).$$

Some basic properties of the mixed cumulants are collected in the next proposition. We omit the simple proofs (Leonov and Shiryaev 1959).

Proposition 6.16. *For any random variables $X, X_1, \dots$ with finite moments (defined on the same probability space), the following holds:*

(i) $\varkappa_k(X) = \varkappa(X, \dots, X)$, *where X is repeated k times.*

(ii) $\varkappa(X_1, \dots, X_k)$ *is symmetric, i.e.,* $\varkappa(X_1, \dots, X_k) = \varkappa(X_{\sigma(1)}, \dots, X_{\sigma(k)})$ *for every permutation σ of* $\{1, \dots, k\}$.

(iii) $\varkappa(aX_1, X_2, \dots, X_k) = a\varkappa(X_1, X_2, \dots, X_k)$, *for any real a.*

(iv) $\varkappa(X_1' + X_1'', X_2, \dots, X_k) = \varkappa(X_1', X_2, \dots, X_k) + \varkappa(X_1'', X_2, \dots, X_k)$.

(v) $\varkappa(X_1, \dots, X_k) = 0$ *if* $\{X_1, \dots, X_k\}$ *can be partitioned into two (or more) non-empty sets of random variables which are independent of each other.*

(vi) $\varkappa(X_1, \dots, X_k) = \sum_{I_1,\dots,I_l}(-1)^{l-1}(l-1)!\prod_{r=1}^{l}\mathbb{E}\prod_{j\in I_r}X_j$, *summing over all partitions of* $\{1, \dots, k\}$ *into non-empty sets* $\{I_1, \dots, I_l\}$, $l \geq 1$.

(vii) $\mathbb{E}(X_1 \cdots X_k) = \sum_{I_1,\dots,I_l}\prod_{r=1}^{l}\varkappa(\{X_i : i \in I_r\})$, *summing as in* (vi). ∎

By Hölder's inequality, $|\mathbb{E}\prod_{j\in I}X_j| \leq \prod_{j\in I}(\mathbb{E}|X_j|^k)^{1/k}$ when $|I| \leq k$, and thus (vi) implies the useful estimate

$$|\varkappa(X_1, \dots, X_k)| \leq C_k\prod_1^k(\mathbb{E}|X_j|^k)^{1/k},$$

where C_k depends on k only. We will use a more refined estimate, which we state as a lemma (Mikhailov 1991); recall the definition of dependency graph in Section 1.2 and let

$$\overline{N}_L(\alpha_1, \ldots, \alpha_r) = \bigcup_{i=1}^{r} \{\beta \in V(L) : \beta = \alpha_i \text{ or } \alpha_i\beta \in E(L)\} \tag{6.8}$$

denote the closed neighborhood of $\{\alpha_1, \ldots, \alpha_r\}$ in L.

Lemma 6.17. *Suppose that $S = \sum_{\alpha\in A} X_\alpha$, where $\{X_\alpha\}_{\alpha\in A}$ is a family of random variables with dependency graph L. Suppose, moreover, that $r \geq 2$ and that M and T are numbers such that*

$$\sum_{\alpha\in A} \mathbb{E}\,|X_\alpha| \leq M$$

and, for every $\alpha_1, \ldots, \alpha_{r-1} \in A$,

$$\sum_{\alpha\in\overline{N}_L(\alpha_1,\ldots,\alpha_{r-1})} \mathbb{E}(|X_\alpha| \mid X_{\alpha_1}, \ldots, X_{\alpha_{r-1}}) \leq T.$$

Then

$$|\varkappa_r(S)| \leq C_r M T^{r-1},$$

where C_r is a constant that depends on r only.

Proof. By Proposition 6.16(i, ii, iv)

$$\varkappa_r(S) = \varkappa(S, \ldots, S) = \sum_{\alpha_1,\ldots,\alpha_r} \varkappa(X_{\alpha_1}, \ldots, X_{\alpha_r}),$$

where, by Proposition 6.16(v), every term in the sum for which $\{\alpha_1, \ldots, \alpha_r\}$ forms a disconnected subgraph of L vanishes. In each of the remaining terms, the indices $\alpha_1, \ldots, \alpha_r$ may be reordered such that $\alpha_s \in \overline{N}_L(\alpha_1, \ldots, \alpha_{s-1})$ when $2 \leq s \leq r$. Hence

$$|\varkappa_r(S)| \leq r! \sum_{\alpha_1,\ldots,\alpha_r}^{*} |\varkappa(X_{\alpha_1}, \ldots, X_{\alpha_r})|, \tag{6.9}$$

where $\sum^*$ denotes the sum over $\alpha_1 \in A$, $\alpha_2 \in \overline{N}_L(\alpha_1)$, $\alpha_3 \in \overline{N}_L(\alpha_1, \alpha_2)$, ..., $\alpha_r \in \overline{N}_L(\alpha_1, \ldots, \alpha_{r-1})$.

Define, for every sequence $\alpha_1, \ldots, \alpha_s$ in A, $s \geq 1$,

$$Y_{\alpha_1,\ldots,\alpha_s} = \sum_{I_1,\ldots,I_k} \prod_{j=1}^{k} \mathbb{E} \prod_{i\in I_j} |X_{\alpha_i}|,$$

summing over all partitions $\{I_1, \ldots, I_k\}$ of $\{1, \ldots, s\}$; by Proposition 6.16(vi),

$$|\varkappa(X_{\alpha_1}, \ldots, X_{\alpha_r})| \leq C'_r Y_{\alpha_1,\ldots,\alpha_r}. \tag{6.10}$$

Now consider a sequence $\alpha_1, \ldots, \alpha_{s-1}$ in A, with $2 \leq s \leq r$. Let $\{I_1, \ldots, I_k\}$ be a partition of $\{1, \ldots, s\}$. We may assume that $s \in I_1$; then let $I'_1 = I_1 \setminus \{s\}$. By assumption (since we may define $\alpha_j = \alpha_{s-1}$ for $s \leq j < r$),

$$\mathbb{E}\Bigl(\sum_{\alpha \in \overline{N}_L(\alpha_1,\ldots,\alpha_{s-1})} |X_\alpha| \Bigm| X_{\alpha_1}, \ldots, X_{\alpha_{s-1}}\Bigr) \leq T$$

and thus

$$\mathbb{E}\Bigl(\prod_{i \in I'_1} |X_{\alpha_i}| \sum_{\alpha \in \overline{N}_L(\alpha_1,\ldots,\alpha_{s-1})} |X_\alpha| \Bigm| X_{\alpha_1}, \ldots, X_{\alpha_{s-1}}\Bigr) \leq T \prod_{i \in I'_1} |X_{\alpha_i}|.$$

Taking the expectation of both sides, we see that

$$\begin{aligned}\sum_{\alpha_s \in \overline{N}_L(\alpha_1,\ldots,\alpha_{s-1})} \mathbb{E} \prod_{i \in I_1} |X_{\alpha_i}| &= \mathbb{E}\Bigl[\prod_{i \in I'_1} |X_{\alpha_i}| \sum_{\alpha_s \in \overline{N}_L(\alpha_1,\ldots,\alpha_{s-1})} |X_{\alpha_s}|\Bigr] \\ &\leq T\,\mathbb{E} \prod_{i \in I'_1} |X_{\alpha_i}|.\end{aligned}$$

Multiplying this inequality by $\prod_{j=2}^k \mathbb{E} \prod_{i \in I_j} |X_{\alpha_i}|$ and summing over all partitions $\{I_1, \ldots, I_k\}$, we obtain on the right-hand side T times a sum, whose terms coincide with the terms in $Y_{\alpha_1,\ldots,\alpha_{s-1}}$, each repeated at most s times. Consequently,

$$\sum_{\alpha_s \in \overline{N}_L(\alpha_1,\ldots,\alpha_{s-1})} Y_{\alpha_1,\ldots,\alpha_s} \leq sTY_{\alpha_1,\ldots,\alpha_{s-1}}.$$

We next sum this inequality over $\alpha_1, \ldots, \alpha_{s-1}$ and obtain

$$\sum_{\alpha_1,\ldots,\alpha_s}^{*} Y_{\alpha_1,\ldots,\alpha_s} \leq sT \sum_{\alpha_1,\ldots,\alpha_{s-1}}^{*} Y_{\alpha_1,\ldots,\alpha_{s-1}}, \qquad 2 \leq s \leq r.$$

Since $\sum_{\alpha_1} Y_{\alpha_1} = \sum_\alpha \mathbb{E}\,|X_\alpha| \leq M$ by assumption, induction now yields

$$\sum\nolimits^{*} Y_{\alpha_1,\ldots,\alpha_s} \leq s!\,MT^{s-1}, \qquad 1 \leq s \leq r. \tag{6.11}$$

The result follows by (6.9), (6.10), and (6.11). ∎

Applications to asymptotic normality

The lemma above leads to the following sufficient condition for asymptotic normality (Mikhailov 1991).

Theorem 6.18. *Suppose that $(S_n)_1^\infty$ is a sequence of random variables such that $S_n = \sum_{\alpha\in A_n} X_{n\alpha}$, where for each n, $\{X_{n\alpha}\}_\alpha$ is a family of random variables with dependency graph L_n. Suppose further that there exist numbers M_n, Q_n and B_r such that*

$$\sum_{\alpha\in A_n} \mathbb{E}\,|X_{n\alpha}| \le M_n \tag{6.12}$$

and, for every n and $r \ge 1$, and $\alpha_1, \ldots, \alpha_r \in A_n$,

$$\sum_{\alpha\in \overline{N}_{L_n}(\alpha_1,\ldots,\alpha_r)} \mathbb{E}\big(|X_{n\alpha}| \bigm| X_{n\alpha_1}, \ldots, X_{n\alpha_r}\big) \le B_r Q_n. \tag{6.13}$$

Let $\sigma_n^2 = \operatorname{Var} S_n$. If, as $n \to \infty$,

$$\frac{M_n Q_n^2}{\sigma_n^3} \to 0 \tag{6.14}$$

then

$$\widetilde{S}_n \xrightarrow{d} \mathrm{N}(0,1).$$

Proof. By Lemma 6.17, with $T = B_{r-1}Q_n$,

$$|\varkappa_r(S_n)| \le C_r B_{r-1}^{r-1} M_n Q_n^{r-1}, \qquad r \ge 2, \tag{6.15}$$

and thus, with $C_r' = C_r B_{r-1}^{r-1}$,

$$|\varkappa_r(\widetilde{S}_n)| \le C_r' \frac{M_n Q_n^{r-1}}{\sigma_n^r} = C_r' \left(\frac{M_n Q_n^2}{\sigma_n^3}\right)^{r-2} \left(\frac{\sigma_n^2}{M_n Q_n}\right)^{r-3}.$$

Since $\sigma_n^2 = \varkappa_2(S_n) \le C_2' M_n Q_n$, by (6.15),

$$|\varkappa_r(\widetilde{S}_n)| \le C_r' (C_2')^{r-3} \left(\frac{M_n Q_n^2}{\sigma_n^3}\right)^{r-2} \to 0, \qquad r \ge 3.$$

The result follows by Theorem 6.14. (Note that $\varkappa_1(\widetilde{S}_n) = 0$ and $\varkappa_2(\widetilde{S}_n) = 1$.) ∎

Example 6.19. Consider again the subgraph count X_G in $\mathbb{G}(n,p)$, where $n \to \infty$ and $p = p(n)$ is a function of n. We will use Theorem 6.18 to give a new proof of Theorem 6.5.

This time we denote the family of subgraphs of K_n that are isomorphic to G by $\{G_\alpha\}_{\alpha\in A_n}$. We let $I_\alpha = \mathbf{1}[G_\alpha \subset \mathbb{G}(n,p)]$ and $X_\alpha = I_\alpha - \mathbb{E}\,I_\alpha$, for simplicity omitting subscripts n. (If we assume that $p(n)$ is bounded away from 1 (for example, $p \le 1/2$), we can also apply the theorem with $X_\alpha = I_\alpha$. We leave this slightly simpler version to the reader. – Exercise!) Define the graph L_n with vertex set A_n by connecting every pair of indices α and β

such that the corresponding graphs G_α and G_β have a common edge; this is evidently a dependency graph for $\{X_\alpha\}$ (Example 1.6).

We verify the conditions of the theorem. First, observe that $\mathbb{E}|X_\alpha| = 2\,\mathbb{E}\,I_\alpha(1-\mathbb{E}\,I_\alpha) \le 2v_G(1-p)\,\mathbb{E}\,I_\alpha$, and thus $\sum_{\alpha\in A_n}\mathbb{E}|X_\alpha| \le 2v_G(1-p)\,\mathbb{E}\,X_G$, that is, (6.12) holds with $M_n = 2v_G(1-p)\,\mathbb{E}\,X_G$.

Next, suppose that $r \ge 1$ and $\alpha_1,\dots,\alpha_r \in A_n$ are given. Define $F = G_{\alpha_1}\cup\cdots\cup G_{\alpha_r}$ and, for every $\alpha\in A_n$, $F_\alpha = G_\alpha \cap F$. Note that $\alpha \in \overline{N}_{L_n}(\alpha_1,\dots,\alpha_r)$ if and only if $e(F_\alpha)\ne 0$. There are fewer than $2^{v_F} \le 2^{rv_G}$ such subgraphs of F, and for each subgraph $H \subseteq F$ there are $O(n^{v_G-v_H})$ choices of α such that $F_\alpha = H$, each with

$$\mathbb{E}\big(|X_\alpha| \mid X_{\alpha_1},\dots,X_{\alpha_{r-1}}\big) \le \mathbb{E}\big(I_\alpha \mid X_{\alpha_1},\dots,X_{\alpha_{r-1}}\big) + \mathbb{E}\,I_\alpha \le 2p^{e_G-e_H}.$$

Since, further, each F_α is isomorphic to a subgraph of G, it follows that

$$\sum_{\alpha\in\overline{N}_{L_n}(\alpha_1,\dots,\alpha_r)} \mathbb{E}\big(|X_\alpha| \mid X_{\alpha_1},\dots,X_{\alpha_{r-1}}\big) \le B_r \sup_{H\subseteq G,\, e_H\ge 1}\frac{\mathbb{E}\,X_G}{\mathbb{E}\,X_H} = B_r\frac{\mathbb{E}\,X_G}{\Phi_G},$$

for a suitable B_r, depending on r and G. Consequently, (6.13) holds with $Q_n = \mathbb{E}\,X_G/\Phi_G$.

Finally, by Lemma 3.5, $\sigma_n^2 = \operatorname{Var}(X_G) \asymp (1-p)(\mathbb{E}\,X_G)^2/\Phi_G$. It follows that

$$\frac{M_nQ_n^2}{\sigma_n^3} = \frac{2v_G(1-p)(\mathbb{E}\,X_G)^3\Phi_G^{-2}}{\sigma_n^3} = O\big((1-p)^{-1/2}\Phi_G^{-1/2}\big). \tag{6.16}$$

This verifies (6.14), since, as shown in the proof of Theorem 6.5, the conditions imply $(1-p)\Phi_G \to \infty$.

Theorem 6.18 thus applies; since $X_G = \sum_\alpha I_\alpha$, and thus $X_G - \mathbb{E}\,X_G = \sum_\alpha X_\alpha$, the result follows.

Remark 6.20. The results above may be improved by weakening the assumptions; in fact, we know of two such improvements in different directions. (We do not know whether they can be combined.)

First, as will be shown later by a different method (Theorem 6.33), it suffices to verify condition (6.13) in Theorem 6.18 for $r = 2$.

Second, by using a theorem by Marcinkiewicz (1939), which states that the normal distributions are the only ones with all but a finite number of cumulants vanishing, it follows (Janson 1988) that in Theorem 6.14 it is enough to assume that the condition $\varkappa_k(X_n) \to \varkappa_k(X)$ holds for $k \ge m$, for any fixed $m < \infty$. (See also Grimmett (1992b).) This improvement is useful in cases where there is a general method to obtain desired estimates for all cumulants of sufficiently large order, although the method fails for a few low-order ones. In particular, it leads to the following strengthening of Theorem 6.18 (Mikhailov 1991). For an application to random graphs, see Janson (1988).

Theorem 6.21. *Suppose that the assumptions of Theorem 6.18 hold, except that* (6.14) *is replaced by*

$$\frac{M_n Q_n^{s-1}}{\sigma_n^s} \to 0 \tag{6.17}$$

for some real $s > 2$. *Then the conclusion still holds, that is,*

$$\widetilde{S}_n \xrightarrow{d} \mathrm{N}(0,1).$$

Proof. If $r \geq s$, then

$$\frac{M_n Q_n^{r-1}}{\sigma_n^r} = \left(\frac{M_n Q_n^{s-1}}{\sigma_n^s}\right)^{\frac{r-2}{s-2}} \left(\frac{\sigma_n^2}{M_n Q_n}\right)^{\frac{r-s}{s-2}} \leq C''_{r,s} \left(\frac{M_n Q_n^{s-1}}{\sigma_n^s}\right)^{\frac{r-2}{s-2}}$$

by (6.15) (with $r = 2$) and, thus, by (6.15) and (6.17), $\varkappa_r(\widetilde{S}_n) \to 0$. The result follows by the improvement of Theorem 6.14 just mentioned. ∎

Note that the proof shows that the assumption (6.17) becomes weaker as s increases. It can be reformulated as

$$\frac{M_n}{\sigma_n}\left(\frac{Q_n}{\sigma_n}\right)^{s-1} \to 0.$$

6.2 STEIN'S METHOD: THE POISSON CASE

A method to show convergence to the normal distribution was given by Stein (1972). The method has later been extended to several other limit distributions; we will here only consider the simplest and most important cases, namely, Stein's original normal case (which is treated in the next section) and Chen's (1975) version for the Poisson case (which is treated below). The method was introduced in the theory of random graphs by Barbour (1982).

Stein's method is well adapted to the type of sums of random variables that appear in many combinatorial applications; it then often leads to calculations very similar to those needed for estimating the second and (for the normal case) third moments when applying the method of moments. Consequently, Stein's method often requires less effort and simpler combinatorial arguments than the method of moments, where we have to estimate moments of arbitrary order.

An important feature of Stein's method is that it does not only give convergence; it actually gives an explicit upper bound for the distance between the distribution of a given random variable and a suitable normal or Poisson distribution. In other words, it is really a method to prove normal or Poisson *approximation* rather than convergence. Hence it leads to estimates of the rate of convergence, which, in practice, often turn out to be of the right order of magnitude.

Here we have talked about the 'distance' between two distributions without explaining what it is. In fact, several possible distances can be defined; for Poisson approximation the most useful is the following.

The *total variation distance* between the distributions of two random variables X and Y is, in general, defined by

$$d_{TV}(X,Y) = \sup_A |\mathbb{P}(X \in A) - \mathbb{P}(Y \in A)|,$$

taking the supremum over all Borel sets A. If X and Y are integer valued, as in the cases we consider below, this is equivalent to

$$d_{TV}(X,Y) = \tfrac{1}{2}\sum_k |\mathbb{P}(X = k) - \mathbb{P}(Y = k)|.$$

We also use hybrid notation, such as $d_{TV}(X, \mathrm{Po}(\lambda))$.

It is easily seen that if $(X_n)_1^\infty$ is a sequence of random variables, and $(\lambda_n)_1^\infty$ is a sequence of positive numbers with $\lambda_n \to \lambda$, then $d_{TV}(X_n, \mathrm{Po}(\lambda_n)) \to 0$ if and only if $X_n \xrightarrow{d} \mathrm{Po}(\lambda)$. Moreover, if $d_{TV}(X_n, \mathrm{Po}(\lambda_n)) \to 0$ and $\lambda_n \to \infty$, the central limit theorem for $\mathrm{Po}(\lambda_n)$ implies that $(X_n - \lambda_n)/\lambda_n^{1/2} \xrightarrow{d} \mathrm{N}(0,1)$. In particular, if further $\lambda_n = \mathbb{E}\, X_n$ and $\mathrm{Var}(X_n) \sim \lambda_n$, then $\widetilde{X}_n \xrightarrow{d} \mathrm{N}(0,1)$. Hence estimates of the total variation distance to a Poisson distribution can imply convergence to both Poisson distributions and normal distributions. (Not all cases of normal convergence are obtained in this way, however; typically we may obtain the cases when the mean and variance are asymptotically equal.)

For the theoretical background for the Stein–Chen method for Poisson approximation we refer to Chen (1975), Stein (1986) and Barbour, Holst and Janson (1992). These references also show how the method leads to explicit results such as the ones below (as well as others). A useful and rather general result obtained by the Stein–Chen method is the following (Barbour, Holst and Janson 1992).

Theorem 6.22. *Suppose that $X = \sum_{\alpha \in A} I_\alpha$, where the I_α are random indicator variables, and suppose that, for each $\alpha \in A$, there exists a family of random indicator variables $J_{\beta\alpha}$, $\beta \in A \setminus \{\alpha\}$, such that*

$$\mathcal{L}(\{J_{\beta\alpha}\}_\beta) = \mathcal{L}(\{I_\beta\}_\beta \mid I_\alpha = 1), \tag{6.18}$$

that is, the joint distribution of $\{J_{\beta\alpha}\}_\beta$ equals the conditional distribution of $\{I_\beta\}_\beta$ given $I_\alpha = 1$. Then, with $\pi_\alpha = \mathbb{E}\, I_\alpha$ and $\lambda = \mathbb{E}\, X = \sum_{\alpha \in A} \pi_\alpha$,

$$d_{TV}(X, \mathrm{Po}(\lambda)) \le \min(\lambda^{-1}, 1) \sum_{\alpha \in A} \pi_\alpha \Big(\pi_\alpha + \sum_{\beta \ne \alpha} \mathbb{E}\, |J_{\beta\alpha} - I_\beta|\Big). \tag{6.19}$$

∎

One way to apply Theorem 6.22 without explicit construction of the variables $J_{\beta\alpha}$ is via a dependency graph. In fact, if the family $\{I_\alpha\}$ has a dependency graph L, then there exist random variables $J_{\beta\alpha}$ with the right distribution (6.18) such that $J_{\beta\alpha} = I_\beta$ when $\alpha\beta \notin E(L)$, so in (6.19) it suffices to consider β that are adjacent to α. For such β we may crudely use $|J_{\beta\alpha} - I_\beta| \le J_{\beta\alpha} + I_\beta$ together with the general relation

$$\pi_\alpha \mathbb{E}\, J_{\beta\alpha} = \pi_\alpha \mathbb{E}(I_\beta \mid I_\alpha = 1) = \mathbb{P}(I_\beta = I_\alpha = 1) = \mathbb{E}(I_\alpha I_\beta), \tag{6.20}$$

which yields

$$\pi_\alpha \mathbb{E}\,|J_{\beta\alpha} - I_\beta| \le \mathbb{E}(I_\alpha I_\beta) + \pi_\alpha \pi_\beta.$$

This leads to the following result.

Theorem 6.23. *Suppose that $X = \sum_{\alpha\in A} I_\alpha$, where the I_α are random indicator variables with a dependency graph L. Then, with $\pi_\alpha = \mathbb{E}\, I_\alpha$ and $\lambda = \mathbb{E}\, X = \sum_{\alpha\in A} \pi_\alpha$ (and with summation over ordered pairs (α, β)),*

$$\begin{aligned} d_{TV}(X, \mathrm{Po}(\lambda)) &\le \min(\lambda^{-1}, 1)\Big(\sum_{\alpha\in A} \pi_\alpha^2 + \sum_{\alpha,\beta:\ \alpha\beta\in E(L)} \big(\mathbb{E}(I_\alpha I_\beta) + \mathbb{E}\, I_\alpha\, \mathbb{E}\, I_\beta\big)\Big) \\ &= \min(\lambda^{-1}, 1)\Big(\mathrm{Var}\, X - \mathbb{E}\, X + 2 \sum_{\alpha,\beta:\ \alpha\beta\in E(L)} \pi_\alpha \pi_\beta + 2 \sum_{\alpha\in A} \pi_\alpha^2\Big). \end{aligned}$$

■

A simple case of Theorem 6.22 is when $J_{\beta\alpha} - I_\beta$ has constant sign. We say that the random indicator variables $(I_\alpha)_{\alpha\in A}$ are *positively related* if, for each $\alpha \in A$, there exist random variables $J_{\beta\alpha}$ with the distribution (6.18), such that $J_{\beta\alpha} \ge I_\beta$ for every $\beta \ne \alpha$; similarly, the variables are *negatively related* if, for each $\alpha \in A$, there exist such $J_{\beta\alpha}$ with $J_{\beta\alpha} \le I_\beta$ for every $\beta \ne \alpha$.

For positively related variables, (6.20) yields

$$\pi_\alpha \mathbb{E}\,|J_{\beta\alpha} - I_\beta| = \pi_\alpha \mathbb{E}(J_{\beta\alpha} - I_\beta) = \mathbb{E}(I_\alpha I_\beta) - \pi_\alpha \pi_\beta, \tag{6.21}$$

which leads to the following consequence of Theorem 6.22; note that the variables $J_{\beta\alpha}$ do not appear explicitly (although their existence is essential). See further Barbour, Holst and Janson (1992), where also a corresponding result for negatively related variables and other similar results are given.

Theorem 6.24. *Suppose that $X = \sum_{\alpha\in A} I_\alpha$, where the I_α are positively related random indicator variables. Then, with $\pi_\alpha = \mathbb{E}\, I_\alpha$ and $\lambda = \mathbb{E}\, X = \sum_{\alpha\in A} \pi_\alpha$,*

$$\begin{aligned} d_{TV}(X, \mathrm{Po}(\lambda)) &\le \min(\lambda^{-1}, 1)\Big(\mathrm{Var}\, X - \mathbb{E}\, X + 2\sum_{\alpha\in A} \pi_\alpha^2\Big) \\ &\le \frac{\mathrm{Var}\, X}{\mathbb{E}\, X} - 1 + 2 \max_{\alpha\in A} \pi_\alpha. \end{aligned}$$

■

Returning to asymptotics, with variables depending on a parameter $n \to \infty$, we thus see that for sums of positively related variables, a sufficient condition for Poisson approximation (with an error tending to 0 as $n \to \infty$) is that the individual probabilities tend to 0 (uniformly) and that the variance is asymptotic to the mean. Since any Poisson distribution has the variance equal to the mean, the latter condition is very natural.

Remark 6.25. Indicator variables that are positively (negatively) related are positively (negatively) correlated, but the converse does not hold. Correlation is only a pairwise property, whereas being positively or negatively related depends on the joint distribution of the whole family.

Example 6.26. Consider again X_G, the number of copies of a fixed graph G in $\mathbb{G}(n,p)$, and suppose that G is strictly balanced. We proved in Theorem 3.19, using the method of moments, that if $np^{m(G)} \to c > 0$, then $X_G \xrightarrow{d} \mathrm{Po}(\lambda)$ with $\lambda = c^{v_G} / \mathrm{aut}(G)$. Here we show how this result follows by the Stein–Chen method. The method further gives an explicit estimate $O(n^{-\beta})$ of the rate of convergence, with $\beta = \min\{v_H - e_H/d(G) : H \subsetneq G,\, e_H > 0\} > 0$, which we, however, leave to the reader (Exercise!).

We write $X_G = \sum_{G'} I_{G'}$ as in the proof of Theorem 6.5 and observe that the sum has $(1+o(1))n^{v_G} / \mathrm{aut}(G)$ terms, each having expectation p^{e_G}. Thus

$$\mathbb{E}\, X_G \sim n^{v_G} p^{e_G} / \mathrm{aut}(G) = (np^{d(G)})^{v_G} / \mathrm{aut}(G) \to \lambda. \tag{6.22}$$

Moreover, since G is strictly balanced, a similar calculation yields $\mathbb{E}\, X_H \to \infty$ for every proper subgraph H of G. It follows as in (3.10), considering the terms with $G' = G''$ and thus $H = G$ separately,

$$\begin{aligned}\mathrm{Var}(X_G) &= \sum_{G'} \mathrm{Var}(I_{G'}) + O\Bigl(\sum_{H \subsetneq G,\, e_H > 0} \frac{(\mathbb{E}\, X_G)^2}{\mathbb{E}\, X_H}\Bigr) \\ &= (1 - p^{e_G})\, \mathbb{E}\, X_G + o(1).\end{aligned}$$

Hence $\mathrm{Var}\, X_G / \mathbb{E}\, X_G \to 1$ and, since further $\max \mathbb{E}\, I_{G'} = p^{e_G} \to 0$, the result follows by Theorem 6.24, provided we can show that the variables $I_{G'}$ are positively related. This can be verified as follows.

Fix a copy G' of G in K_n. The conditional distribution of $\mathbb{G}(n,p)$ given $I_{G'} = 1$ is the same as the distribution of the union $\mathbb{G}(n,p) \cup G'$, obtained by adding the edges in G' to $\mathbb{G}(n,p)$. Consequently, we may define $J_{G''G'} = \mathbf{1}[G'' \subset \mathbb{G}(n,p) \cup G']$ with G'' ranging over the copies of G in K_n; these variables have the correct joint distribution (6.18) and evidently $J_{G''G'} \geq I_{G''}$.

In this example we may alternatively apply Theorem 6.23 with L as in the proof of Theorem 6.5; this yields the same result (without having to verify that the variables are positively related), with only a slightly worse bound for $d_{TV}(X_G, \mathrm{Po}(\lambda))$.

In the preceding example, as in many others, it is easy to construct explicitly variables $J_{\beta\alpha}$ to show that the variables I_α are positively related. An alternative is to deduce the existence of suitable $J_{\beta\alpha}$ by the following abstract result. (Recall that we do not really care what the variables $J_{\beta\alpha}$ are, once we know that they exist with (6.18) and $J_{\beta\alpha} \geq I_\beta$.) For a proof, using the FKG inequality (Theorem 2.12), see Barbour, Holst and Janson (1992, Section 2.2) and the references given there.

Theorem 6.27. *Suppose that the indicator variables $\{I_\alpha\}_{\alpha\in A}$ all are increasing functions of some underlying independent random variables $\{Y_j\}$. Then the variables $\{I_\alpha\}_{\alpha\in A}$ are positively related; in particular, Theorem 6.24 applies to their sum.* ∎

For example, the subgraph counts in Example 6.26 are increasing funcions of the edge indicators, and we see immediately that they are positively related.

Example 6.28. Let X be the number of isolated vertices in $\mathbb{G}(n,p)$. Clearly, $X = \sum_1^n I_i$, where $I_i = \mathbf{1}[\text{vertex } i \text{ is isolated}]$. In this case, the indicator variables I_i are *decreasing* functions of the edge indicators in $\mathbb{G}(n,p)$, but that is just as good; we can apply Theorem 6.27 with Y_j being the edge indicators in the complement of $\mathbb{G}(n,p)$, or use the explicit construction $J_{ji} = \mathbf{1}[j \text{ is isolated in } \mathbb{G}_i]$, where $\mathbb{G}_i$ equals $\mathbb{G}(n,p)$ with all edges from vertex i removed. Either way, it follows that the I_i are positively related. (In this example, Theorem 6.23 is not useful.)

For example, if $p = \log n/n + c/n$ for some fixed real constant c, then

$$\mathbb{E}\,X = n(1-p)^{n-1} \sim ne^{-np} = e^{-c}$$

and

$$\begin{aligned}\operatorname{Var} X &= \bigl(1-(1-p)^{n-1}\bigr)\,\mathbb{E}\,X + n(n-1)\bigl((1-p)^{2n-3} - (1-p)^{2(n-1)}\bigr)\\ &= \mathbb{E}\,X + o(1),\end{aligned}$$

and Theorem 6.24 shows that $X \xrightarrow{d} \operatorname{Po}(e^{-c})$. In particular,

$$\mathbb{P}\bigl(\mathbb{G}(n,p) \text{ has no isolated vertices}\bigr) = \mathbb{P}(X=0) \to e^{-e^{-c}}.$$

(This yields another proof of Corollary 3.31.)

The number of vertices of degree at most a given number $d > 0$ can be treated in the same way. A similar argument works also for the number S_d of vertices of degree exactly d, but this time the corresponding indicators I_i are not positively related, and we use Theorem 6.22. This yields the result, first proved by Erdős and Rényi (1960) by the method of moments, that $S_d \xrightarrow{d} \operatorname{Po}(c)$ if $\mathbb{E}\,S_d \to c < \infty$ (which, for $d \geq 1$, happens in two ranges of p). For details, see Karoński and Ruciński (1987) and Barbour, Holst and Janson (1992), where also further examples are given.

Example 6.29. Let, as in Section 3.6, T_G be the number of isolated copies of G in $\mathbb{G}(n,p)$. Then $X_G = \sum_{G'} I_{G'}$, where, as in Example 6.26, G' ranges over the copies of G in K_n, but now $I_{G'}$ is the indicator that G' is an isolated subgraph of $\mathbb{G}(n,p)$.

In order to define suitable random variables $J_{G''G'}$, we first define, for a given G', a modification $\widetilde{\mathbb{G}}$ of $\mathbb{G}(n,p)$ by adding all edges in G' and deleting every other edge in $\mathbb{G}(n,p)$ incident with a vertex in G'. We then define $J_{G''G'}$ to be the indicator that G'' is an isolated subgraph of $\widetilde{\mathbb{G}}$. Then $\widetilde{\mathbb{G}}$ is a random graph distributed as $\mathbb{G}(n,p)$ conditioned on $I_{G'} = 1$, and thus the variables $J_{G''G'}$ have the correct joint distribution (6.18). Moreover, $J_{G''G'} = 0$ if $G' \cap G'' \neq \emptyset$, but $G' \neq G''$, while $J_{G''G'} \geq I_{G''}$ if $G' \cap G'' = \emptyset$.

It is easily seen that if G is connected and unicyclic and $np \to c > 0$, or if G is a tree of order v and either $n^v p^{v-1} \to c > 0$ or $vnp - \log n - (v-1)\log\log n \to c \in (-\infty, \infty)$, then $\mathbb{E}\, T_G \to \lambda < \infty$. In these cases, Theorem 6.22 yields $T_G \xrightarrow{d} \mathrm{Po}(\lambda)$ by straightforward calculations (Exercise!). (Compare with Section 3.6.)

The same argument applies also to the random varible T_v counting all isolated trees of order v (and not just copies of a specific trees), which proves the final part of Theorem 3.30.

(These results were originally proved by Erdős and Rényi (1960) by the method of moments.)

Example 6.30. We have so far applied the Stein–Chen method to the random graph $\mathbb{G}(n,p)$, but it applies also to $\mathbb{G}(n,M)$. For example, consider the subgraph count X_G as in Example 6.26, but now for $\mathbb{G}(n,M)$. The main difference from the $\mathbb{G}(n,p)$ case is that for $\mathbb{G}(n,M)$, we can use neither Theorem 6.23 (because there does not exist a sparse dependency graph) nor Theorem 6.24 (because the indicator variables $I_{G'}$ are not positively related); instead we use Theorem 6.22 with the following construction.

The conditional distribution of $\mathbb{G}(n,M)$ given $I_{G'} = 1$ is the same as the distribution of the random graph $\widetilde{\mathbb{G}}$ obtained from $\mathbb{G}(n,M)$ by first adding all edges in G' that are not already present, and then deleting the same number of edges, randomly chosen among the edges outside G'. We may thus define $J_{G''G'} = \mathbf{1}[G'' \subset \widetilde{\mathbb{G}}]$ and $V_{G'} = \sum_{G'' \neq G'} J_{G''G'}$.

It is straightforward to estimate $\mathbb{E}\,|J_{G''G'} - I_{G'}|$, which, by (6.19), yields an estimate of $d_{TV}(X_G, \mathrm{Po}(\mathbb{E}\, X_G))$ for $\mathbb{G}(n,M)$, but we omit the details (Exercise!).

Similar constructions apply to the other examples above.

6.3 STEIN'S METHOD: THE NORMAL CASE

The original version of Stein's method yields normal approximation; see Stein (1972, 1986) for a general description. The method was applied to random graphs by Barbour (1982) and Barbour, Karoński and Ruciński (1989), to which we refer for further details. In particular, Stein's method yields the following rather general result (Barbour, Karoński and Ruciński 1989), based on constructing a suitable decomposition of the studied random variable. We let d_1 denote the distance between distributions defined by

$$d_1(X,Y) = \sup\{|\mathbb{E}\,h(X) - \mathbb{E}\,h(Y)| : \sup_{x\in\mathbb{R}}|h(x)| + \sup_{x\in\mathbb{R}}|h'(x)| \le 1\};$$

note that $d_1(X_n,Y) \to 0$ implies $X_n \xrightarrow{d} Y$. (This distance is well adapted to Stein's method, although other distances are more commonly used in other contexts; *cf.* Barbour, Karoński and Ruciński (1989).)

Theorem 6.31. *Suppose that W is a random variable which can be decomposed as follows: For some finite index sets A and B_α, $\alpha \in A$, and square integrable random variables X_α, W_α, Z_α, $Z_{\alpha\beta}$, $W_{\alpha\beta}$ and $V_{\alpha\beta}$, $\alpha \in A$, $\beta \in B_\alpha$:*

$$\begin{aligned} W &= \sum_{\alpha\in A} X_\alpha; \\ W &= W_\alpha + Z_\alpha, && \alpha \in A; \\ Z_\alpha &= \sum_{\beta\in B_\alpha} Z_{\alpha\beta}, && \alpha \in A; \\ W_\alpha &= W_{\alpha\beta} + V_{\alpha\beta}, && \alpha \in A,\ \beta \in B_\alpha; \end{aligned}$$

where, further, $\mathbb{E}\,X_\alpha = 0$, W_α is independent of X_α and $W_{\alpha\beta}$ is independent of the pair $(X_\alpha, Z_{\alpha\beta})$. Then, for some universal constant C, $\sigma^2 = \operatorname{Var} W$ and $\widetilde{W} = \sigma^{-1}W$,

$$\begin{aligned} d_1\big(\widetilde{W}, \mathrm{N}(0,1)\big) \le C\sigma^{-3}\Big(&\sum_{\alpha\in A} \mathbb{E}(|X_\alpha| Z_\alpha^2) \\ &+ \sum_{\alpha\in A}\sum_{\beta\in B_\alpha} \big(\mathbb{E}\,|X_\alpha Z_{\alpha\beta} V_{\alpha\beta}| + \mathbb{E}\,|X_\alpha Z_{\alpha\beta}|\,\mathbb{E}\,|Z_\alpha + V_{\alpha\beta}|\big)\Big). \end{aligned} \tag{6.23}$$

■

In applications, one has to construct decompositions as above of a given variable, keeping the right-hand side of (6.23) small. For sums of random variables with a lot of independence, as measured by a suitably sparse dependency graph, there is a straightforward construction, which leads to the following result.

Theorem 6.32. *Suppose that $W = \sum_{\alpha\in A} X_\alpha$, where $\{X_\alpha\}_{\alpha\in A}$ is a family of random variables with dependency graph L and, further, $\mathbb{E}\,X_\alpha = 0$, $\alpha \in A$. Let $\sigma^2 = \operatorname{Var} W$ and assume that $0 < \sigma^2 < \infty$. Then, for some universal constant C and with $\overline{N}_\Gamma(\alpha)$ the closed neighborhood of α as in (6.8),*

$$d_1\big(\widetilde{W}, \mathrm{N}(0,1)\big) \le C\sigma^{-3} \sum_{\alpha\in A} \sum_{\beta,\gamma\in\overline{N}_\Gamma(\alpha)} \big(\mathbb{E}\,|X_\alpha X_\beta X_\gamma| + \mathbb{E}\,|X_\alpha X_\beta|\,\mathbb{E}\,|X_\gamma|\big). \tag{6.24}$$

Proof. We apply Theorem 6.31 with $B_\alpha = \overline{N}_\Gamma(\alpha)$, $W_\alpha = \sum_{\beta\notin\overline{N}_\Gamma(\alpha)} X_\beta$, $Z_\alpha = \sum_{\beta\in\overline{N}_\Gamma(\alpha)} X_\beta$, $Z_{\alpha\beta} = X_\beta$, $W_{\alpha\beta} = \sum_{\gamma\notin\overline{N}_\Gamma(\alpha)\cup\overline{N}_\Gamma(\beta)} X_\gamma$ and $V_{\alpha\beta} = \sum_{\gamma\in\overline{N}_\Gamma(\beta)\setminus\overline{N}_\Gamma(\alpha)} X_\gamma$. It is then easily seen that

$$\sum_{\alpha\in A}\mathbb{E}(|X_\alpha|Z_\alpha^2) + \sum_{\alpha\in A}\sum_{\beta\in B_\alpha}(\mathbb{E}\,|X_\alpha Z_{\alpha\beta}V_{\alpha\beta}| + \mathbb{E}\,|X_\alpha Z_{\alpha\beta}|\,\mathbb{E}\,|Z_\alpha + V_{\alpha\beta}|)$$
$$\le 2\sum_{\alpha\in A}\sum_{\beta,\gamma\in\overline{N}_L(\alpha)}(\mathbb{E}\,|X_\alpha X_\beta X_\gamma| + \mathbb{E}\,|X_\alpha X_\beta|\,\mathbb{E}\,|X_\gamma|)$$

and the result follows. (We may assume that $\mathbb{E}\,X_\alpha^2 < \infty$, since otherwise the right-hand side of (6.24) is infinite.) ∎

In particular, this yields an improvement of Theorem 6.18.

Theorem 6.33. *Suppose that $(S_n)_1^\infty$ is a sequence of random variables such that $S_n = \sum_{\alpha\in A_n} X_{n\alpha}$, where for each n, $\{X_{n\alpha}\}_\alpha$ is a family of random variables with dependency graph L_n. Suppose further that there exist numbers M_n and Q_n such that*

$$\sum_{\alpha\in A_n}\mathbb{E}\,|X_{n\alpha}| \le M_n \tag{6.25}$$

and for every $\alpha_1, \alpha_2 \in A_n$,

$$\sum_{\alpha\in\overline{N}_{L_n}(\alpha_1,\alpha_2)}\mathbb{E}\big(|X_{n\alpha}| \mid X_{n\alpha_1}, X_{n\alpha_2}\big) \le Q_n. \tag{6.26}$$

Let $\sigma_n^2 = \operatorname{Var} S_n$. Then

$$d_1(\widetilde{S}_n, \mathrm{N}(0,1)) = O\Big(\frac{M_n Q_n^2}{\sigma_n^3}\Big).$$

In particular, if

$$\frac{M_n Q_n^2}{\sigma_n^3} \to 0, \tag{6.27}$$

then

$$\widetilde{S}_n \xrightarrow{d} \mathrm{N}(0,1).$$

Proof. By replacing $X_{n\alpha}$ by $X_{n\alpha} - \mathbb{E}\,X_{n\alpha}$ (and S_n by $S_n - \mathbb{E}\,S_n$), we may assume that $\mathbb{E}\,X_{n\alpha} = 0$; note that (6.25) and (6.26) still hold if we replace M_n and Q_n by $2M_n$ and $2Q_n$.

It follows, arguing as in Lemma 6.17, that

$$\sum_{\alpha\in A}\sum_{\beta,\gamma\in\overline{N}_L(\alpha)}(\mathbb{E}\,|X_\alpha X_\beta X_\gamma| + \mathbb{E}\,|X_\alpha X_\beta|\,\mathbb{E}\,|X_\gamma|) \le 2M_nQ_n^2$$

and the result follows by Theorem 6.32. ∎

Example 6.34. Consider again the subgraph count X_G in $\mathbb{G}(n,p)$, where $n \to \infty$ and $p = p(n)$ is a function of n.

We have $X_G - \mathbb{E}\,X_G = \sum_{\alpha \in A_n} X_\alpha$, where A_n and X_α are as in Example 6.19. As already shown in Example 6.19, with the dependency graph L_n defined there, (6.25) and (6.26) hold with $M_n = O\big((1-p)\,\mathbb{E}\,X_G)\big)$ and $Q_n = O(\mathbb{E}\,X_G/\Phi_G)$, which yields, as in (6.16),

$$M_n Q_n^2/\sigma_n^3 = O\big((1-p)^{-1/2}\Phi_G^{-1/2}\big).$$

Consequently, Theorem 6.33 yields a new proof of Theorem 6.5, with the additional information that

$$d_1\big(\widetilde{X}_G, \mathrm{N}(0,1)\big) = O\big((1-p)^{-1/2}\Phi_G^{-1/2}\big).$$

Theorem 6.31 is more flexible than Theorem 6.32 and can be applied also to sums where all summands X_α are dependent. One such case is when counting the number of subsets of (the vertex set of) $\mathbb{G}(n,p)$ that satisfy a given *semi-induced* property, that is, a property that depends only on the edges with at least one endpoint in the set. We begin with an example.

Example 6.35. Denote the number of vertices of degree d in $\mathbb{G}(n,p)$ by S_d. We consider a fixed $d \geq 0$ and let $n \to \infty$ with $p = p(n)$. Then (assuming for simplicity $np^2 = o(1)$; for larger p, $\mathbb{E}\,S_d \to 0$ rapidly),

$$\mathbb{E}\,S_d = n\binom{n-1}{d}p^d(1-p)^{n-1-d} \sim \tfrac{1}{d!}n^{d+1}p^d e^{-np}.$$

Consequently,

$$\mathbb{E}\,S_d \to \infty \iff n^{d+1}p^d \to \infty \text{ and } np - \log n - d\log\log n \to -\infty.$$

Let $I_i = \mathbf{1}[i \text{ has degree } d \text{ in } \mathbb{G}(n,p)]$, and $X_i = I_i - \mathbb{E}\,I_i$; then $S_d - \mathbb{E}\,S_d = \sum_1^n X_i$. In this case, there is (in general) no independence between any two X_i and X_j, and we use the following construction.

Let, for a set $F \subseteq [n]$ and $i \in [n]$,

$$I_i^F = \begin{cases} \mathbf{1}[i \text{ has degree } d \text{ in } \mathbb{G}(n,p) \setminus F], & i \notin F, \\ 0, & i \in F, \end{cases}$$

and let $X_i^F = I_i^F - \mathbb{E}\,I_i^F$. Moreover, for $i, j \in [n]$, let $B_i = A = [n]$ and

$$\begin{aligned} Z_{ij} &= I_j - I_j^{\{i\}}, \\ Z_i &= \sum_{j=1}^{n} Z_{ij}, \\ W_i &= \sum_{j=1}^{n} X_j^{\{i\}} - \mathbb{E}\, Z_i, \\ V_{ij} &= \sum_{k=1}^{n} (I_k^{\{i\}} - I_k^{\{i,j\}}), \\ W_{ij} &= \sum_{k=1}^{n} X_k^{\{i,j\}} - \mathbb{E}\, V_{ij} - \mathbb{E}\, Z_i. \end{aligned}$$

The conditions of Theorem 6.31 then are satisfied, and it is not difficult to show that both sums in (6.23) are $O(\mathbb{E}\, S_d)$; see Barbour, Karoński and Ruciński (1989) for details. Consequently,

$$d_1(\widetilde{S}_d, \mathrm{N}(0,1)) = O\big((\operatorname{Var} S_d)^{-3/2}\, \mathbb{E}\, S_d\big).$$

Moreover, a simple calculation yields

$$\operatorname{Var} S_d = \frac{n}{n-1}\binom{n-1}{d}^2 \big((n-1)p-d\big)^2 p^{2d-1}(1-p)^{2n-2d-3} + \mathbb{E}\, S_d - n^{-1}(\mathbb{E}\, S_d)^2,$$

and it is easily seen that if $d \geq 1$ or $d = 0$ and $np = \Omega(1)$, then

$$\operatorname{Var} S_d \asymp \mathbb{E}\, S_d,$$

and consequently

$$d_1(\widetilde{S}_d, \mathrm{N}(0,1)) = O\big((\mathbb{E}\, S_d)^{-1/2}\big);$$

in particular, $\widetilde{S}_d \xrightarrow{d} \mathrm{N}(0,1)$ if furthermore $\mathbb{E}\, S_d \to \infty$.

In the remaining case $d = 0$ (isolated vertices) and $np \to 0$, we have $\mathbb{E}\, S_0 \sim n$ and $\operatorname{Var} S_0 \sim 2n^2 p$, whence we only obtain $d_1(\widetilde{S}_0, \mathrm{N}(0,1)) = O(n^{-2}p^{-3/2})$; this proves $\widetilde{S}_0 \xrightarrow{d} \mathrm{N}(0,1)$, provided $n^{-4/3} \ll p \ll n^{-1}$. This result can be extended to the full range where $\operatorname{Var} S_0 \to \infty$, that is, when $\mathbb{E}\, S_0 \to \infty$ and $n^2 p \to \infty$. In fact, when $np \to 0$, $\operatorname{Var} S_0 \sim \operatorname{Var} S_1 \sim -\operatorname{Cov}(S_0, S_1) \sim 2n^2 p$, and thus $\operatorname{Var}(S_0 + S_1) = o(n^2 p) = o(\operatorname{Var} S_0)$; consequently, $\widetilde{S}_0 = -\widetilde{S}_1 + o_p(1)$, and the result follows since, as just shown, $\widetilde{S}_1 \xrightarrow{d} \mathrm{N}(0,1)$ provided furthermore $n^2 p \to \infty$. (See, further, Barbour, Karoński and Ruciński (1989) and Kordecki (1990).)

Since $\operatorname{Var} S_d \to \infty$ is necessary for asymptotic normality, we have proved the following result.

Theorem 6.36. *If $d \geq 0$, then $\widetilde{S}_d \xrightarrow{d} \mathrm{N}(0,1)$ if and only if $\operatorname{Var} S_d \to \infty$. For $d \geq 1$, this is equivalent to $\mathbb{E}\, S_d \to \infty$, i.e., $n^{d+1}p^d \to \infty$ and $np - \log n -$*

$d \log\log n \to -\infty$; *for* $d = 0$ *it is equivalent to* $n^2 p \to \infty$ *and* $np - \log n \to -\infty$. ∎

Remark 6.37. As remarked in Section 6.2, S_d has an asymptotic Poisson distribution when $\mathbb{E}\, S_d \to c < \infty$. In fact (Karoński and Ruciński 1987), the Stein–Chen method yields Poisson approximation also when $\mathbb{E}\, S_d \to \infty$, provided $np \to 0$ and $d \geq 2$, or $np \to \infty$ (in these cases $\operatorname{Var} S_d \sim \mathbb{E}\, S_d$), which gives another proof (historically the first) of Theorem 6.36 in these cases.

For a general semi-induced property $\mathcal{P}$, we similarly construct a decomposition of $W = \sum I_\alpha$, where α ranges over the subsets of $[n]$ of a given size and I_α is the indicator that α has the property $\mathcal{P}$, by defining

$$I_\alpha^F = \begin{cases} \mathbf{1}[\alpha \text{ has the property } \mathcal{P} \text{ in } \mathbb{G}(n,p) \setminus F], & \alpha \cap F = \emptyset, \\ 0, & \alpha \cap F \neq \emptyset, \end{cases}$$

and then proceeding as above. Another example of this is the following (Barbour 1982, Barbour, Karoński and Ruciński 1989).

Theorem 6.38. *Let* T_k *be the number of isolated trees of order* k *in* $\mathbb{G}(n,p)$, *where* $k \geq 2$ *is fixed. Then* $\widetilde{T}_k \xrightarrow{d} \mathrm{N}(0,1)$ *if and only if* $\mathbb{E}\, T_k \to \infty$, *that is, when* $n^k p^{k-1} \to \infty$ *and* $knp - \log n - (k-1)\log\log n \to -\infty$.

Sketch of proof. We apply Theorem 6.31 with the construction just indicated. The sums in (6.23) are both $O(\mathbb{E}\, T_k)$ and $\operatorname{Var} T_k \asymp \mathbb{E}\, T_k$; thus

$$d_1(\widetilde{T}_k, \mathrm{N}(0,1)) = O\big((\mathbb{E}\, T_k)^{-1/2}\big).$$ ∎

Again, as stated in Theorem 3.30 and Example 6.29, Erdős and Rényi (1960) proved that $T_k \xrightarrow{d} \mathrm{Po}(c)$ when $\mathbb{E}\, T_k \to c < \infty$ (in both ranges of p). Note further that the first (threshold) part of Theorem 3.30 follows easily from Theorem 6.38.

6.4 PROJECTIONS AND DECOMPOSITIONS

A standard method when studying asymptotic distributions is to approximate the studied random variable by another one, which is simpler in some sense. Indeed, by Cramér's theorem (see Section 1.2), if $X_n - Y_n \xrightarrow{p} 0$ and $Y_n \xrightarrow{d} Z$, then $X_n \xrightarrow{d} Z$ too.

The first projection

Consider a random variable X which is a *graph functional*, that is, a (real-valued) variable that depends only on the isomorphism type of $\mathbb{G}(n,p)$. Then, the simplest choice of an approximating variable is a linear function $aL + b$

of the number of edges $L = e(\mathbb{G}(n,p)) = X_{K_2}$. (Here a and b are constants that may depend on n and $p = p(n)$.) Since $L \in \mathrm{Bi}\big(\binom{n}{2}, p\big)$, the central limit theorem yields $\widetilde{L} \overset{d}{\to} \mathrm{N}(0,1)$, provided that $n^2 p \to \infty$ and $n^2(1-p) \to \infty$, so it remains only to study the error $X - aL - b$.

We choose the coefficients a and b such as to minimize the L^2-norm of the error, that is, $Y = aL + b$ is the projection in $L^2(\mathbb{P})$ of X onto the two-dimensional subspace of linear functions of L. This is the usual linear regression, and as is well known, then $X - Y$ is orthogonal to 1 and L, which leads to

$$a = \frac{\mathrm{Cov}(X,L)}{\mathrm{Var}\, L} = \frac{\mathrm{Cov}(X,L)}{\binom{n}{2}p(1-p)},$$
$$b = \mathbb{E}\, X - a\,\mathbb{E}\, L.$$

Moreover,

$$\mathbb{E}(X-Y)^2 = \mathrm{Var}(X-Y) = \mathrm{Var}(X-aL) = \mathrm{Var}\, X - a^2\,\mathrm{Var}\, L.$$

Now $L = \sum_e I_e$, summing over all edges e in K_n, where I_e is the indicator $\mathbf{1}[e \in \mathbb{G}(n,p)]$. By symmetry, $\mathrm{Cov}(X, I_e)$ is independent of e and thus, for any edge e,

$$\begin{aligned} a &= \mathrm{Cov}(X,L)/\tbinom{n}{2}p(1-p) = \mathrm{Cov}(X,I_e)/p(1-p) \\ &= \big(\mathbb{E}(XI_e) - p\,\mathbb{E}\, X\big)\,/\,p(1-p) \\ &= \big(\mathbb{E}(X \mid e \in \mathbb{G}(n,p)) - \mathbb{E}\, X\big)\,/\,(1-p). \end{aligned}$$

The approximating variable $aL + b$ is known as *the first projection* of X, and this approach to proving normality is called *the first projection method.* It can be summarized as the following theorem.

Theorem 6.39. *Suppose that X_n is a graph functional of $\mathbb{G}(n,p)$, with $p = p(n)$, and let*

$$\alpha_n = \mathbb{E}\big(X_n \mid e \in \mathbb{G}(n,p)\big) - \mathbb{E}\, X_n$$

for an edge $e \in K_n$. If $n^2 p \to \infty$, $n^2(1-p) \to \infty$ and

$$\mathrm{Var}\, X_n \sim \binom{n}{2} p(1-p)^{-1}\alpha_n^2, \tag{6.28}$$

then $\widetilde{X}_n \overset{d}{\to} N(0,1)$.

Proof. By the discussion above, we let $Y_n = a_n L_n + b_n$, with $a_n = (1-p)^{-1}\alpha_n$ and $b_n = \mathbb{E}\, X_n - a_n\,\mathbb{E}\, L_n$, and find

$$\mathbb{E}(X_n - Y_n)^2 = \mathrm{Var}\, X_n - a_n^2\tbinom{n}{2}p(1-p) = o(\mathrm{Var}\, X_n).$$

Hence, if $\sigma_n^2 = \operatorname{Var} X_n$, then $\mathbb{E}|(X_n - Y_n)/\sigma_n|^2 \to 0$ and, furthermore, $\operatorname{Var} Y_n \sim \operatorname{Var} X_n$. Consequently,

$$\frac{X_n - \mathbb{E} X_n}{\sigma_n} = \frac{Y_n - \mathbb{E} Y_n}{\sigma_n} + \frac{X_n - Y_n}{\sigma_n},$$

where $(Y_n - \mathbb{E} Y_n)/\sigma_n = (\operatorname{Var} Y_n / \operatorname{Var} X_n)^{1/2} \widetilde{L}_n \xrightarrow{d} \mathrm{N}(0,1)$ and $(X_n - Y_n)/\sigma_n \xrightarrow{p} 0$. The result follows by Cramér's theorem (Section 1.2). ∎

Example 6.40. Consider again the subgraph count X_G. Given an edge $e \in K_n$, there are $e_G (n)_{v_G} / \binom{n}{2} \operatorname{aut}(G)$ copies of G in K_n that contain e, and thus

$$\begin{aligned}\alpha_n &= \mathbb{E}(X \mid I_e = 1) - \mathbb{E} X = e_G (n)_{v_G} \left(\tbinom{n}{2} \operatorname{aut}(G)\right)^{-1} (p^{e_G - 1} - p^{e_G}) \\ &\sim 2 e_G n^{v_G - 2} \left(\operatorname{aut}(G)\right)^{-1} p^{e_G - 1}(1 - p).\end{aligned}$$

It is easily seen that $\binom{n}{2} p(1-p)^{-1} \alpha_n^2$ asymptotically equals the contribution to $\operatorname{Var}(X_G)$ from the terms in (3.10) corresponding to two copies of G that intersect in a single edge (Exercise!). Consequently, the condition (6.28) is equivalent to K_2 being the only leading overlap of G, that is, to $np^{m^{(2)}(G)} \to \infty$; see Section 3.2.

This example is typical; the method of the first projection is (usually) very easy to apply, but it works only sometimes and often does not give the full result.

Higher projections

It is natural to try to extend the range of the first projection method by projecting onto a larger space of variables, thus reducing the error in the approximation.

The first projection uses only information on subgraphs of $\mathbb{G}(n,p)$ with two vertices; the next step (sometimes called the *second projection*) is to use information on the subgraphs with three vertices. Each such subgraph has 0, 1, 2 or 3 edges, and is determined up to isomorphism by its number of edges, so the second projection is a linear combination of the random variables τ_0, τ_1, τ_2 and τ_3, where τ_j counts the number of triples of vertices in $\mathbb{G}(n,p)$ with j edges between them. Equivalently, by simple algebra, the second projection can be expressed as a linear combination of the constant 1 and the three subgraph counts $X_{K_2} = L$, X_{P_2} and X_{K_3}, where P_2 is the path of length 2.

In neither of these representations, however, are the four basis variables orthogonal, so it will be more convenient to use a third representation. It can be constructed from the subgraph counts above by the usual orthogonalization procedure, but we prefer to define it directly in the next subsection, at the same time generalizing it to larger subgraphs.

A general decomposition

Let H be a graph. Consider the $(n)_{v_H}$ different injective mappings from the vertices of H into $\{1, \ldots, n\}$. Each such mapping φ maps H onto a copy $\varphi(H)$ of H in K_n, and we define

$$S_n(H) = S_{n,p}(H) = \sum_{\varphi} \prod_{e \in \varphi(H)} (I_e - p), \tag{6.29}$$

where, as above, $I_e = \mathbf{1}[e \in \mathbb{G}(n,p)]$.

In other words, we sum $\prod_{e \in H'}(I_e - p)$ over all copies H' of H in K_n, counted with multiplicities $\operatorname{aut}(H)$. Note that if we replace $(I_e - p)$ by I_e in (6.29), we obtain $\operatorname{aut}(H) X_H$.

It is obvious that $S_n(H)$ depends only on the isomorphism type of H. Hence we may regard $\{S_n(H)\}_H$ as a family of random variables, indexed by *unlabelled* graphs H.

The simplest examples are

$$\begin{aligned} S_n(\emptyset) &= 1 \qquad \text{(trivial but useful)} \\ S_n(K_1) &= n \qquad \text{(trivial and less useful)} \\ S_n(K_2) &= 2 \sum_{e \in K_n} (I_e - p) = 2\big(e(\mathbb{G}(n,p)) - \tbinom{n}{2}p\big). \end{aligned}$$

It is easily seen that if H has any isolated vertices, removing them changes $S_n(H)$ only by a non-random factor (depending on n). Hence we may restrict attention to H without isolated vertices.

Since the variables $I_e - p$ are independent and have mean 0, two products $\prod(I_e - p)$ are orthogonal unless they coincide, and the following results are easily obtained.

Lemma 6.41. *Suppose that H and K are graphs without isolated vertices.*

(i) *If $H \neq \emptyset$, then $\mathbb{E}\, S_n(H) = 0$.*

(ii) *If $H \neq \emptyset$, then*

$$\operatorname{Var} S_n(H) = \mathbb{E}\, S_n(H)^2 = \operatorname{aut}(H)(n)_{v_H}(p(1-p))^{e_H}.$$

(iii) *If H and K are non-isomorphic, then $S_n(H)$ and $S_n(K)$ are orthogonal:*

$$\operatorname{Cov}(S_n(H), S_n(K)) = \mathbb{E}[S_n(H)S_n(K)] = 0. \qquad \blacksquare$$

We next show that the variables $S_n(H)$ can be used to decompose any graph functional.

Lemma 6.42. *Every graph functional X of $\mathbb{G}(n,p)$ has a unique expansion*

$$X = \sum_H \hat{X}_n(H) S_n(H) \tag{6.30}$$

for some real coefficients $\hat{X}_n(H) = \hat{X}_{n,p}(H)$, *where* H *ranges over the unlabelled graphs with no isolated vertices and at most* n *vertices. Furthermore, the terms in* (6.30) *are orthogonal and*

$$\operatorname{Var} X = \sum_{H\neq\emptyset} \hat{X}_n(H)^2 \operatorname{Var} S_n(H) = \sum_{H\neq\emptyset} \operatorname{aut}(H)(n)_{v_H}(p(1-p))^{e_H}\hat{X}_n(H)^2. \tag{6.31}$$

Proof. Trivially,

$$X = \sum_G X(G) \prod_{e\in G} I_e \prod_{e\notin G}(1 - I_e),$$

where we sum over all graphs G with vertex set $\{1,\dots,n\}$. If we substitute $I_e = (I_e - p) + p$ and expand, X will be expressed as a linear combination of terms $\prod_{e\in H}(I_e - p)$, $H \subseteq K_n$, and (6.30) is obtained by collecting terms with isomorphic H together.

Lemma 6.41 implies that the terms in (6.30) are orthogonal, and that (6.31) holds. Moreover, Lemma 6.41 and (6.30) yield also

$$\hat{X}_n(H) = \mathbb{E}(XS_n(H))/\,\mathbb{E}\,S_n(H)^2,$$

and thus the term $\hat{X}_n(H)$ is uniquely determined. ∎

Here $\hat{X}_n(\emptyset)S_n(\emptyset) = \hat{X}_n(\emptyset) = \mathbb{E}\,X$, so the randomness enters through the terms in (6.30) with H non-null.

The first projection equals $\hat{X}_n(\emptyset)S_n(\emptyset) + \hat{X}_n(K_2)S_n(K_2)$, and is thus obtained by ignoring all terms in (6.30) with $v_H \geq 3$. Similarly, the second projection equals $\hat{X}_n(\emptyset)S_n(\emptyset)+\hat{X}_n(K_2)S_n(K_2)+\hat{X}_n(P_2)S_n(P_2)+\hat{X}_n(K_3)S_n(K_3)$, that is, the sum of the four terms with $v_H \leq 3$. More generally, we can select any set of graphs H and consider only the sum of the corresponding terms in (6.30) as an approximation of X.

In order to use this idea to obtain asymptotic distributions from the decomposition (6.30), we have to know asymptotic distributions of the basis variables $S_n(H)$. We already know that $S_n(K_2) = 2(L - \mathbb{E}\,L)$ is asymptotically normal, provided p is not too close to 0 or 1; this extends to every $S_n(H)$ with H connected, while disconnented H give other limits. More precisely, we have the following theorem, proved using a continuous time martingale limit theorem in Janson (1994a). (For fixed p, it was earlier proved by the method of moments (Janson 1990a, Janson and Nowicki 1991).)

Theorem 6.43. *Suppose that* $p = p(n) \to p_0$ *as* $n \to \infty$, *with* $0 \leq p_0 \leq 1$. *Then there exist random variables* $U(H)$, *where* H *ranges over unlabelled graphs, such that if* H *is any graph without isolated vertices for which*

$$np^{m(H)} \to \infty, \tag{6.32}$$

then, as $n \to \infty$,

$$n^{-v_H/2}p^{-e_H/2}S_{n,p}(H) \xrightarrow{d} U(H). \tag{6.33}$$

The convergence in (6.33) *holds jointly for any finite number of graphs H that satisfy* (6.32). *The limit variables $U(H)$ are determined by the following properties:*

(i) *If H is connected and $e_H > 0$, then $U(H)$ has a normal distribution with mean $\mathbb{E}\, U(H) = 0$ and variance*

$$\mathbb{E}\, U(H)^2 = \operatorname{aut}(H)(1-p_0)^{e_H}. \tag{6.34}$$

(ii) *If $H_1, \dots, H_m$ are different (i.e., non-isomorphic) connected unlabelled graphs, then $U(H_1), \dots, U(H_m)$ are independent.*

(iii) *If H has connected components $H_1, \dots, H_m$, each having at least one edge, then $U(H)$ is a polynomial in $U(H_1), \dots, U(H_m)$, known as the Wick product $:U(H_1) \cdots U(H_m):$, see, e.g., Janson* (1997) *for a definition. In particular, for $m = 2$,*

$$U(H) = \;:U(H_1)U(H_2):\; = U(H_1)U(H_2) - \mathbb{E}\big(U(H_1)U(H_2)\big). \tag{6.35}$$

Furthermore, (6.34) *holds for every H, and $\mathbb{E}\, U(H_1)U(H_2) = 0$ if H_1 and H_2 are two different unlabelled graphs without isolated vertices.* ∎

We return to the study of a general graph functional X, or more formally, a sequence X_n of functionals of $\mathbb{G}(n,p)$, where $p = p(n)$ is a given sequence. Then X_n has a (unique) expansion (6.30).

The simplest case is when only a finite set of graphs H, independent of n, is needed in the expansion (6.30). Assume further that (6.32) holds for these graphs H. The asymptotic behavior of X_n then follows from Theorem 6.43, provided we know the asymptotic behavior of the coefficients $\hat{X}_n(H)$. Both normal and non-normal limits may be obtained by this procedure. In fact, X_n is asymptotically normal if and only if the terms with H connected dominate the decomposition (6.30).

Even if no finite set of graphs suffices for the expansion of every X_n, it is frequently the case that a finite set gives a good approximation. In general, let $\mathcal{H}$ be a family of non-null unlabelled graphs without isolated vertices. We say that X_n is *dominated* by $\mathcal{H}$ (for the given sequence $p(n)$) if, as $n \to \infty$,

$$\operatorname{Var} X_n \sim \sum_{H \in \mathcal{H}} \hat{X}_n(H)^2 \operatorname{Var} S_n(H).$$

In this case, X_n has the same asymptotic distribution (if any) as the projection $\sum_{H\in\mathcal{H}} \hat{X}_n(H) S_n(H)$. In particular, if there exists a finite dominating family $\mathcal{H}$, we may apply Theorem 6.43 to this projection and obtain limit results just as for the case of a finite expansion.

Remark 6.44. The first and second projection methods can now be recognized as the special cases of this procedure with the dominating families

$\mathcal{H} = \{K_2\}$ and $\mathcal{H} = \{K_2, P_2, K_3\}$, respectively. Note that in these cases, all graphs $H \in \mathcal{H}$ are connected, and thus X_n is always asymptotically normal when these methods apply.

The method extends through a truncation argument to a more general situation, called here *asymptotically finitely dominated*, where $\mathcal{H}$ is infinite, but for every $\varepsilon > 0$ a finite subset (independent of n) of $\mathcal{H}$ suffices to yield at least $(1-\varepsilon)\operatorname{Var} X_n$ in (6.31).

This method yields the following rather general result, where we define

$$\hat{X}_n^*(H) = n^{v_H/2} p^{e_H/2} \hat{X}_n(H).$$

Note that by Lemma 6.41 we have, for $H \neq \emptyset$,

$$\operatorname{Var}\big(\hat{X}_n(H) S_n(H)\big) \sim (1-p)^{e_H} \operatorname{aut}(H) \hat{X}_n^*(H)^2. \tag{6.36}$$

Theorem 6.45. *Let X_n be a graph functional of $\mathbb{G}(n,p)$, where $p = p(n)$. Suppose that $p \to p_0 \in [0,1]$, that β_n is a sequence of positive numbers, and that $\mathcal{H}$ is a family of non-null graphs without isolated vertices such that, for every $H \in \mathcal{H}$,*

$$np^{m(H)} \to \infty \tag{6.37}$$

and

$$a(H) = \lim_{n\to\infty} \hat{X}_n^*(H)/\beta_n \quad \textit{exists.} \tag{6.38}$$

Suppose further that

$$\operatorname{Var}(X_n)/\beta_n^2 \to \sum_{H\in\mathcal{H}} a(H)^2 \operatorname{aut}(H)(1-p_0)^{e_H} < \infty. \tag{6.39}$$

Then,

$$\frac{X_n - \mathbb{E}\, X_n}{\beta_n} \xrightarrow{d} \sum_{H\in\mathcal{H}} a(H) U(H), \tag{6.40}$$

where $U(H)$ is as in Theorem 6.43. (If $\mathcal{H}$ is infinite, the sum is interpreted in L^2.)

Sketch of proof. The case $p_0 = 1$ is trivial (and is better handled by studying the complementary graph, which is $\mathbb{G}(n, 1-p)$).

Thus assume $0 \le p_0 < 1$. Given $\varepsilon > 0$, we may choose a finite subset $\mathcal{H}_\varepsilon$ of $\mathcal{H}$ such that $\sum_{H\in\mathcal{H}_\varepsilon} a(H)^2 \operatorname{aut}(H)(1-p_0)^{e_H} > \sum_{H\in\mathcal{H}} a(H)^2 \operatorname{aut}(H)(1-p_0)^{e_H} - \varepsilon$. It follows by Lemma 6.41, (6.36) and (6.38) that if Y_n^ε is the projection $\sum_{H\in\mathcal{H}_\varepsilon} \hat{X}_n(H) S_n(H)$, then $\operatorname{Var}(Y_n^\varepsilon/\beta_n) \to \sum_{H\in\mathcal{H}_\varepsilon} a(H)^2 \operatorname{aut}(H)(1-p_0)^{e_H}$ and, consequently, by (6.39), for n large,

$$\mathbb{E}\Big(\frac{X_n - \mathbb{E}\, X_n}{\beta_n} - \frac{Y_n^\varepsilon}{\beta_n}\Big)^2 = \operatorname{Var}\frac{X_n}{\beta_n} - \operatorname{Var}\frac{Y_n^\varepsilon}{\beta_n} < \varepsilon.$$

Since further Theorem 6.43 implies that $Y_n^\varepsilon/\beta_n \xrightarrow{d} \sum_{H\in\mathcal{H}_\varepsilon} a(H)U(H)$ as $n \to \infty$, (6.40) follows by Billingsley (1968, Theorem 4.2). ∎

Corollary 6.46. *Under the assumptions of Theorem 6.45, if further, every graph $H \in \mathcal{H}$ is connected and at least one $a(H) \neq 0$, then X_n is asymptotically normal,*

$$\widetilde{X}_n \xrightarrow{d} \mathrm{N}(0,1). \tag{6.41}$$

Proof. If every H is connected, then each $U(H)$ is normal and so is the sum $\sum a(H)U(H)$; (6.41) follows by (6.34) and (6.39). ∎

Corollary 6.47. *Under the assumptions of Theorem 6.45, if further, $p_0 < 1$, $\mathcal{H}$ is finite and $a(H) \neq 0$ for some disconnected $H \in \mathcal{H}$, then X_n has a non-normal asymptotic distribution.*

Proof. If some H is disconnected (and $p_0 \neq 1$), then $U(H)$ is a polynomial of degree ≥ 2 in normal variables and it is easily seen that the sum in (6.40) is such a polynomial too. Such a polynomial has a distribution with too large tails to be normal, see, for example, Janson (1997, Theorem 6.12). ∎

Remark 6.48. Suppose that $p_0 < 1$ and that not all $a(H) = 0$. If $\mathcal{H}$ is finite, then (6.39) is, by (6.38) and Lemma 6.41, equivalent to the condition that X_n is dominated by $\mathcal{H}$. If $\mathcal{H}$ is infinite, (6.39) is stronger and, in fact, equivalent to X_n being asymptotically finitely dominated by $\mathcal{H}$.

In the normal case, we may replace the assumption (6.38) on convergence of the coefficients by a suitable upper bound.

Theorem 6.49. *Let X_n be a graph functional of $\mathbb{G}(n,p)$, where $p = p(n)$. Suppose that $p \to p_0 \in [0,1]$, and that X_n is dominated by a family $\mathcal{H}$ of connected graphs such that, for every $H \in \mathcal{H}$, $np^{m(H)} \to \infty$ and the numbers*

$$b(H) = \sup_n \hat{X}_n^*(H)/(\operatorname{Var} X_n)^{1/2} \tag{6.42}$$

are finite and satisfy

$$\sum_{H\in\mathcal{H}} b(H)^2 \operatorname{aut}(H) < \infty. \tag{6.43}$$

Then,

$$\widetilde{X}_n \xrightarrow{d} \mathrm{N}(0,1).$$

Proof. Let $\beta_n = (\operatorname{Var} X_n)^{1/2}$. Then, since X_n is dominated by $\mathcal{H}$,

$$\begin{aligned} 1 = \operatorname{Var} X_n/\beta_n^2 &\sim \sum_{H\in\mathcal{H}} \operatorname{Var}\bigl(\hat{X}_n(H)S_n(H)/\beta_n\bigr) \\ &= \sum_{H\in\mathcal{H}} n^{-v_H} p^{-e_H} \hat{X}_n^*(H)^2 \operatorname{Var} S_n(H)/\beta_n^2. \end{aligned} \tag{6.44}$$

Since $\hat{X}_n^*(H)/\beta_n = O(1)$ for $H \in \mathcal{H}$ by (6.42), there exists a subsequence along which $\hat{X}_n^*(H)/\beta_n \to a(H)$ for every $H \in \mathcal{H}$ and some $a(H)$. Along this subsequence, each term in the sum in (6.44) converges, by Lemma 6.41, to $a(H)^2 \operatorname{aut}(H)(1-p_0)^{e_H}$ and is bounded, using also (6.42), by $b(H)^2 \operatorname{aut}(H)$. Consequently, we may by (6.43) apply the dominated convergence theorem to (6.44), obtaining

$$1 = \operatorname{Var} X_n/\beta_n^2 \to \sum_{H\in\mathcal{H}} a(H)^2 \operatorname{aut}(H)(1-p_0)^{e_H}$$

and thus (6.39) holds. By Theorem 6.45, applied along the subsequence, $\widetilde{X}_n \xrightarrow{d} \mathrm{N}(0,1)$ along the subsequence.

Moreover, given any subsequence, this argument shows that there is a subsubsequence with $\widetilde{X}_n \xrightarrow{d} \mathrm{N}(0,1)$, and the result follows by the subsubsequence principle. ∎

Applications to subgraph counts in $\mathbb{G}(n,p)$

We use the decomposition method to obtain old and new results for subgraph counts.

Example 6.50. Consider again the subgraph count X_G in $\mathbb{G}(n,p)$, where $n\to\infty$ and $p = p(n)$ is a function of n. We will use the results above to give yet another proof (our last) of Theorem 6.5.

We can write, *cf.* (6.29) and the discussion there,

$$X_G = \frac{1}{\operatorname{aut}(G)} \sum_{\varphi} \prod_{e\in\varphi(G)} I_e, \tag{6.45}$$

summing over all injective mappings from the vertices of G into $\{1,\dots,n\}$. As in the proof of Lemma 6.42, if we substitute $I_e = (I_e - p) + p$, expand the product in (6.45) as a sum of terms of the type $\prod_{e\in\varphi(H)}(I_e-p)p^{e_G-e_H}$ with $H \subseteq G$ and rearrange the terms, we obtain an orthogonal expansion (6.30) with only $H \subseteq G$ (and without isolated vertices) appearing. Furthermore, it follows from this argument that, for such H,

$$\hat{X}_{Gn}(H) \asymp n^{v_G - v_H} p^{e_G - e_H}$$

and thus (omitting the subscript n)

$$\hat{X}_G^*(H) \asymp n^{v_G - v_H/2} p^{e_G - e_H/2}. \tag{6.46}$$

Now consider $p = p(n)$. If $p \to p_0$, $0 < p_0 < 1$, then $\hat{X}_G^*(H) \asymp n^{v_G - v_H/2}$, and thus X_G is dominated by the $H \neq \emptyset$ with smallest v_H, that is, by $\{K_2\}$. By a simple application of Theorem 6.49, X_G is asymptotically normal. (This is essentially the first projection method; see Remark 6.44 and Example 6.40.)

If $p \to 1$, X_G still is dominated by $\{K_2\}$, and, assuming $n^2(1-p) \to \infty$, asymptotic normality follows similarly.

If $p \to 0$, then, by (6.46), X_G is dominated by the set of non-null $H \subseteq G$ for which $n^{-v_H/2}p^{-e_H/2}$ is of largest order, that is, of the leading overlaps defined in Section 3.2. The assumption $np^{m(G)} \to \infty$ implies by Lemma 3.15 that all leading overlaps are connected. Since also $m(H) \le m(G)$ for $H \subseteq G$, Theorem 6.49 yields $\widetilde{X}_G \xrightarrow{d} \mathrm{N}(0,1)$.

Example 6.51. We next consider *induced* subgraph counts. Let $Y_G(\mathbb{G}(n,p))$ be the number of induced subgraphs of $\mathbb{G}(n,p)$ that are isomorphic to G. We see, as before, that Y_G has a finite decomposition (6.30), but now the sum has to be taken over all graphs H (without isolated vertices) with $v_H \le v_G$.

In the case $p \to 0$, it is easy to see that the decomposition (6.30) is dominated by the same terms as in the decomposition for X_G, that is, by the terms corresponding to leading overlaps of G, except in the case when G has no edges. Hence, by the argument in Example 6.50, Y_G is asymptotically normal provided $np^{m(G)} \to \infty$. (In the exceptional case $e_G = 0$, X_G is degenerate but not Y_G; Y_G is dominated by $\{K_2\}$ and is thus asymptotically normal provided $n^2p \to \infty$.) We leave the details to the reader (Exercise!).

The case $p \to 1$ is reduced to the case $p \to 0$ by considering complements; $Y_G(\mathbb{G}(n,p))$ equals the number of induced copies of the complement of G in the complement of $\mathbb{G}(n,p)$, that is, in $\mathbb{G}(n,1-p)$.

The case $p \to p_0 \in (0,1)$ is more interesting. Let us, for simplicity, assume that $p(n) = p$ is constant. In this case, a detailed calculation shows that, with $\alpha_n = (n)_{v_G} / \operatorname{aut}(G)$,

$$\hat{Y}_G(K_2) = \frac{\alpha_n}{n(n-1)}\left(e_G - \tbinom{v_G}{2}p\right)p^{e_G-1}(1-p)^{\binom{v_G}{2}-e_G-1}.$$

More generally, for every H there is a polynomial Q_H such that

$$\hat{Y}_G(H) = \frac{(n)_{v_G}}{(n)_{v_H}} Q_H(p).$$

In particular, $\hat{Y}_G(H) = O(n^{v_G - v_H})$ and thus $\hat{Y}^*_G(H) = O(n^{v_G - v_H/2})$; moreover, O can here be replaced by Θ if $Q_H(p) \ne 0$, while $\hat{Y}_G(H) = \hat{Y}^*_G(H) = 0$ if $Q_H(p) = 0$. Consequently, Y_G is dominated by the smallest non-null graphs H with $Q_H(p) \ne 0$.

If $p \ne e_G/\binom{v_G}{2}$, then $Q_{K_2}(p) \ne 0$ and thus Y_G is dominated by $\{K_2\}$ (just as X_G is); consequently, Y_G is asymptotically normal by Theorem 6.49. (This is essentially the first projection method; see Remark 6.44.) In this case, $\operatorname{Var} Y_G \asymp \hat{Y}^*_G(K_2)^2 \asymp n^{2v_G-2}$.

If, however, $p = e_G/\binom{v_G}{2}$, then $\hat{Y}_G(K_2) = 0$ and we have to study larger H. The next possibility is $v_H = 3$, which holds for two graphs H, namely, P_2 and K_3. (Recall that we only consider graphs H without isolated vertices.) Hence, if further $\hat{Y}_G(P_2) \ne 0$ or $\hat{Y}_G(K_3) \ne 0$, then Y_G is dominated by $\{P_2, K_3\}$.

Since both P_2 and K_3 are connected, Y_G is asymptotically normal in this case too by Theorem 6.49. (This is essentially the second projection method; see Remark 6.44.) In this case, however, $\operatorname{Var} Y_G \asymp n^{2v_G-3}$.

The remaining possibility is that $\hat{Y}_G(K_2) = \hat{Y}_G(P_2) = \hat{Y}_G(K_3) = 0$. Graphs G that satisfy these conditions are known as *p-proportional*. For such G and p, it may be shown that $\hat{Y}_G(2K_2) \neq 0$, where $2K_2$ is the graph consisting of two disjoint edges. Hence the list of possible cases ends here: Y_G is dominated by $\{H : v_H = 4\}$ and $\operatorname{Var}(Y_G) \asymp n^{2v_G-4}$. Furthermore, since $2K_2$ is disconnected, Corollary 6.47 shows that Y_G is *not* asymptotically normal.

More precisely, this argument yields the following result; see Barbour, Karoński and Ruciński (1989) and Janson (1990a, 1995a) for further details.

Theorem 6.52. *Consider* $\mathbb{G}(n,p)$ *where* p *is fixed,* $0 < p < 1$.

(i) *We have*

$$n^{1-v_G}(Y_G - \mathbb{E}\, Y_G) \xrightarrow{d} \mathrm{N}(0, \sigma_2^2),$$

for some $\sigma_2^2 \geq 0$;

(ii) $\sigma_2^2 = 0$ *if and only if* $p = e_G / \binom{v_G}{2}$, *and then*

$$n^{3/2-v_G}(Y_G - \mathbb{E}\, Y_G) \xrightarrow{d} \mathrm{N}(0, \sigma_3^2),$$

for some $\sigma_3^2 \geq 0$;

(iii) $\sigma_2^2 = \sigma_3^2 = 0$ *if and only if* G *is p-proportional, and then*

$$n^{2-v_G}(Y_G - \mathbb{E}\, Y_G) \xrightarrow{d} a(Z_1^2 - 1) + bZ_2,$$

where $Z_1, Z_2 \in \mathrm{N}(0,1)$ *are independent,* a, b *are constants, and* $a < 0$. *This limit is non-degenerate and not normal.* ∎

The parameters in the limit distributions may be given explicitly; here we only remark that a calculation shows that, for (iii),

$$a = 2p(1-p)Q_{2K_2}(p) = -\frac{1}{\operatorname{aut}(G)}\binom{v_G}{2}p^{e_G}(1-p)^{\binom{v_G}{2}-e_G}.$$

Remark 6.53. It is not trivial to construct proportional graphs, and not even obvious that any exist at all. The smallest proportional graphs have eight vertices; the first example was given in Barbour, Karoński and Ruciński (1989), and another example is the wheel consisting of a cycle of seven vertices, all joined to a central vertex. Deterministic and probabilistic constructions showing that p-proportional graphs exist for every rational $p \in (0,1)$ are given by Janson and Kratochvíl (1991), Kärrman (1993), and Janson and Spencer (1992).

Kärrman (1994) has further constructed a graph (with 64 vertices) that is proportional (with $p = 1/2$) and, furthermore, such that $\hat{Y}_G(H) = 0$ for every H with $v_H = 4$ except $2K_2$; in this case the constant b vanishes.

Applications to subgraph counts in $\mathbb{G}(n, M)$

We have so far treated the decomposition method for $\mathbb{G}(n,p)$. In fact, the results extend to the continuous time random graph process $\{\mathbb{G}(n,t)\}_t$ defined in Chapter 1, with a random graph evolving in time. The method then yields results on convergence of graph functionals as stochastic processes. See Janson (1994a) for details; an example treated there is the asymptotic normality of the maximum number of isolated edges during the evolution of a random graph.

The process version of the method also leads to results for $\mathbb{G}(n,M)$ by considering the random time when the evolving graph has exactly M edges. We let $p = M/\binom{n}{2}$ and define the variables $S_n(H)$ by (6.29), now letting I_e be the edge indicators for $\mathbb{G}(n,M)$. The expansion (6.30) is still valid, with the same $\hat{X}_n(H)$ as for $\mathbb{G}(n,p)$, but the terms are no longer orthogonal. Note that $S_n(K_2) = 0$ for $\mathbb{G}(n,M)$, so clearly the term with $H = K_2$ disappears from (6.30). Moreover, a further analysis shows that also terms where H contains an isolated edge (i.e., a component K_2) require special treatment. This leads to the following result; for the proof we again refer to Janson (1994a), which also contains some extensions.

Theorem 6.54. *Let X_n be a graph functional of $\mathbb{G}(n,M)$, where $M = M(n) \to \infty$, such that there is a finite family $\mathcal{H}$ with $\hat{X}_n(H) = 0$ for $H \notin \mathcal{H}$ (for every p). Let $p = M(n)/\binom{n}{2}$, and let $\mathcal{H}'$ be the subfamily of all graphs in $\mathcal{H}$ with at least three vertices. Suppose that $p \to p_0 < 1$, and that β_n is a sequence of positive numbers such that for every $H \in \mathcal{H}'$,*

$$a(H) = \lim_{n\to\infty} \hat{X}_n^*(H)/\beta_n \quad \textit{exists.}$$

Suppose further that $a(H) = 0$ for every $H \in \mathcal{H}'$ with two or more isolated edges, and that if $H \in \mathcal{H}'$ with $a(H) \neq 0$, then $np^{m(H)} \to \infty$. Then,

$$\frac{X_n - \alpha_n}{\beta_n} \xrightarrow{d} \sum_{H\in\widetilde{\mathcal{H}}} a(H)U(H), \tag{6.47}$$

where $\widetilde{\mathcal{H}} = \{H \in \mathcal{H}' :$ every component of H has at least three vertices$\}$, $U(H)$ is as in Theorem 6.43 and α_n equals the expectation of X_n calculated for $\mathbb{G}(n,p)$.

In particular, if $a(H) = 0$ for every $H \in \mathcal{H}$ with two or more components with at least three vertices, but $a(H) \neq 0$ for some $H \in \mathcal{H}'$, then X_n is asymptotically normal. ∎

Note that the smallest H that gives a non-normal term in (6.47) is the union of two copies of P_2, which has six vertices.

Comparing the limits in Theorems 6.45 and Theorem 6.54, we see that the following holds, at least provided both theorems apply (and, presumably, more generally): If a graph functional is dominated by a family of graphs that have no isolated edges, then it has the same asymptotic distribution for $\mathbb{G}(n,p)$ and $\mathbb{G}(n,M)$, with $M \sim p\binom{n}{2}$. On the other hand, if the functional is dominated by K_2 (i.e., the situation when the first projection applies), then the asymptotic distributions may be completely different; moreover, the variance for $\mathbb{G}(n,p)$ is of a larger magnitude than the variance for $\mathbb{G}(n,M)$. (See Pittel (1990) for a different approach to the relation between asymptotic distributions for the two models.)

Example 6.55. We may now study the subgraph counts X_G (arbitrary subgraphs) and Y_G (induced subgraphs), where G is fixed, for the random graph $\mathbb{G}(n,M)$, $M = M(n)$; see Examples 6.50 and 6.51 for the corresponding results for $\mathbb{G}(n,p)$. We define $p = M/\binom{n}{2}$, and begin by studying X_G.

Since the term with $H = K_2$ in the decomposition (6.30) plays no role for $\mathbb{G}(n,M)$, we consider the graphs $H \neq \emptyset, K_2$ and find the ones with largest order of $\hat{X}^*_G(H)$; by (6.46), these are the subgraphs $H \subseteq G$ with $v_H \geq 3$ and smallest order of $n^{v_H}p^{e_H}$.

Let us first assume that G has a component with three or more vertices, that is, G has a subgraph P_2, and assume that $np^{m(G)} \to \infty$. Then the argument in the proof of Lemma 3.15, together with $n^{v(2K_2)}p^{e(2K_2)} \gg n^{v(P_2)}p^{e(P_2)}$, shows that every such extremal H is connected, and it follows by Theorem 6.54 that $X_G(\mathbb{G}(n,M))$ is asymptotically normal. We see further that the asymptotics of X_G are the same for $\mathbb{G}(n,M)$ as for $\mathbb{G}(n,p)$ when M is so small that K_2 is not a leading overlap, but not for larger M. In the case $M \approx p\binom{n}{2}$ with $p \in (0,1)$ fixed, $\operatorname{Var} X_G \asymp n^{2v_G-2}$ for $\mathbb{G}(n,p)$ but $\operatorname{Var} X_G \asymp n^{2v_G-3}$ for $\mathbb{G}(n,M)$ (dominated by P_2 and possibly K_3, when K_2 is ignored).

In the exceptional case when every component of G has at most two vertices, that is, G consists of isolated vertices and edges, X_G is deterministic if $e_G \leq 1$. If $e_G \geq 2$, a modification of the argument above shows that X_G is asymptotically normal in this case too, provided $n^{3/2}p \to \infty$ and $n^{3/2}(1-p) \to \infty$; we omit the details.

The normalization used in Theorem 6.54 is not the natural one by the mean and variance of X_G, but it may be shown that in this case (and many others), all moments of the normalized variable converge to the corresponding moments of $\mathrm{N}(0,1)$, and hence convergence holds with the natural normalization too. We may summarize the result as follows.

Theorem 6.56. *If $e_G > 1$, $M = M(n) \gg n^{1/2}$, $\binom{n}{2} - M \gg n^{1/2}$ and $np^{m(G)} \to \infty$, where $p = M/\binom{n}{2}$, then $\widetilde{X}_G(\mathbb{G}(n,M)) \xrightarrow{d} \mathrm{N}(0,1)$.* ∎

Turning to the induced subgraph count Y_G, we see by similar arguments that when $p \to 0$, we have (just as for $\mathbb{G}(n,p)$) the same dominating graphs H as for X_G, provided $P_2 \subseteq G$, and thus Y_G is asymptotically normal provided M is not too small. Again, this may be shown also in the exceptional case when G consists of isolated edges and vertices, provided $v_G \geq 3$ (otherwise Y_G is constant).

Theorem 6.57. *If $v_G \geq 3$, $M = M(n) \gg n^{1/2}$, $p = M/\binom{n}{2} \to 0$ and $np^{m(G)} \to \infty$, then $\widetilde{Y}_G(\mathbb{G}(n, M)) \xrightarrow{d} \mathrm{N}(0, 1)$.* ∎

Again the case $M/\binom{n}{2} \to 1$ can be handled by considering the complements, and we obtain asymptotic normality in this case too, provided $\binom{n}{2} - M$ is not too small.

Finally, in the case $M \approx p\binom{n}{2}$, p fixed, Theorem 6.54 yields the following analogue of Theorem 6.52; see Janson (1994a, 1995a) for details. We let $\mathcal{U}_k^0$ denote the set of unlabelled graphs with k vertices, none of them isolated, and let $\mathcal{U}_k^c$ denote the subset of connected graphs with k vertices.

Theorem 6.58. *Let $0 < p < 1$ be fixed and consider $\mathbb{G}(n, M)$ where (for simplicity) $M = M(n) = \lfloor \binom{n}{2} p \rfloor$.*

(i) *We have*

$$n^{3/2-v_G}(Y_G - \mathbb{E}\, Y_G) \xrightarrow{d} \mathrm{N}(0, \sigma_3^2)$$

for some $\sigma_3^2 \geq 0$;

(ii) *$\sigma_3^2 = 0$ if and only if $\hat{Y}_G(H) = 0$ for $H \in \mathcal{U}_3^0 = \{P_2, K_3\}$, and then*

$$n^{2-v_G}(Y_G - \mathbb{E}\, Y_G) \xrightarrow{d} \mathrm{N}(0, \sigma_4^2)$$

for some $\sigma_4^2 \geq 0$;

(iii) *$\sigma_3^2 = \sigma_4^2 = 0$ if and only if $\hat{Y}_G(H) = 0$ for $H \in \mathcal{U}_3^0 \cup \mathcal{U}_4^c$, and then*

$$n^{5/2-v_G}(Y_G - \mathbb{E}\, Y_G) \xrightarrow{d} \mathrm{N}(0, \sigma_5^2)$$

for some $\sigma_5^2 \geq 0$;

(iv) *$\sigma_3^2 = \sigma_4^2 = \sigma_5^2 = 0$ if and only if $\hat{Y}_G(H) = 0$ for $H \in \mathcal{U}_3^0 \cup \mathcal{U}_4^0 \cup \mathcal{U}_5^c$, and then*

$$n^{3-v_G}(Y_G - \mathbb{E}\, Y_G) \xrightarrow{d} a(Z_1^2 - 1) + b(Z_2^2 - 1) + cZ_3,$$

where $Z_1, Z_2, Z_3 \in \mathrm{N}(0, 1)$ are independent, a, b, c are constants, and $a \neq 0$. This limit is non-degenerate and not normal. ∎

Kärrman's (1994) example shows that there exists a graph G such that case (iii) occurs. We do not, however, know if there exists any graph such that case (iv) happens; thus we do not know whether $Y_G(\mathbb{G}(n, M))$ is always asymptotically normal.

In any case, as remarked in connection with Theorem 6.52, a p-proportional graph G has $\hat{Y}_G(2K_2) \neq 0$, and thus is not an example of (iv) in the present theorem. In other words, the classes of graphs that yield non-normal limits in Theorems 6.52 and 6.58 are disjoint.

Other projections

We have in this section studied projections using the variables $S_n(H)$ only. Other projections are useful in other situations; we mention a few examples and references.

A similar decomposition has been used by de Jong (1996) to extend Theorem 6.5 to random hypergraphs.

Andersson (1998) has studied (directed) subgraph counts in a random tournament, obtaining results similar to Theorem 6.52. In that case, however, there is an infinite list of possible cases.

Janson (1994b) studied the numbers of spanning trees, Hamilton cycles or perfect matchings in $\mathbb{G}(n,p)$ and $\mathbb{G}(n,M)$. For $\mathbb{G}(n,M)$, with $M \gg n^{3/2}$ and $\binom{n}{2} - M \gg n$, these random numbers are shown to be asymptotically normal, by approximating with a linear function of X_{P_2}; a kind of "first projection" for $\mathbb{G}(n,M)$. For $M \asymp n^{3/2}$, as well as for $\mathbb{G}(n,p)$ with $p = \Omega(n^{-1/2})$ and $1-p \gg n^{-2}$, the numbers are shown to be asymptotically log-normal.

Furthermore, in Chapter 9 we will prove results on asymptotic distributions for random regular graphs by projecting onto functions of the cycle counts Z_i; in that case the basic variables Z_i have asymptotic Poisson distributions and the resulting asymptotic distributions are quite different from the ones obtained here, see Theorem 9.12.

6.5 FURTHER METHODS

Finally, we briefly mention a couple of other methods.

Martingales

As mentioned above, the proof of Theorem 6.43 is based on a martingale limit theorem; another martingale limit theorem is used by de Jong (1996). Such theorems may also be used directly in situations where the methods described above fail. We will not go into details, which are rather technical, and mention only one example.

Example 6.59. Barbour, Janson, Karoński and Ruciński (1990) studied the number X_d of cliques of a given fixed size $d \geq 2$ in $\mathbb{G}(n,p)$, where a clique is defined as a maximal complete subgraph, that is, a K_d that is not contained in a K_{d+1}. It was shown that if $p = p(n)$ is such that $\mathbb{E}\, X_d \to \lambda < \infty$, then $X_d \xrightarrow{d} \mathrm{Po}(\lambda)$, and if $\mathbb{E}\, X_d \to \infty$, then $\widetilde{X}_d \xrightarrow{d} \mathrm{N}(0,1)$.

The Poisson part was proved using the Stein–Chen method. For the normal part, different methods were used for different ranges of p; for certain p the first projection method works, and for a larger range it is possible to use Corollary 6.46 (with the family $\mathcal{H}$ consisting of K_2, K_3, K_d and the 'multistars' $M_{d,r}$ obtained by adding $r \geq 1$ vertices to K_d, joining them to all vertices of K_d). For p close to the upper threshold, this fails, since then (6.39) does not hold; instead a martingale limit theorem was invoked directly.

Generating functions

A method that is widely used to find asymptotics, including asymptotic distributions, for combinatorial problems is to define a suitable generating function and obtain results through a study of its asymptotics. This method has, however, been used rather sparsely for random graphs. We refer to Pittel (1990) and to Janson, Knuth, Łuczak and Pittel (1993) for examples.

7

The Chromatic Number

In this chapter we present results on the chromatic number which, due to their elegance and importance, range among the very best in the theory of random graphs. We begin with Frieze's beautiful method, which combines the second moment method with large deviation inequalities to estimate the independence number of random graphs. In Section 7.4 we describe Bollobás's ingenious argument for determining the chromatic number of dense random graphs – probably the most important and celebrated result on random graphs for the last years. Then we analyze an expose-and-merge algorithm, based on Matula's original idea, which is used for estimating the chromatic number in the sparse case. Finding the chromatic number can be viewed as a vertex partition problem. In the last part of the chapter we discuss this problem in a more general form.

7.1 THE STABILITY NUMBER

Let us recall that a set of vertices of a graph G is *independent*, or *stable*, if it contains no edges of G. The size of the largest among such sets, denoted by $\alpha(G)$, is called the *independence* (or *stability*) *number* of G. In this section we study the behavior of $\alpha(\mathbb{G}(n,p))$ – a random variable closely related to the chromatic number $\chi(\mathbb{G}(n,p))$. Note that the independence number of a graph is the same as the clique number of the complementary graph, so the results below can also be stated in terms of the clique number of $\mathbb{G}(n, 1-p)$.

again. Thus, a_i achieves its largest value either at $i_1 = \min\{i \geq 1 : b_i < 1\}$, or at $i_2 = \max\{i < k : b_i > 1\} + 1$. The reader can easily verify (Exercise!) that i_1 is much smaller than k, so the factor $(1-p)^{-i}$ is almost negligible and $i_1 = O(1 + k^2/n)$; more precisely, if $k^2/n \leq 1$, then $i_1 = 1$, and otherwise $i_1 \asymp k^2/n$ (in fact, $i_1 \sim k^2/n$ if $k^2/n \to \infty$ and $np \to \infty$). In order to find a lower bound for i_2, set $i_2' = \lceil k(1 - 1/\log^5 np)\rceil$. Then

$$b_{i_2'-1} \geq \frac{k}{2n\log^{10} np}\exp(i_2'p + O(i_2'p^2)) \geq \frac{np}{\log^{12} np} > 1,$$

and so $i_2 \geq i_2'$.

Let us estimate the value of a_{i_2}. The inequalities $i_2 \geq i_2' = \lceil k(1 - 1/\log^5 np)\rceil$ and (7.7) imply that for $i_2' < k$ we have

$$\begin{aligned} a_{i_2} &= \frac{\binom{k}{i_2}\binom{n-k}{k-i_2}}{\binom{n}{k}(1-p)^{\binom{k}{2}}}(1-p)^{\binom{k}{2}-\binom{i_2}{2}} \leq \frac{\binom{k}{k-i_2}\binom{n-k}{k-i_2}}{\mathbb{E}X} \\ &\leq \Big(\frac{e^2nk}{(k-i_2)^2}\Big)^{k-i_2}\exp\Big(-\frac{\varepsilon k}{2}\Big) \\ &\leq \Big(\frac{e^2nk}{(k-i_2')^2}\exp\Big(-\frac{\varepsilon k}{2(k-i_2')}\Big)\Big)^{k-i_2'} \\ &\leq \Big(e^2np\log^{10} np\exp\Big(-\frac{\varepsilon}{2}\log^5 np\Big)\Big)^{1/p\log^5 np} = o(n^{-1}), \end{aligned}$$

while for $i_2' = k$, directly from (7.7), we get $a_{i_2} = a_k = 1/\mathbb{E}X = o(n^{-1})$. Thus, the contribution to $\sum_i a_i$ coming from the terms with large indices is negligible, and the sum is dominated by terms with indices close to i_1.

Let us consider two cases. If $p \geq \log^2 n/\sqrt{n}$, then $k < 2\sqrt{n}/\log n$, $i_1 = 1$ and, furthermore, for every $i \leq i_3 = \lceil \log np\rceil$ and n large enough,

$$a_i = \frac{\binom{k}{i}\binom{n-k}{k-i}}{\binom{n}{k}}(1-p)^{-\binom{i}{2}} \leq k^{2i}\frac{(n-k)_{k-i}}{(n)_k}\exp(i^2p/2) \leq \Big(\frac{2k^2}{n}\Big)^i,$$

so $a_{i_3} = o(n^{-1})$. Hence,

$$\sum_{i=1}^{k} a_i \leq \sum_{i=1}^{i_3-1}\Big(\frac{2k^2}{n}\Big)^i + k\max\{a_{i_3}, a_{i_2}\} \leq \frac{4k^2}{n} + o(1) = o(1). \tag{7.9}$$

Next, suppose that $C_\varepsilon/n \leq p \leq \log^2 n/\sqrt{n}$ for some large constant $C_\varepsilon > 0$. Then, rather crudely,

$$a_{i_1} \leq (1-p)^{-\binom{i_1}{2}} \leq \exp((1/2 + o(1))pi_1^2) \leq \exp\Big(\frac{k}{2\log^3 np}\Big)$$

and

$$\sum_{i=1}^{k} a_i \leq k\max\{a_{i_1}, a_{i_2}\} \leq k\exp\Big(\frac{k}{2\log^3 np}\Big) \leq \exp\Big(\frac{k}{\log^3 np}\Big) - 1. \tag{7.10}$$

Now, in order to get (7.4) and (7.5), it is enough to put the values of $\mathbb{E}\,X$ and $\mathbb{E}\,X^2$ given by (7.7), (7.8), (7.9) and (7.10) into (3.3) – a stronger form of the second moment method. ■

Let us remark that, although inequality (3.2) is slightly weaker than (3.3), for most random graph problems it works just as well. Here, however, it is not the case – for small p the inequality (3.2) gives a worthless negative lower bound for the probability $\mathbb{P}(X(k_\varepsilon) > 0)$.

On the other hand, at first sight, the estimate (7.5) does not look terribly useful either – although positive it tends to 0 faster than $\exp(-\sqrt{n}/2\log^4 n)$. One might hope that this is because our estimates were too crude. Indeed, one can bound the variance of X more carefully and show that (3.3) yields (7.4) also for some p which tends to 0 faster than $\log^2 n/\sqrt{n}$. However, for $p = p(n)$ which tends to 0 very quickly the second moment method utterly fails. It is not hard to understand why it does so poorly in this case: if, say, $p = n^{-3/4}$, then the largest stable sets are larger than $n^{3/4}$, and the majority of pairs of such sets share a substantial amount of elements which makes $\mathbb{E}(X^2)$ much larger than $(\mathbb{E}\,X)^2$. Quite surprisingly, (7.5) can still be used for the evaluation of $\alpha(\mathbb{G}(n,p))$, provided it is supplemented with a large deviation inequality of a martingale or Talagrand type. We owe this profound observation to Frieze (1990), who showed that the estimates for $\alpha(\mathbb{G}(n,p))$ given by (7.6) remain valid also for $p \ll 1/\sqrt{n}$.

Theorem 7.4. *Let $\varepsilon > 0$ and let $k_{\pm\varepsilon}$ be defined as in* (7.2). *Then there exists a constant C_ε such that for $C_\varepsilon/n \le p = p(n) \le \log^{-2} n$* a.a.s.

$$k_{-\varepsilon} \le \alpha(\mathbb{G}(n,p)) \le k_\varepsilon .$$

Proof. By Lemma 7.3, the assertion holds for $\log^2 n/\sqrt{n} \le p \le \log^{-2} n$, and for the whole range of $p = p(n)$ *a.a.s.* $\alpha(\mathbb{G}(n,p)) \le k_\varepsilon$. Thus, it is enough to show that if $C_\varepsilon/n \le p \le \log^2 n/\sqrt{n}$, then *a.a.s.* $\alpha(\mathbb{G}(n,p)) \ge k_{-\varepsilon}$.

Note that Talagrand's large deviation inequality, Theorem 2.29, can be applied to $\alpha(\mathbb{G}(n,p))$ with $c_i = 1$ and $\psi(r) = \lceil r \rceil$ (see Examples 2.35 and 2.33), and thus by (2.35) we get

$$\begin{aligned}\mathbb{P}(\alpha(\mathbb{G}(n,p)) \le k_{-\varepsilon} - 1)\,\mathbb{P}(\alpha(\mathbb{G}(n,p)) \ge k_{-\varepsilon/2})& \\ \le \exp\Big(-\frac{(k_{-\varepsilon/2} - k_{-\varepsilon} + 1)^2}{4k_{-\varepsilon/2}}\Big)& \\ \le \exp\Big(-\frac{(\varepsilon/p)^2}{8\log np/p}\Big) = \exp\Big(-\frac{\varepsilon^2}{8p\log np}\Big)&.\end{aligned} \tag{7.11}$$

Thus, combining (7.11) and (7.5) (for $\varepsilon/2$), it follows that, for large n,

$$\begin{aligned}\mathbb{P}(\alpha(\mathbb{G}(n,p)) < k_{-\varepsilon}) &\le \exp\Big(-\frac{\varepsilon^2}{8p\log np} + \frac{2}{p\log^2 np}\Big) \\ &\le \exp\Big(-\frac{\varepsilon^2}{16p\log n}\Big) = o(1)\ .\end{aligned}$$

■

7.2 THE CHROMATIC NUMBER: A GREEDY APPROACH

Let us recall that the chromatic number $\chi(G)$ of a graph G is the smallest integer ℓ such that the vertex set of G can be partitioned into ℓ stable sets. The problem of computing the value of the chromatic number of a graph has drawn much attention in graph theory, but in this chapter we will not use any sophisticated results from this area, relying mainly on known elementary facts about $\chi(G)$ (see, e.g., Bollobás (1998) or Diestel (1996)).

We begin our study of $\chi(\mathbb{G}(n,p))$ with the simple observation that for any graph G with n vertices and stability number $\alpha(G)$, the chromatic number $\chi(G)$ is bounded from below by $\lceil n/\alpha(G)\rceil$. Thus, the upper bounds for $\alpha(\mathbb{G}(n,p))$ given by Theorem 7.1 and 7.4 yield immediately that *a.a.s.* $\chi(\mathbb{G}(n,p)) \geq n/\hat{k}_\varepsilon$ if p is a constant (or $p = p(n)$ tends to 0 slowly enough), and $\chi(\mathbb{G}(n,p)) \geq n/k_\varepsilon$ if $C_\varepsilon/n \leq p = p(n) \leq \log^{-2} n$ for a large enough constant $C = C_\varepsilon$. Replacing these bounds by simpler, slightly smaller expressions we arrive at the following result.

Corollary 7.5.

(i) *If $p = p(n) > n^{-\delta}$ for every $\delta > 0$ but $p \leq c$ for some $c < 1$, then* a.a.s.

$$\chi(\mathbb{G}(n,p)) \geq \frac{n}{2\log_b n - \log_b \log_b n},$$

where $b = 1/(1-p)$.

(ii) *There exists a constant C_0 such that if $C_0/n \leq p = p(n) \leq \log^{-2} n$, then* a.a.s.

$$\chi(\mathbb{G}(n,p)) \geq \frac{np}{2\log np - 2\log\log np + 1}. \qquad \blacksquare$$

The main question about $\chi(\mathbb{G}(n,p))$ is whether the vertex set of $\mathbb{G}(n,p)$ can be partitioned into stable sets of nearly maximum size, that is, whether

$$\chi(\mathbb{G}(n,p)) = (1 + o_p(1))n/\alpha(\mathbb{G}(n,p)). \tag{7.12}$$

In this section we examine an algorithmic approach to this problem, and describe a simple algorithm coloring the vertices of $\mathbb{G}(n,p)$ which *a.a.s.* uses only twice as many colors as anticipated in (7.12).

Let $D(G) = \max_{H\subseteq G} \delta(H)$ be the *degeneracy number* of a graph G and, as in Section 3.1, let $m(G) = \max_{H\subseteq G} |E(H)|/|V(H)|$. Much of our argument will rely on the following well-known simple upper bound on $\chi(G)$.

Lemma 7.6. *There exists a polynomial time algorithm which colors the vertices of every graph using at most $1 + D(G)$ colors. In particular,*

$$\chi(G) \leq 1 + D(G) \leq 1 + 2m(G). \qquad \blacksquare$$

The above fact is particularly well suited for small subgraphs of a random graph, which are quite sparse and thus can be effectively colored with only a few colors. Throughout this chapter we will use the following estimates of the density of small subgraphs of $\mathbb{G}(n,p)$. Statements (i)–(iii) below can be easily verified using the first moment method (Exercise!). In order to prove (iv) it is enough to compute the expected number of subgraphs of $\mathbb{G}(n,p)$ with $m(F) \geq 1.45$ and fewer than $0.05n$ vertices, and use the fact that if $np \leq 1.001$ then *a.a.s.* the size of the largest component of $\mathbb{G}(n,p)$ is smaller than $0.05n$ (see Theorem 5.4).

Lemma 7.7.

(i) *There exists a constant C_0 such that for $np \geq C_0$* a.a.s. *every subgraph F of $\mathbb{G}(n,p)$ with fewer than $n/\log^2 np$ vertices satisfies $m(F) \leq np/\log^2 np$.*

(ii) *If $p \leq \log^2 n/\sqrt{n}$, then* a.a.s. *for every subgraph F of $\mathbb{G}(n,p)$ with fewer than $2\sqrt{n \log n}$ vertices we have $m(F) \leq \log^3 n$.*

(iii) *If $p \leq n^{-6/7}$, then* a.a.s. *for every subgraph F of $\mathbb{G}(n,p)$ with fewer than $70\sqrt{n \log n}$ vertices $m(F) \leq 1.45$ holds.*

(iv) *If $np \leq 1.001$, then* a.a.s. *$m(F) \leq 1.45$ for every subgraph of $\mathbb{G}(n,p)$. In particular,* a.a.s. *$\chi(\mathbb{G}(n,p)) \leq 3$.* ■

Note that although it follows from the proof of Theorem 7.4 that the expected number of stable sets of size $(1 - o(1))\alpha(\mathbb{G}(n,p))$ is quite large, such sets have a natural tendency to cluster together and so, possibly, they do not cover all vertices of the random graph. However, in the first attempt to estimate $\chi(\mathbb{G}(n,p))$ from above, we will defer this problem for a while, and instead look more closely at the stable sets which are about half the size of the largest one. More specifically, the following fact can be shown using the first moment method (Exercise!). A stable set is called *maximal* if it is not contained in any other stable set.

Lemma 7.8.

(i) *There exists a constant C_0 such that if $C_0/n \leq p \leq \log^{-2} n$, then with probability $1 - o(n^{-2})$, $\mathbb{G}(n,p)$ contains no maximal stable set smaller than $(\log np - 3 \log\log np)/p$.*

(ii) *If $p \geq \log^{-3} n$ but $p \leq c$ for some $c < 1$, then with probability $1 - o(n^{-2})$ every maximal stable set of $\mathbb{G}(n,p)$ is larger than $\log_b n - 3\log_b \log_b n$, where $b = 1/(1-p)$.* ■

Lemma 7.8 tells us that *a.a.s.* each stable set much smaller than $\frac{1}{2}\alpha(\mathbb{G}(n,p))$ can be extended to a bigger one. In particular, every vertex belongs to a stable set of size about $\alpha(\mathbb{G}(n,p))/2$. As was observed by Grimmett and McDiarmid

(1975), one can use this fact to describe an algorithm which colors the vertices of $\mathbb{G}(n,p)$ with $(2+o_p(1))n/\alpha(\mathbb{G}(n,p))$ colors.

Theorem 7.9. *There exists a polynomial time algorithm* CHR *for which the following hold:*

(i) *there exists a constant C_0 such that if $C_0/n \le p \le 1/\log^2 n$, then* a.a.s. CHR *uses no more than $np/(\log np - 6\log\log np)$ colors to properly color the vertices of $\mathbb{G}(n,p)$;*

(ii) *if $p \ge \log^{-2} n$ but $p \le c$ for some $c < 1$, then* a.a.s. *the number of colors used by* CHR *to color the vertices of $\mathbb{G}(n,p)$ is bounded from above by $n/(\log_b n - 6\log_b \log_b n)$, where $b = 1/(1-p)$.*

Proof. We will show only the first part of the assertion; the proof of (ii) is similar. Note first that in a random graph $\mathbb{G}(n,p)$ one can find a maximal stable subset S by examining only those pairs of vertices of $\mathbb{G}(n,p)$ which have at least one end in S. Indeed, to construct S greedily start from any vertex, put it into S, then check for every other vertex if it has neighbors among the vertices already in S; if this is not the case add such a vertex to S. Now color the vertices of S using the first color. Then, the graph obtained from $\mathbb{G}(n,p)$ by deleting the vertices of S can be viewed as the random graph $\mathbb{G}(n-|S|,p)$, so we can repeat the above greedy procedure over and over again, until the number of vertices in the graph drops below $n/\log^2 np$. If C_0 is sufficiently large, then, due to Lemma 7.8, the number of colors used so far is *a.a.s.* bounded from above by

$$\frac{np}{\log(np/\log^2 np) - 3\log\log(np/\log^2 np)} \le \frac{np}{\log np - 5\log\log np}.$$

Furthermore, Lemmas 7.6 and 7.7(i) imply that *a.a.s.* the remaining vertices of the graph can be effectively colored by at most $2np/\log^2 np + 1$ colors. Consequently, *a.a.s.*

$$\chi(\mathbb{G}(n,p)) \le \frac{np}{\log np - 5\log\log np} + \frac{2np}{\log^2 np} + 1 \le \frac{np}{\log np - 6\log\log np}. \quad \blacksquare$$

Can we do better than the algorithm CHR and effectively color $\mathbb{G}(n,p)$ using a substantially smaller number of colors? Clearly, in order to reduce the number of colors by a constant factor, we need to describe a fast procedure IND(δ) which *a.a.s.* finds in $\mathbb{G}(n,p)$ a stable set larger than $(1/2+\delta)\alpha(\mathbb{G}(n,p))$ for some $\delta > 0$. (The problem of the existence of IND(δ) was posed by Karp (1976) and, as was observed by Juels and Peinado (1998), has some interesting cryptographic consequences.) In fact, Matula (1987) showed that, having such a procedure as a subroutine, one could devise an algorithm which *a.a.s.* colors the vertices of $\mathbb{G}(n,p)$ with fewer than $(2-\delta')n/\alpha(\mathbb{G}(n,p))$ colors, for some positive constant δ'.

However, at this moment neither do we know what IND(δ) should look like, nor do we have any idea how to show that such a procedure does not exist. Although it is possible to find in $\mathbb{G}(n,p)$ stable sets slightly larger than those whose existence is assured by Lemma 7.8, and describe algorithms which *a.a.s.* color the vertices of $\mathbb{G}(n,p)$ with fewer colors than CHR does, all improvements are only with respect to the second-order terms: all these procedures produce stable sets of size $(1/2+o_p(1))\alpha(\mathbb{G}(n,p))$. (Examples of such algorithms were given, e.g., by Bollobás and Thomason (1985), Pittel and Weishaar (1997), and Jerrum (1992); a survey on the algorithmic theory of random graphs was published by Frieze and McDiarmid (1997).) Thus, the problem of existence of IND(δ) continues to be the most important open question of algorithmic flavor in the theory of random graphs. To illustrate our ignorance concerning this subject let us mention that even the following problem, posed independently, in various variants, by Jerrum (1992) and Kučera (1995), has not been settled so far.

Problem 7.10. *Let $0 < a < 1/2$, $0 < p < 1$, and let $\mathbb{G}_{n,p,a}$ be a random graph obtained from $\mathbb{G}(n,p)$ by choosing randomly a subset S of size $\lfloor n^a \rfloor$ of the vertex set of $\mathbb{G}(n,p)$ and removing from $\mathbb{G}(n,p)$ all the edges with both ends in S. Describe a polynomial time algorithm which* a.a.s. *finds a maximum stable set in $\mathbb{G}_{n,p,a}$ for every $a > 0$.*

Note that if $a > 0$, then *a.a.s.* S is the unique maximum stable set of $\mathbb{G}_{n,p,a}$ (Exercise!). Alon, Krivelevich and Sudakov (1998) provided an algorithm which *a.a.s.* finds S in a polynomial time for $a = 1/2$. Moreover, as observed by Kučera (1995), if $a > 1/2$, then the problem has an immediate solution: in this case all vertices of S *a.a.s.* can be identified just by inspecting their degrees (Exercise!).

7.3 THE CONCENTRATION OF THE CHROMATIC NUMBER

From Theorems 7.1, 7.4 and 7.9 we know that for every $\delta > 0$ and $c < 1$, and for every function p such that $C_\delta/n \leq p \leq c$ for some constant C_δ, the chromatic number $\chi(\mathbb{G}(n,p))$ *a.a.s.* lies between $n/\alpha(\mathbb{G}(n,p))$ and $(2+\delta)n/\alpha(\mathbb{G}(n,p))$. But is it sharply concentrated, in other words, is there some function $h(n) = h(n,p)$ such that *a.a.s.* $\chi(\mathbb{G}(n,p)) = (1+o_p(1))h(n)$?

One can immediately see using the vertex exposure martingale and Corollary 2.27 that $\chi(\mathbb{G}(n,p))$ is concentrated in an interval of length $O_p(\sqrt{n})$. This fact, however, does not answer our question in the case of a sparse random graph, when the chromatic number is of an order smaller than $\sqrt{n}$. Nonetheless, Shamir and Spencer (1987) proved a sharp concentration of $\chi(\mathbb{G}(n,p))$ throughout the entire evolution of $\mathbb{G}(n,p)$. Here we present a somewhat stronger version of their result given by Łuczak (1991c), based on an idea of Frieze.

Theorem 7.11. *For every sequence $p = p(n)$ there exists a function $h(n)$ such that the following hold:*

(i) *if* $p \geq n^{-6/7}$, *then* $\chi(\mathbb{G}(n,p)) = (1 + o_p(1))h(n)$;

(ii) *if* $p \leq n^{-6/7}$, *then* a.a.s. $h(n) \leq \chi(\mathbb{G}(n,p)) \leq h(n) + 1$.

Proof. As we indicated earlier, statement (i) is easy for a sufficiently dense random graph. If $p \geq p_0 = \log^2 n/\sqrt{n}$, then, by Corollary 7.5, *a.a.s.* $\chi(\mathbb{G}(n,p)) \geq \chi(\mathbb{G}(n,p_0)) \geq \sqrt{n}\log n/2$. Corollary 2.27 applied to the vertex exposure martingale, gives

$$\mathbb{P}\big(|\chi(\mathbb{G}(n,p)) - \mathbb{E}\,\chi(\mathbb{G}(n,p))| \geq \sqrt{n}\log\log n\big) = o(1),$$

that is, in this case the assertion (i) follows with $h(n) = \mathbb{E}\,\chi(\mathbb{G}(n,p))$.

Assume now that $p \leq \log^2 n/\sqrt{n}$ and let $h = h(n)$ be the smallest natural number for which

$$\mathbb{P}(\chi(\mathbb{G}(n,p)) \leq h) \geq 1/\log n.$$

Consequently, $\mathbb{P}(\chi(\mathbb{G}(n,p)) < h) < 1/\log n$, and so *a.a.s.* $\chi(\mathbb{G}(n,p)) \geq h$. Denote by $\bar{Y}$ the number of vertices in the largest induced subgraph of $\mathbb{G}(n,p)$ which can be colored by at most $h(n)$ colors, and set $Y = n - \bar{Y}$. We first prove that $\mathbb{E}\,Y < \sqrt{n\log n}$. Indeed, suppose that $\mathbb{E}\,Y \geq \sqrt{n\log n}$. Since altering the presence of the edges incident to a single vertex cannot affect the value of Y by more than one, Corollary 2.27 applies to the random variable Y with vertex exposure. This gives

$$\begin{aligned}\mathbb{P}(\chi(\mathbb{G}(n,p) \leq h) = \mathbb{P}(Y = 0) &\leq \mathbb{P}(Y \leq \mathbb{E}\,Y - \sqrt{n\log n}) \\ &\leq \exp(-\log n/2) < 1/\log n\,,\end{aligned}$$

contradicting the choice of h. Thus $\mathbb{E}\,Y < \sqrt{n\log n}$ and, once again using Corollary 2.27, we get

$$\mathbb{P}(Y \geq 2\sqrt{n\log n}) \leq \mathbb{P}(Y \geq \mathbb{E}\,Y + \sqrt{n\log n}) \leq \exp(-\log n/2) < 1/\log n.$$

Hence *a.a.s.* all except at most $2\sqrt{n\log n}$ vertices of $\mathbb{G}(n,p)$ can be colored using at most h colors. Moreover, Lemmas 7.6 and 7.7(ii) imply that for such a function p *a.a.s.* each subgraph of $\mathbb{G}(n,p)$ with at most $2\sqrt{n\log n}$ vertices can be colored using at most $2\log^3 n + 1$ colors. Thus, in this case *a.a.s.*

$$h \leq \chi(\mathbb{G}(n,p)) \leq h + 2\log^3 n + 1. \tag{7.13}$$

Note now that if $p \geq n^{-6/7}$, then, by Corollary 7.5, *a.a.s.* $\chi(\mathbb{G}(n,p)) \geq n^{1/7}/\log n$. Thus, (7.13) implies that $2\log^3 n + 1 = o(h)$ and, again by (7.13), the assertion (i) follows.

The argument for $p < n^{-6/7}$ is slightly more involved. Let us start with some comments on the rather uninteresting case $np < 1.001$. If $np \to 0$, then *a.a.s.* $\mathbb{G}(n,p)$ is a forest (see Section 5.1) and so *a.a.s.* $1 \le \chi(\mathbb{G}(n,p)) \le 2$. If $n^{-1/10} \le np \le 1.001$, then *a.a.s.* $\mathbb{G}(n,p)$ contains at least one edge, and thus, by Lemma 7.7(iv), *a.a.s.* $2 \le \chi(\mathbb{G}(n,p)) \le 3$.

Thus, we may and will assume that $1.001/n \le p \le n^{-6/7}$. Then, with probability $1 - o(1/\log n)$, $\mathbb{G}(n,p)$ contains an odd cycle of size $2\lfloor(\log\log n)^2\rfloor + 1$, (Exercise!) and so $h \ge 3$. Hence, Lemma 7.7(iii) and the argument presented above imply that there exists $h = h(n) \ge 3$ such that *a.a.s.* $\mathbb{G}(n,p)$ has the following properties:

(i) all except at most $2\sqrt{n\log n}$ vertices of the graph can be colored with at most h colors;

(ii) for every subgraph F of the graph with fewer than $70\sqrt{n\log n}$ vertices we have $m(F) \le 1.45$.

Now to complete the proof it is enough to show that every graph $G = (V, E)$ with the above two properties can be colored using at most $h + 1$ colors.

Let S be a subset of V such that $|S| \le 2\sqrt{n\log n}$ and the vertices of $V \backslash S$ can be colored with at most h colors. We recursively define an ascending sequence of sets $S = S_0 \subseteq S_1 \subseteq \cdots \subseteq S_t \subseteq V$ in such a way that $|S_t| \le 62\sqrt{n\log n}$ and the neighborhood of S_t is a stable set in G. The recursive step is simple: if the set S_i has already been found and the neighborhood of S_i contains an edge $\{v, w\}$, we put $S_{i+1} = S_i \cup \{v, w\}$. Note that $|S_{i+1}| = |S_i| + 2$ and that $e(S_{i+1}) \ge e(S_i) + 3$, where $e(S_i)$ is the number of edges of G contained in S_i. Hence, after i steps,

$$|S_i| = |S| + 2i \quad \text{and} \quad e(S_i) \ge 3i.$$

But, due to property (ii), as long as $i < 34\sqrt{n\log n}$ we have

$$3i \le e(S_i) \le 1.45|S_i| = 1.45|S| + 2.9i \le 3\sqrt{n\log n} + 2.9i$$

and so the procedure must end after at most $t \le 30\sqrt{n\log n}$ steps. Consequently, G contains a set S_t of size smaller than $|S| + 2t \le 62\sqrt{n\log n}$ such that its neighborhood $N(S_t)$ is stable in G and all vertices outside S_t can be colored by at most h colors.

Now one can color the vertices of G with at most $h+1$ colors in the following way. All the vertices not belonging to $S_t \cup N(S_t)$ are colored with the first h colors, while the vertices of $N(S_t)$ are colored by the $(h+1)$st color. Finally, the set S_t, which due to Lemma 7.6 spans a subgraph with chromatic number at most three, can be colored by any three of the first h colors. ■

Recently Alon and Krivelevich (1997), adding to the above argument one more ingredient, the Lovász Local Lemma, showed that $\chi(\mathbb{G}(n,p))$ is asymptotically concentrated on at most two points as long as $p \le n^{-1/2-\varepsilon}$. They also observed the following consequence of the two-point distribution, leading, by a suitable choice of $p(n)$, to a one-point distribution.

Corollary 7.12. *For every $\varepsilon > 0$ and every positive integer sequence $r = r(n) \le n^{1/2-\varepsilon}$ there exists a probability sequence $p = p(n)$ such that for sufficiently large n*

$$\mathbb{P}(\chi(\mathbb{G}(n,p)) = r) \ge 1 - \varepsilon.$$

Proof. We may assume that $\varepsilon < 1$ and $r > 1$. Define $p = p(n)$ as the infimum of all real numbers $\rho \in (0,1)$ such that

$$\mathbb{P}(\chi(\mathbb{G}(n,\rho)) < r) \le \varepsilon/2.$$

Corollary 7.5 implies that $p < n^{-1/2-\varepsilon/2}$, and thus we may apply Theorem 7.11(ii) (in the extended range). It follows that $h(n) = r - 1$, but $\mathbb{P}(\chi(\mathbb{G}(n,p)) = r - 1) \le \varepsilon/2$. ∎

Let us compare the obtained concentration of $\chi(\mathbb{G}(n,p))$ with that of $\alpha(\mathbb{G}(n,p))$. From Theorem 7.11 we know that $\chi(\mathbb{G}(n,p))$ is concentrated on at most two points when $p(n)$ tends to 0 quickly enough, while Remark 7.2 states that a similar two-point concentration holds for $\alpha(\mathbb{G}(n,p))$ when p tends to 0 very slowly with n. This connection is probably due to the relation $\chi(G) \ge n/\alpha(G)$ which, for random graphs, tends to become an asymptotic equation. The fact that Theorem 7.11 does not specify the function $h(n)$, while the concentration function of $\alpha(\mathbb{G}(n,p))$ is explicit, indicates that dealing with the chromatic number is more difficult. The reason is not hard to understand: the stability number is a local parameter of a random graph, while the chromatic number characterizes its overall structure. Nevertheless, in the next two sections we will find the asymptotic value of $h(n)$.

7.4 THE CHROMATIC NUMBER OF DENSE RANDOM GRAPHS

When Shamir and Spencer (1987) used Azuma's inequality (Theorem 2.25) to show that both the stability number and, more importantly, the chromatic number of $\mathbb{G}(n,p)$, are sharply concentrated around their expectations, it was not expected that martingales could help in finding the asymptotic value of $\chi(\mathbb{G}(n,p))$. As we saw in the previous section, Shamir and Spencer proved that, in particular, for every constant p there exists a sequence $h(n)$ such that $\chi(\mathbb{G}(n,p)) = (1 + o_p(1))h(n)$. In view of Corollary 7.5(i) and Theorem 7.9(ii) we have, with $b = 1/(1-p)$,

$$\frac{n}{2\log_b n} \le h(n) \le \frac{n}{\log_b n}.$$

It came as a great surprise when Bollobás (1988a) used martingales to show that the truth lies at the left endpoint of the above interval. His paper estimating $\chi(\mathbb{G}(n,p))$ for dense random graphs was based on a beautiful, insightful and, at the same time, very simple argument. By the second moment method it was known (see Theorem 7.1) that if $0 < p < 1$ is a constant, then *a.a.s.*

$\mathbb{G}(n,p)$ contains a stable set of size $\hat{k} \sim 2\log_b n$. Bollobás observed that one can use large deviation inequalities to show that *a.a.s. each* subgraph of $\mathbb{G}(n,p)$ on at least $n/\log^2 n$ vertices contains a stable set of size k not far from $\hat{k}$. But then we are done! Indeed, we can color greedily disjoint, k-element stable sets of $\mathbb{G}(n,p)$, until the number of uncolored vertices drops below $n/\log^2 n$. Finally, we color the uncolored vertices by new, distinct colors (since $n/\log^2 n$ is much smaller than the anticipated value of $\chi(\mathbb{G}(n,p))$ this does not increase significantly the number of colors used in the procedure).

A crucial ingredient of the above argument is an exponential upper bound for the probability that $\mathbb{G}(n,p)$ contains no large stable set; Bollobás derived it from the martingale inequalities in Corollary 2.27 using an elegant thinning argument. We will instead deduce this bound from Theorem 2.18.

Lemma 7.13. *Let $0 < p < 1$ be a constant and $b = 1/(1-p)$. Then the probability that $\mathbb{G}(n,p)$ contains no stable set of size $\lceil 2\log_b n - 2.1\log_b\log_b n\rceil$ is bounded from above by* $\exp(-(1-p)n^2/33\log_b^5 n)$.

Proof. Recall that $X(k)$ stands for the number of stable k-element sets of the random graph $\mathbb{G}(n,p)$. By Theorem 2.18(ii), applied to the cliques in the complement $\mathbb{G}(n,1-p)$ of $\mathbb{G}(n,p)$, we have

$$\mathbb{P}(X(k)=0) \le \exp\left(-\frac{[\mathbb{E}\,X(k)]^2}{\sum\sum_{A',A''}\mathbb{E}\,X_{A'}X_{A''}}\right), \tag{7.14}$$

where

$$X_A = \begin{cases} 1 & \text{if } A \text{ is stable in } \mathbb{G}(n,p) \\ 0 & \text{otherwise,} \end{cases}$$

and the sum in the denominator is taken over all pairs A', A'' such that $|A'| = |A''| = k$ and $|A' \cap A''| \ge 2$. As in Section 7.1, setting $k = \lceil 2\log_b n - 2.1\log_b\log_b n\rceil$, one obtains

$$\frac{\sum\sum \mathbb{E}\,X_{A'}X_{A''}}{[\mathbb{E}\,X(k)]^2} = \frac{\sum_{i=2}^{k}\binom{k}{i}\binom{n-k}{k-i}(1-p)^{\binom{k}{2}-\binom{i}{2}}}{\mathbb{E}\,X(k)}.$$

Elementary, though tedious calculations (Exercise!) show that in the above sum the first term is the largest one. Thus

$$\begin{aligned}\frac{\sum\sum \mathbb{E}\,X_{A'}X_{A''}}{[\mathbb{E}\,X(k)]^2} &\le \frac{(k-1)\binom{k}{2}\binom{n-k}{k-2}(1-p)^{\binom{k}{2}-1}}{\mathbb{E}\,X(k)} \\ &\le \frac{\frac{k^5}{(1-p)(n-2k)^2}\binom{n}{k}(1-p)^{\binom{k}{2}}}{\mathbb{E}\,X(k)} \le \frac{33\log_b^5 n}{(1-p)n^2}.\end{aligned}$$ ∎

Bollobás's result on the chromatic number of dense random graphs follows from the above lemma almost immediately (we present it here in a slightly weaker version).

Theorem 7.14. *Let* $0 < p < 1$ *be a constant and* $b = 1/(1-p)$. *Then* a.a.s.

$$\frac{n}{2\log_b n - \log_b \log_b n} \le \chi(\mathbb{G}(n,p)) \le \frac{n}{2\log_b n - 8\log_b \log_b n}.$$

Proof. Since the lower bound for $\chi(\mathbb{G}(n,p))$ is given by Corollary 7.5, we need only to prove that *a.a.s.* the vertices of $\mathbb{G}(n,p)$ can be properly colored by no more than $n/(2\log_b n - 8\log_b \log_b n)$ colors. Note first that *a.a.s.* each subgraph of $\mathbb{G}(n,p)$ on at least $n/\log^2 n$ vertices contains a stable set of size at least

$$2\log_b\Big(\frac{n}{\log^2 n}\Big) - 2.1\log_b \log_b\Big(\frac{n}{\log^2 n}\Big) \ge 2\log_b n - 7\log_b \log_b n.$$

Indeed, due to Lemma 7.13, the expected number of subgraphs of $\mathbb{G}(n,p)$ induced by at least $n/\log^2 n$ vertices containing no stable sets of this size is bounded from above by

$$2^n \exp\Big(-\frac{(1-p)n^2/\log^4 n}{33\log_b^5(n/\log^2 n)}\Big) \le 2^n \exp\Big(-\frac{n^2}{\log^{10} n}\Big) \to 0.$$

Thus, *a.a.s.* one can greedily color all except at most $n/\log^2 n$ vertices of $\mathbb{G}(n,p)$ with at most $n/(2\log_b n - 7\log_b \log_b n)$ colors. Clearly, the remaining vertices can be generously colored each by a new color and so *a.a.s.*

$$\begin{aligned}\chi(\mathbb{G}(n,p)) &\le \frac{n}{2\log_b n - 7\log_b \log_b n} + \frac{n}{\log^2 n}\\ &\le \frac{n}{2\log_b n - 8\log_b \log_b n}.\end{aligned}$$

■

Remark 7.15. With a little more work, Bollobás's method yields the sharper estimate (McDiarmid 1989)

$$\chi(\mathbb{G}(n,p)) = \frac{n}{2\log_b n - 2\log_b \log_b n + O_C(1)}$$

for p constant; moreover, with the error term replaced by $O_C(1/p)$, this holds for $p \to 0$ too, provided $p > n^{-\delta}$ for every $\delta > 0$.

7.5 THE CHROMATIC NUMBER OF SPARSE RANDOM GRAPHS

Unfortunately, Bollobás's ingenious argument cannot be used to determine the chromatic number of $\mathbb{G}(n,p)$ in the whole range of $p = p(n)$. Although it works very well for $p \ge n^{-\delta}$, where δ is a small positive constant, we are in deep trouble when the probability $p = p(n)$ tends to zero very quickly. Then, the left-hand side of the inequality (7.14) tends to one as $n \to \infty$ and, as we have already seen when proving Theorem 7.4, finding the correct asymptotic size of the largest stable set in $\mathbb{G}(n,p)$ requires a combination of a large deviation inequality and the second moment method. One may still hope that Frieze's approach can be used to show that the probability of $\mathbb{G}(n,p)$ containing no large stable sets of size $k \sim \alpha(\mathbb{G}(n,p))$ tends to 0 much faster than $\binom{n}{k}^{-1}$, as required in Bollobás's method. However, a quick inspection of the proof of Theorem 7.4 reveals that it is based on the large deviation inequality (7.11), which cannot "capture" probabilities smaller than $\exp(-k)$, where $k \sim \alpha(\mathbb{G}(n,p))$.

Thus, in order to deal with small edge probability p we will need a new idea: Matula's *expose-and-merge approach* (see Matula (1987) and Matula and Kučera (1990), where it was used to determine, independently from Bollobás, the correct size of the chromatic number of $\mathbb{G}(n,1/2)$). We follow Łuczak (1991b) and combine this method with Frieze's argument used in the proof of Theorem 7.4 to find the correct asymptotic order of the chromatic number of a random graph basically for all values of $p \geq C/n$, where C is a sufficiently large constant. Here we consider only $p = p(n)$ for which $p \leq \log^{-7} n$; if $p \geq \log^{-7} n$ but $p \leq c$ for some $c < 1$, then one can get better estimates for $\chi(\mathbb{G}(n,p))$ using Bollobás's argument presented in the previous section (see Remark 7.15).

Theorem 7.16. *There exists C_0 such that for every $p = p(n)$ satisfying $C_0/n \leq p \leq \log^{-7} n$ a.a.s.*

$$\frac{np}{2\log np - 2\log\log np + 1} \leq \chi(\mathbb{G}(n,p)) \leq \frac{np}{2\log np - 40\log\log np}.$$

The proof of Theorem 7.16 is based on Lemma 7.18 below, which, in turn, relies on a strengthening of Theorem 7.4 given by Łuczak (1991b). This result, stated as Lemma 7.17, can be verified by following closely Frieze's original argument, but since its computational part is much more involved, we state it without proof.

Lemma 7.17. *Let $\varepsilon > 0$ and $k = \lfloor 2(\log np - \log\log np + 1 - \log 2 - \varepsilon)/p \rfloor$. Then there exists a constant C_ε such that for $C_\varepsilon/n \leq p(n) \leq \log^{-7} n$, with probability at least $1 - o(n^{-1})$, $\mathbb{G}(n,p)$ contains $\lceil n \log^{-5}(np)/k \rceil$ disjoint stable sets, each of k vertices.* ∎

Lemma 7.18. *There is a constant C_0 such that for $C_0/n \leq p \leq \log^{-7} n$, and n large enough, with probability greater than $1 - \log^{-1} np$, more than $n - 2n\log^{-3} np$ vertices of $\mathbb{G}(n,p)$ can be properly colored with fewer than $np/(2\log np - 38\log\log np)$ colors.*

Proof. We will show the statement using Matula's "expose-and-merge" technique. Let $C_0/n \leq p \leq \log^{-7} n$, where C_0 is assumed to be large enough for later estimates. Furthermore, set

$$\bar{k} = \lceil (2 \log np - 37 \log \log np)/p \rceil,$$
$$\ell = \lceil n/(\bar{k} \log^{22} np) \rceil,$$
$$m = \lceil n \log^{-17} np \rceil.$$

Choose a subset A_1 of $[n]$ uniformly at random among all subsets of $[n]$ with m vertices. Since the subgraph H_1 induced in $\mathbb{G}(n,p)$ by A_1 can be viewed as $\mathbb{G}(m,p)$, by Lemma 7.17, with probability at least $1 - o(m^{-1})$ there are ℓ disjoint stable sets in H_1, each of size $\bar{k}$. Let us choose uniformly at random one such family, $I_1^1, \ldots, I_\ell^1$. This extra randomization gives each $\bar{k}$-element subset of A_1 the same overall chance of being chosen as one of these stable sets. We mark all vertices from the set $W = \bigcup_{i=1}^{\ell} I_i^1$ as *used* and all pairs of vertices $\{v, w\}$ with $v, w \in A_1$ as *exposed.*

Now choose another set A_2, uniformly at random among all subsets of $[n] \setminus W$ of m vertices, and let H_2' be the graph induced by A_2 in $\mathbb{G}(n,p)$. We would like to apply Lemma 7.17 to H_2' but, although H_2' can be viewed as $\mathbb{G}(m,p)$, its structure may depend on the structure of H_1, because some pairs of vertices from A_2 could be already exposed. Here comes Matula's ingenious recipe. Let us ignore all exposed pairs, at least for a moment, and for each exposed pair $\{v, w\}$ perform another random experiment in which the probability of success is p, and connect v and w by an edge according to its outcome. The graph obtained this way from H_2' is denoted by H_2. Note that H_2 can no longer be viewed as a subgraph of $\mathbb{G}(n,p)$ – when we expose the exposed pairs for the second time, we might have drawn an edge between two vertices which are not adjacent in $\mathbb{G}(n,p)$, and vice versa. However, H_2 has one great advantage: it can be identified with a random graph $\mathbb{G}(m,p)$ which is independent of H_1, because H_1 and H_2 were generated in separate sequences of random experiments. Thus, we can apply Lemma 7.17 and choose in H_2, again randomly, a family of ℓ disjoint sets $I_1^2, \ldots, I_\ell^2$, which are stable in H_2 (but not necessarily in $\mathbb{G}(n,p)$). Finally, we include all vertices of $\bigcup_{i=1}^{\ell} I_i^2$ in the set W containing the used vertices, and all pairs from A_2 in the set of exposed pairs (some of them may have been already marked as exposed).

Let us repeat this procedure $r = \lceil \log^{22} np - \log^{19} np \rceil$ times. According to Lemma 7.17, the probability that for some graph H_i, $1 \le i \le r$, we have not succeeded with the choice of the family $I_1^i, \ldots, I_\ell^i$ is $o(r/m) = o(\log^{-1} np)$. Thus, let us assume that during the procedure we have produced a family of $r\ell$ disjoint sets I_j^i, where for each i, $1 \le i \le r$, all sets I_j^i, $1 \le j \le \ell$, are stable in H_i. We will use them to define a proper coloring of all except at most $2n \log^{-3} np$ vertices of $\mathbb{G}(n,p)$.

As we have already noticed, a set I_j^t may not be stable in $\mathbb{G}(n,p)$, because when generating H_t, we could have included in I_j^t a pair $\{v, w\}$ which was an edge of $\mathbb{G}(n,p)$, but which was exposed at one of the earlier stages of the algorithm. Let s be the smallest index for which $\{v, w\} \in H_s$. We denote the number of such *troublesome* edges by Y.

Now we estimate the expectation of Y. Since $1 \le s < t \le r$, there exist $\binom{r}{2}$ choices for s and t. At the s-th stage, when we picked a set A_s of size m, the ends $v, w \in A_s$ of a troublesome edge could be chosen in one of $\binom{m}{2}$ ways; the probability that the pair $\{v, w\}$ appeared as an edge of H_s and thus of $\mathbb{G}(n,p)$ is, of course, p. Now it remains to bound the probability that both v and w belonged to one of the sets $I_1^t, I_2^t \dots, I_\ell^t$. Note first that at the t-th stage of the procedure the set A_t, and thus also the sets $I_1^t, I_2^t \dots, I_\ell^t$, were chosen from at least $n - r\ell\bar{k} \ge 0.5n \log^{-3} np$ vertices. Since $|I_i^t| = \bar{k}$ for all $i = 1, 2, \dots, \ell$, the probability that v belonged to $\bigcup_{i=1}^{\ell} I_i^t$ is smaller than $2\ell\bar{k}/n \log^{-3} np$. Similarly, the probability that w was contained in the set I_i^t which contained v is bounded from above by $2\bar{k}/n \log^{-3} np$. Thus,

$$\begin{aligned}
\mathbb{E}Y &\le \binom{r}{2}\binom{m}{2} p \frac{4(\bar{k})^2 \ell}{n^2 \log^{-6} np} \\
&\le \frac{\log^{44} np}{2} \frac{n^2}{2\log^{34} np} \frac{8n \log^7 np}{n^2 \log^{22} np} = 2n \log^{-5} np.
\end{aligned}$$

Hence, from Markov's inequality, we get

$$\mathbb{P}(Y > n \log^{-3} np) \le 2\log^{-2} np \le 0.5 \log^{-1} np\,.$$

Consequently, with probability at least $1 - \log^{-1} np$, the procedure described above generates a system of disjoint sets I_j^i, $1 \le i \le r$, $1 \le j \le \ell$, which contain not more than $n \log^{-3} np$ troublesome edges. Let us delete one end of each such edge from $\bigcup_{i=1}^{r} \bigcup_{j=1}^{\ell} I_j^i$. Then, the resulting set has at least

$$r\ell\bar{k} - n\log^{-3} np \ge n - 2n \log^{-3} np$$

vertices and, on the other hand, it can be colored using not more than

$$r\ell \le \frac{n}{\bar{k}} + r < \frac{np}{2\log np - 38 \log\log np}$$

colors. ■

Proof of Theorem 7.16. The lower bound for $\chi(\mathbb{G}(n,p))$ is given by Corollary 7.5. In order to get the upper bound for the chromatic number of $\mathbb{G}(n,p)$ observe that, due to Lemma 7.18, with probability at least $1 - \log^{-1} np$ all except at most $2n\log^{-3} np$ vertices of $\mathbb{G}(n,p)$ can be colored using at most $np/(2\log np - 38 \log\log np)$ colors. Furthermore, Lemmas 7.6 and 7.7(i) imply that with probability $1 - o(1)$ the subgraph induced in $\mathbb{G}(n,p)$ by the uncolored vertices can be colored using at most $2np \log^{-2} np + 1$ additional colors. Consequently, with probability at least $1 - o(1) - \log^{-1}(np)$,

$$\begin{aligned}
\chi(\mathbb{G}(n,p)) &\le \frac{np}{2\log np - 38\log\log np} + \frac{2np}{\log^2 np} + 1 \\
&\le \frac{np}{2\log np - 39\log\log np}.
\end{aligned} \tag{7.15}$$

Thus, for say, $np \geq \log n$, *a.a.s.* the chromatic number is bounded from above by $np/(2\log np - 39\log\log np)$. Finally, if $C_0 \leq np \leq \log n$, then (7.15) together with Theorem 7.11 implies that *a.a.s.*

$$\chi(\mathbb{G}(n,p)) \leq \frac{np}{2\log np - 39\log\log np} + 1 \leq \frac{np}{2\log np - 40\log\log np}. \qquad \blacksquare$$

The above theorem states that for np being a large enough constant the chromatic number of $\mathbb{G}(n,p)$ is *a.a.s.* about $np/(2\log np)$. But how large must this constant be to guarantee that the chromatic number is at least k, for a given natural number $k \geq 4$? Recently, Achlioptas and Friedgut (1999) have shown the existence of a sequence $d_k(n)$ such that *a.a.s.* $\mathbb{G}(n,(d_k(n)-\varepsilon)/n)$ has chromatic number at most k, while $\mathbb{G}(n,(d_k(n)+\varepsilon)/n)$ does not. The sequence $d_k(n)$ is certainly bounded, but it is not known (although widely believed) that it converges to a limit. To avoid this problem one can define the threshold constant c_k, setting

$$c_k = \inf\{d: \ a.a.s.\ \chi(\mathbb{G}(n,d/n)) \geq k\}.$$

Theorem 7.16 states that for large k we have

$$c_k = (2+o(1))k\log k, \tag{7.16}$$

where the $o(1)$ stands for a quantity which tends to 0 as $k \to \infty$. For small values of k the constants c_k have been estimated by Chvatál (1991), Molloy (1996), and Achlioptas and Molloy (1999). In particular, it turns out that for $k \geq 3$, during the evolution of $\mathbb{G}(n,p)$, the first subgraph of $\mathbb{G}(n,p)$ of minimum degree k appears before the chromatic number of the random graph jumps to $k+1$ (see Molloy (1996) and Achlioptas and Molloy (1997)). Note also that if k is large then the non-empty k-core appears in the random graph when its expected average degree is about k (see Section 5.1). Thus, for large k, (7.16) implies that the k-core, at the moment it emerges in $\mathbb{G}(n,p)$, *a.a.s.* has chromatic number smaller than k.

7.6 VERTEX PARTITION PROPERTIES

The concept of chromatic number can be modified in various ways, and quite a few of its variants have been studied in the theory of random graphs (see, e.g., Bollobás and Thomason (1995, 1997)). In this section we consider one such generalization of the chromatic number, closely related to Ramsey theory. Note that having chromatic number greater than r is equivalent to the property that every r-coloring results in an edge with both endpoints of the same color. This is nothing else but a special case of a general Ramsey property for graphs, often depicted by the following Erdős–Rado arrow notation. Given two graphs F and G, we write

$$F \to (G)_r^1$$

if for every r-coloring of the *vertices* of F there is a monochromatic copy of G in F. (A similar notion but with respect to *edge*-coloring will be thoroughly studied in the next chapter.) Thus, $\chi(F) > r$ if and only if $F \to (K_2)^1_r$. With this extent of generality in mind, we will refocus our interest as compared to the case $G = K_2$ studied in the previous sections of this chapter, and instead of asking for an analogue of the chromatic number, that is, for the smallest number of colors $r = r(n,p)$ for which

$$\mathbb{G}(n,p) \to (G)^1_r,$$

we will fix r and look for a threshold probability function for the above property.

Intuitively, the threshold should be determined by the requirement that for each subgraph H of G, the number of copies of H in the random graph $\mathbb{G}(n,p)$ is of the order of the magnitude of n. Let us explain the reason behind this heuristic. The copies of G contained in $\mathbb{G}(n,p)$ are fairly uniformly distributed and so, when the number of copies of G is much smaller than n, most of these copies are almost disjoint (see Section 3.5). Thus, it seems plausible that one can color the vertices of $\mathbb{G}(n,p)$, even with just two colors, and not create a monochromatic copy of G. On the other hand, if a vertex of $\mathbb{G}(n,p)$ belongs on average to many copies of G, then coloring all vertices of $\mathbb{G}(n,p)$ one by one we may expect that sooner or later we will put ourselves in a position in which coloring a new vertex inevitably leads to a monochromatic copy of G.

As for every three graphs $H \subset G$, and F, the property $F \to (G)^1_r$ implies $F \to (H)^1_r$, the same heuristic applies to every subgraph H of G. Note now that for each $H \subseteq G$,

$$\mathbb{E}(X_H) = \Theta(n^{v_H} p^{e_H}) = \Theta\big(n(np^{e_H/(v_H-1)})^{v_H-1}\big). \tag{7.17}$$

Hence, we anticipate that the threshold for the property $\mathbb{G}(n,p) \to (G)^1_r$ should be of the form $n^{-1/m^{(1)}(G)}$ where, let us recall, for a graph G with at least two vertices,

$$m^{(1)}(G) = \max_{H \subseteq G,\, v_H \geq 2} \frac{e_H}{v_H - 1}.$$

Note that, not surprisingly, the same function appeared in Theorem 4.9 as the threshold function for the property $\mathsf{F}_G(\varepsilon)$ that all but at most εn vertices of $\mathbb{G}(n,p)$ can be covered by vertex disjoint copies of G.

Theorem 7.19. *For every integer r, $r \geq 2$, and for every graph G which contains at least one edge and, if $r = 2$, satisfies $\Delta(G) \geq 2$, there exist positive constants c and C such that*

$$\lim_{n\to\infty} \mathbb{P}\big(\mathbb{G}(n,p) \to (G)^1_r\big) = \begin{cases} 0 & \text{if } p \leq cn^{-1/m^{(1)}(G)}, \\ 1 & \text{if } p \geq Cn^{-1/m^{(1)}(G)}. \end{cases}$$

Remark 7.20. Note that the number of colors r does not appear in the exponent of the threshold function, but is hidden in the constants.

Remark 7.21. The case in which G is a matching and $r = 2$ is somewhat different. One can show using the second moment method (Exercise!) that if $pn = c \geq 1$ then *a.a.s.* $\mathbb{G}(n,p)$ contains at least $\log\log n$ vertex disjoint cycles of odd length and thus *a.a.s.* $\mathbb{G}(n,p) \to (G)^1_2$. However, if $np = c$ and $0 < c < 1$, then one can use the method of moments (see Section 6.1) to show that the number of odd cycles converges in distribution to Poisson distribution $\mathrm{Po}(\lambda)$ for some positive constant $\lambda = \lambda(c)$ (Exercise!). Furthermore, in this case *a.a.s.* $\mathbb{G}(n,p)$ consists of trees and unicyclic components (Theorem 5.5), and for such a graph F and a matching G we have $F \to (G)^1_2$ if and only if F contains at least $2e(G) - 1$ odd cycles (Exercise!). Thus,

$$\lim_{n\to\infty} \mathbb{P}(\mathbb{G}(n,p) \to (G)^1_2) = a(c),$$

where

$$0 < a(c) = \sum_{i=2e(G)-1}^{\infty} \frac{\lambda^i}{i!} e^{-\lambda} < 1.$$

Hence, in order to have *a.a.s.* $\mathbb{G}(n,p) \not\to (G)^1_2$, we need $p \ll n^{-1}$.

Remark 7.22. It is tempting to conjecture that for given G and r the two constants c and C can be chosen arbitrarily close to each other, that is, there is a single constant $\bar{C}$ such that we can take $c = \bar{C} - \varepsilon$ and $C = \bar{C} + \varepsilon$ for every $\varepsilon > 0$. There is at present not much hope, however, to determine or even show the existence of such a $\bar{C}$. As we mentioned in the previous section, this problem remains open even in the simplest case $G = K_2$.

A partial result has recently been obtained by Friedgut and Krivelevich (2000), who proved that if G is strictly K_1-balanced, then the threshold is sharp in the sense defined in Section 1.6; in this case it means that there exists such a $\bar{C} = C(n) = \Theta(1)$, which, however, possibly depends on n.

Proof of Theorem 7.19. Suppose that $\mathbb{G}(n,p) \not\to (G)^1_r$. Then the largest color class of any coloring with no monochromatic G spans a G-free subgraph of size at least n/r. The probability that this happens is, by Theorem 3.9, smaller than

$$2^n\, \mathbb{P}(\mathbb{G}(n/r,p) \not\supset G) < 2^n \exp\{-c_G \Phi_G(n/r,p)\},$$

where $\Phi_G(n,p) = \min\{\mathbb{E}\, X_H : H \subseteq G,\ e_H \geq 1\}$ and the constant c_G depends on G only. It follows from (7.17) that for $p \geq Cn^{-m^{(1)}(G)}$ we have $\Phi_G(n,p) \geq C'n$, where C' grows to infinity as a function of C, so that the probability of $\mathbb{G}(n,p) \not\to (G)^1_r$ tends to 0 for C sufficiently large.

For the proof of the 0-statement of Theorem 7.19 we assume that $p \leq cn^{-1/m^{(1)}(G)}$, where c is a sufficiently small positive constant. The reader can easily check that if $m^{(1)}(G) = 1$, that is, if G is a forest with at least one edge, then Theorem 5.5 implies that the assertion holds whenever $c < 1$ (Exercise!). Thus, without loss of generality we may assume that $v_G \geq 3$ and that for every proper subgraph H of G with at least two vertices

$$\frac{e_H}{v_H - 1} < \frac{e_G}{v_G - 1},$$

that is, G is strictly K_1-balanced (see Section 3.2). (If this was not the case, one could replace G with its smallest subgraph H for which $e_H/(v_H - 1) = m^{(1)}(G)$.) This assumption implies that there are no isolated vertices in G and that, for each proper subgraph H of G with at least two vertices,

$$n^{v_H - 1} p^{e_H} = \Omega(n^{\varepsilon}) \tag{7.18}$$

for some $\varepsilon > 0$.

Our proof will consist of two parts: a deterministic one, where we show that every graph F with $F \to (G)^1_2$ contains a dense subgraph of a special type, and a probabilistic one, where we prove that *a.a.s.* such dense subgraphs do not appear in $\mathbb{G}(n,p)$.

In order to show the first part of the statement we will need a number of definitions. Let us recall that a hypergraph $\mathcal{H}$ is a pair $(\mathcal{V}, \mathcal{E})$, where $\mathcal{V}$ denotes the set of vertices and $\mathcal{E}$ is a family of subsets of $\mathcal{V}$, called hyperedges. A hypergraph $\mathcal{H}$ has chromatic number $\chi(\mathcal{H})$ at least three if every 2-coloring of vertices of $\mathcal{H}$ leads to at least one monochromatic hyperedge. We say that $\mathcal{H}$ is 3-edge-critical if $\chi(\mathcal{H}) \geq 3$ but the deletion of any hyperedge results in losing this property. For graphs F and G, let $\mathcal{H}(F, G)$ be the hypergraph with vertex set $V(F)$, whose hyperedges are the vertex sets of all copies of G contained in F. Note that for each hyperedge A of $\mathcal{H}(F, G)$, we have $|A| = v_G$. We denote by $G(A)$ a copy of G in F which corresponds to the hyperedge A of $\mathcal{H}(F, G)$, and by $G(\mathcal{H}_0)$ the graph $\bigcup_{A \in \mathcal{H}_0} G(A)$, where $\mathcal{H}_0$ is a subhypergraph of $\mathcal{H}$. Note that $F \to (G)^1_2$ if and only if the chromatic number of $\mathcal{H}(F, G)$ is at least three. Furthermore, we may assume that $\mathcal{H}(F, G)$ is 3-edge-critical, since otherwise we could replace $\mathcal{H}(F, G)$ with a 3-edge-critical subgraph, ignoring some copies of G in F. In our further considerations we will use the following result about 3-edge-critical hypergraphs, the simple proof of which is left to the reader (Exercise!).

Proposition 7.23. *If $\mathcal{H}$ is a 3-edge-critical hypergraph, then for every hyperedge A of $\mathcal{H}$ and every vertex $v \in A$ there is a hyperedge A' such that $A \cap A' = \{v\}$.* ∎

A *linear path* $(A_1, \ldots, A_\ell)$ is a hypergraph with hyperedges $A_1, \ldots, A_\ell$, $\ell \geq 1$, such that

$$|A_i \cap A_j| = \begin{cases} 1 & \text{if } j = i \pm 1, \\ 0 & \text{otherwise.} \end{cases}$$

A *linear [quasi-linear] cycle* $(A_0, A_1, \ldots, A_\ell)$ is a hypergraph which consists of a linear path $(A_1, \ldots, A_\ell)$, $\ell \geq 2$, and a hyperedge A_0 such that

$$|A_0 \cap A_i| = \begin{cases} 1 & \text{if } i = 1, \\ 0 & \text{for } i = 2, \ldots, \ell - 1, \\ s & \text{if } i = \ell, \end{cases}$$

where $s = 1$ [$s \geq 1$], respectively). A cycle which is quasi-linear but not linear we will call *spoiled.*

Let $\mathcal{P}$ be the longest linear path in $\mathcal{H} = \mathcal{H}(F, G)$. By Proposition 7.23, $\mathcal{P}$ contains at least two hyperedges. Let x and y be two vertices which belong only to the first hyperedge of $\mathcal{P}$, and let A_x and A_y be two hyperedges of $\mathcal{H}$ (which, of course, correspond to two copies of G contained in F) whose existence is guaranteed by Proposition 7.23, that is, $A_z \cap A_1 = \{z\}$, $z = x, y$.

By the maximality of $\mathcal{P}$, $|V(\mathcal{P}) \cap A_z| \geq 2$, $z = x, y$. Let $i_z = \min\{i \geq 2 : A_z \cap A_i \neq \emptyset\}$, $z = x, y$, and assume that, say, $i_y \leq i_x$. The hyperedges $A_1, \ldots, A_{i_x}, A_x$ form a quasi-linear cycle $\mathcal{C}$, which is linear if and only if $|A_x \cap A_{i_x}| = 1$. Otherwise $\mathcal{C}$ is spoiled. We also have $|A_y \cap V(\mathcal{C})| \geq 2$. Moreover, there is an edge in $G(A_y)$ which does not belong to $G(\mathcal{C})$. Indeed, as $\delta(G) \geq 1$, take any edge of $G(A_y)$ incident to y. We call the pair $(\mathcal{C}, A_y)$ a *cycle with handle.* So, we have just proved a deterministic statement that if $F \to (G)^1_2$ then the hypergraph $\mathcal{H}(F, G)$ contains a quasi-linear cycle with handle.

Now we will show that *a.a.s.* no such structure does exist in $\mathcal{H}(\mathbb{G}(n,p), G)$. Let X, Y and Z be random variables counting, respectively, linear paths of length at least $B \log n$, spoiled cycles $\mathcal{C} = (A_0, \ldots, A_t)$, and linear cycles with handles $(\mathcal{C}, A)$ of length less than $B \log n + 1$, in the random hypergraph $\mathcal{H}(\mathbb{G}(n,p), G)$, where $B = B(c, G)$ is a big enough constant. Straightforward estimates show that their expectations all converge to 0 as $n \to \infty$. Indeed,

$$\mathbb{E}(X) \leq \sum_{t > B \log n} n^{t(v_G - 1) + 1} p^{t e_G} < n \sum_{t > B \log n} (c^{e_G})^t = o(1),$$

$$\mathbb{E}(Y) \leq \sum_{t > 2} \sum_{H \subset G} n^{t(v_G - 1) - (v_H - 1)} p^{t e_G - e_H} = o(1)$$

and

$$\mathbb{E}(Z) = O\left(\sum_{t=3}^{B \log n} \sum_{H \subset G} n^{(t+1)(v_G - 1) - (v_H - 1)} (\log n)^{v_H} p^{(t+1)e_G - e_H} \right) = o(1),$$

where the inner sums extend over all proper subgraphs H of G with at least two vertices and correspond, in case of Y, to all possible shapes of the intersection $G(A_0) \cap G(A_t)$ and, in the case of Z, to all possible shapes of the intersection $G(A) \cap G(\mathcal{C})$. The index t stands for the number of hyperedges in a path or cycle. The logarithmic factor in the last estimate represents the number of choices of the vertices at which the handle A is attached to the cycle. Finally, we made use of formula (7.18).

Thus, by Markov's inequality, $\mathbb{P}(X = Y = Z = 0) \to 1$ as $n \to \infty$, which completes the proof of Theorem 7.19. ∎

8

Extremal and Ramsey Properties

As the reader has undoubtedly noticed, statements of many results in this book could well begin with the phrase: "Let G be a fixed graph and let $\mathbb{G}(n,p)$ be the random graph such that $p(n) = \ldots$". Let us recall that in Section 3.1 we studied the probability that $\mathbb{G}(n,p)$ contains a copy of G for different values of $p = p(n)$, while in Section 3.3 and Chapter 6 the asymptotic behavior of the random variable X_G which counts copies of G in $\mathbb{G}(n,p)$ was thoroughly analyzed. In Section 3.4 we dealt with the property that every vertex of $\mathbb{G}(n,p)$ belongs to a copy of G and in Section 4.2 a connection between this property and the property that $\mathbb{G}(n,p)$ has a G-factor was considered. Finally, in Section 7.6, by estimating the order of the largest G-free subset of $V(\mathbb{G}(n,p))$, we proved that $n^{-1/m^{(1)}(G)}$ is the threshold for the property that every coloring of the vertices of $\mathbb{G}(n,p)$ with a fixed number of colors leads to a monochromatic copy of G.

In this chapter we investigate further variations of this familiar theme which are the edge versions of the questions from Section 7.6. Specifically, we will study the size of the largest subgraph of $\mathbb{G}(n,p)$ that contains no copy of G and look for the threshold function $p = p(n)$ which guarantees that each coloring of the edges of $\mathbb{G}(n,p)$ with a fixed number of colors creates a monochromatic copy of G.

It turns out that dealing with edges rather than vertices makes the above two problems much different from their vertex counterparts. This is apparent already when $G = K_3$. From the proof of Theorem 4.9 it follows that if $n^2p^3 \to \infty$ then *a.a.s.* every *induced* subgraph of $\mathbb{G}(n,p)$ of order εn contains a triangle, while, on the other hand, it is very well known that every graph contains a triangle-free subgraph with at least half of the edges. Thus, before we proceed any further, we will examine some heuristics leading to the main results of this chapter.

8.1 HEURISTICS AND RESULTS

An edge partition problem

Following the Erdős–Rado arrow notation, for two graphs F and G and a natural number r, $r \geq 2$, we write $F \to (G)_r^2$ if every coloring of the edges of F with r colors creates a monochromatic copy of G. Since in the whole chapter we always color the edges, not vertices, we will omit the superscript 2, writing just $F \to (G)_r$. Our first goal is to determine for what $p = p(n)$ a random graph has the Ramsey property $\mathbb{G}(n,p) \to (G)_r$.

Let us notice that for given G and r the property $F \to (G)_r$ is increasing, so it has a threshold function (Theorem 1.24). In the vertex case the threshold was determined by the requirement that the number of copies of G should be of the order of n, the number of vertices in $\mathbb{G}(n,p)$. When coloring the edges we are facing a similar situation. If the number of copies of G is much smaller than the number of edges of $\mathbb{G}(n,p)$, most copies of G are edge disjoint, and the edges can be colored so that no copy of G is monochromatic. Although this fact is hardly surprising, its proof is rather tedious. It is similar to but more involved than the proof of the 0-statement of Theorem 7.19 in Section 7.6. Copies of G may locally cluster together and a substantial amount of work is needed to show that a proper coloring does exist (Rödl and Ruciński 1993). On the other hand, if the number of copies of G is much larger than the number of edges, one may expect that they are so uniformly distributed around the graph that each coloring leads to many monochromatic copies of G. The quantities X_G and X_{K_2} are of the same order of magnitude when, setting as usual $v_G = |V(G)|$ and $e_G = |E(G)|$,

$$n^{v_G} p^{e_G} = \Theta(n^2 p).$$

But the property $F \to (G)_r$ is hereditary with respect to taking subgraphs of G and, just as in studying the containment problem in Chapter 3 or the vertex coloring problem in Section 7.6, one has to consider all (non-empty) subgraphs of G, which leads to the condition

$$\min_{G' \subseteq G,\, e(G')>0} \{n^{v(G')} p^{e(G')}\} = \Theta(n^2 p),$$

or, using the notation of Chapter 3, $\Phi_G = \Theta(n^2 p)$.

Hence, the threshold function for the property $\mathbb{G}(n,p) \to (G)_r$ should be $p = p(n) = n^{-1/m^{(2)}(G)}$, where $m^{(2)}(G)$ was defined in (3.18).

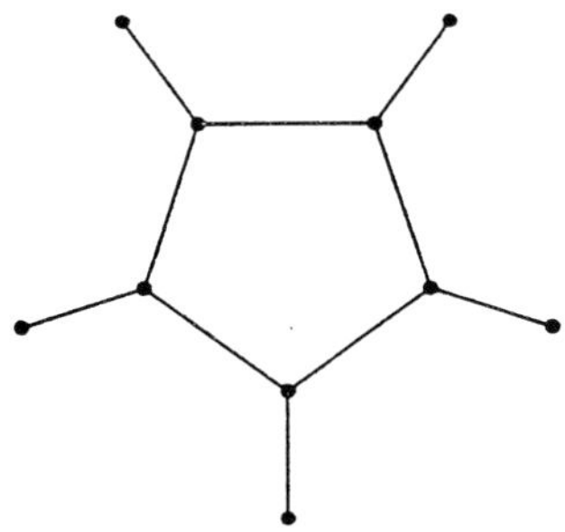

Fig. 8.1 A sunshine graph.

The following result proved by Rödl and Ruciński (1995) states that this is indeed the case, except for star forests (i.e., forests where every component is a star). Recall that P_3 is the path with 3 *edges.*

Theorem 8.1. *Let $r \geq 2$, and let G be a graph with at least one edge.*

(i) *If G is a star forest, then the threshold function for $\mathbb{G}(n,p) \to (G)_r$ is $n^{-1-1/((\Delta(G)-1)r+1)}$.*

(ii) *If $r = 2$ and G is a forest consisting of stars and P_3's, with at least one P_3, then there exists a constant C such that*

$$\lim_{n\to\infty} \mathbb{P}\big(\mathbb{G}(n,p) \to (G)_2\big) = \begin{cases} 0, & \text{if } p \ll n^{-1/m^{(2)}(G)} = 1/n, \\ 1, & \text{if } p \geq Cn^{-1/m^{(2)}(G)} = C/n. \end{cases}$$

(iii) *In every other case, there exist constants $c = c(G,r)$ and $C = C(G,r)$ such that*

$$\lim_{n\to\infty} \mathbb{P}\big(\mathbb{G}(n,p) \to (G)_r\big) = \begin{cases} 0, & \text{if } p \leq cn^{-1/m^{(2)}(G)}, \\ 1, & \text{if } p \geq Cn^{-1/m^{(2)}(G)}. \end{cases}$$

Proof of (i) and (ii). Case (i), a star forest, is easy. If, for example, G is a single star $K_{1,k}$, then $F \to (G)_r$ for every F with $\Delta(F) \geq (k-1)r+1$, while $F \not\to (G)_r$ if F is a forest with smaller $\Delta(F)$ (Exercise!). It then follows that the threshold for $\mathbb{G}(n,p) \to (G)_r$ coincides with the threshold for the existence of a $K_{1,(k-1)r+1}$ given by Theorem 3.4. Note that by Theorem 3.19, this is a coarse threshold which cannot be tightened as in Cases (ii) or (iii).

For (ii), the other special case, assume for simplicity that $G = P_3$. As was first observed by Friedgut (personal communication), the 0-statement in (ii) cannot be improved as in (iii), since $F \to (P_3)_2$ if F is the sunshine graph in Figure 8.1 (or any other graph obtained by adding pendant edges at every vertex of an odd cycle of length at least 5). This observation also yields the 1-statement in (ii), with $C \approx 1.35$, the solution of $C(1-e^{-C}) = 1$ (Advanced

Exercise!). The 0-statement follows because $\mathbb{G}(n,p)$ is, for $p \ll 1/n$, *a.a.s.* a forest (see Section 5.1), and if F is a forest then $F \not\to (P_3)_2$ (Exercise!). ■

An extended outline of the proof of the 1-statement in (iii), which utilizes the Szemerédi Regularity Lemma, will be presented in Section 8.4. The special case $G = K_3$ was first proved for an arbitrary number of colors in Rödl and Ruciński (1994). For the proof of the 0-statement, see Łuczak, Ruciński and Voigt (1992) and Rödl and Ruciński (1993).

Remark 8.2. Note the different types of dependency on r for the different cases. If G is a star forest, then the threshold function is a power of n where the power depends on r. For other G, the rate does not depend on r, which affects the result only through the constants $c(G,r)$ and $C(G,r)$; if G is as in (ii), we have a coarser type of threshold for $r = 2$ than for $r \geq 3$.

Note also that in Case (iii) we necessarily have $C(G,r) \to \infty$ as $r \to \infty$. In fact, with $3r$ colors arranged in r groups of three colors each, let $p = c(G,3)rn^{-1/m^{(2)}(G)}$ and assign randomly one of the color groups to each edge of $\mathbb{G}(n,p)$. This exhibits $\mathbb{G}(n,p)$ as the union of r disjoint copies of $\mathbb{G}(n,p/r)$, each of which *a.a.s.* can be properly colored with the three corresponding colors; hence *a.a.s.* $\mathbb{G}(n,p) \not\to (G)_{3r}$, and thus $C(G,3r) > c(G,3)r$. (We do not know the true order of growth of $C(G,r)$; the upper bound given by the proof is enormous.)

Remark 8.3. Just as for the vertex case in Theorem 7.19, it is tempting to conjecture that in Case (iii), for given G and r there is a single constant $\bar{C} = \bar{C}(G,r)$ such that we can take $c = \bar{C} - \varepsilon$ and $C = \bar{C} + \varepsilon$ for every $\varepsilon > 0$. A proof of this statement seems difficult even for the simplest choices of G and r. A partial result has recently been obtained by Friedgut and Krivelevich (2000), who proved that if $r = 2$ and G is a tree other than a star or P_3, then the threshold is sharp, in the sense defined in Section 1.6, which means that there exists such a quantity $\bar{C} = \bar{C}(n)$, which, however, possibly depends on n.

A Turán-type problem

Another problem which continues to stimulate research in the theory of random graphs has the flavor of extremal graph theory, rather than Ramsey theory. Instead of partitioning the edges of $\mathbb{G}(n,p)$ into several classes, one tries here to determine the minimum size of a subgraph of $\mathbb{G}(n,p)$ which guarantees the containment of a copy of G. This problem, unlike the partition problem, in general remains open.

Let us introduce some notation. For two graphs F and G, we denote by $\mathrm{ex}(F,G)$ the number of edges in the largest subgraph of F containing no copy of G and set $\overline{\mathrm{ex}}(F,G)$ for the fraction of the number of edges of F in such a subgraph, that is,

$$\mathrm{ex}(F,G) = \max\{e_H : G \not\subseteq H \subseteq F\}$$

and

$$\overline{\text{ex}}(F,G) = \text{ex}(F,G)/e_F.$$

When $e_F = 0$, we define $\overline{\text{ex}}(F,G) = 1$ for every G.

The function $\text{ex}(K_n, G)$ has been studied extensively in extremal graph theory (see Bollobás's (1978) monograph). A celebrated result of Erdős, Stone and Simonovits (Erdős and Stone 1946, Erdős and Simonovits 1966) states that for each G with at least one edge

$$\text{ex}(K_n, G) = \left(1 - \frac{1}{\chi(G) - 1} + o(1)\right)\binom{n}{2}, \tag{8.1}$$

or, equivalently, $\lim_{n\to\infty} \overline{\text{ex}}(K_n, G) = \frac{\chi(G)-2}{\chi(G)-1}$. Thus, the asymptotic behavior of $\overline{\text{ex}}(K_n, G)$ depends exclusively on the chromatic number $\chi(G)$ of the graph G.

If $F \neq K_n$, the function $\overline{\text{ex}}(F,G)$ is much harder to study. Obviously, one always has $\overline{\text{ex}}(F,G) \leq 1$, and equality holds if and only if $G \nsubseteq F$ (unless F and G are both empty). Is it true that for a given graph G one can make $\overline{\text{ex}}(F,G)$ arbitrarily small by a suitable choice of F? As was already noticed, this is not the case when $G = K_3$, since for every graph F we have $\overline{\text{ex}}(F, K_3) \geq 1/2$. In general, an easy probabilistic argument (Exercise!) yields $\overline{\text{ex}}(F,G) \geq \lim_{n\to\infty} \overline{\text{ex}}(K_n, G)$, provided $\chi(G) \geq 3$. This asymptotic inequality turns out to be exact and holds for bipartite graphs too.

Proposition 8.4. *For every graph F on n vertices and for every graph G, $\overline{\text{ex}}(F,G) \geq \overline{\text{ex}}(K_n, G)$.*

Proof. Let F be any graph with the vertex set $[n]$ and let H denote a graph on $[n]$ such that $G \nsubseteq H$ and $e_H = \text{ex}(K_n, G)$. For a given permutation $\sigma : [n] \to [n]$ let $H(\sigma)$ be the graph obtained from H by relabelling the vertices according to σ and let $F(\sigma) = F \cap H(\sigma)$. Finally, let σ^{rand} be a random permutation of $[n]$. The expected number of edges in $F(\sigma^{\text{rand}})$ is

$$e_F e_H \Big/ \binom{n}{2} = e_F \,\text{ex}(K_n, G) \Big/ \binom{n}{2} = e_F \,\overline{\text{ex}}(K_n, G),$$

so there exists a permutation σ_0 such that $e(F(\sigma_0)) \geq e_F \,\overline{\text{ex}}(K_n, G)$. Furthermore $F(\sigma_0)$, as a subgraph of $H(\sigma_0)$, contains no copy of G. Thus $\overline{\text{ex}}(F,G) \geq \overline{\text{ex}}(K_n, G)$ and the assertion follows. ∎

Corollary 8.5. *For every graph G and all $m < n$, we have $\overline{\text{ex}}(K_m, G) \geq \overline{\text{ex}}(K_n, G)$.*

Proof. Apply Proposition 8.4 with F being the union of a complete graph K_m and $n - m$ isolated vertices. ∎

Proposition 8.4 suggests that the right "Turán-type" question to ask about random graphs is the following: For which functions $p = p(n)$ do we have

$$\overline{\text{ex}}(\mathbb{G}(n,p), G) = (1 + o_p(1))\,\overline{\text{ex}}(K_n, G)? \tag{8.2}$$

It will follow from the next result that if (8.2) holds for some p_1, then it does so for each $p_2 \geq p_1$. In this result we technically assume that $\mathbb{G}(n, p_1)$ and $\mathbb{G}(n, p_2)$ are related in the natural way, which is to say that they are two stages of the same random graph process $\{\mathbb{G}(n,t)\}_t$ (just as, e.g., when applying the two-round-exposure technique).

Proposition 8.6. *Let G be a graph with $\Delta(G) \geq 2$, and let $p_1 = p_1(n)$ and $p_2 = p_2(n)$ be such that $p_1 \leq p_2$. Then*

$$\overline{\mathrm{ex}}(\mathbb{G}(n,p_1), G) \geq (1 + o_p(1))\,\overline{\mathrm{ex}}(\mathbb{G}(n,p_2), G).$$

Proof. We will actually prove the corresponding statement for the uniform model $\mathbb{G}(n, M)$: *If $\Delta(G) \geq 2$ and $M_1 = M_1(n) \leq M_2(n) = M_2$, then*

$$\overline{\mathrm{ex}}(\mathbb{G}(n,M_1), G) \geq (1 + o_p(1))\,\overline{\mathrm{ex}}(\mathbb{G}(n,M_2), G). \tag{8.3}$$

The proposition follows by conditioning on the number of edges in both $\mathbb{G}(n,p_1)$ and $\mathbb{G}(n,p_2)$.

Again, we regard $\mathbb{G}(n,M_1)$ and $\mathbb{G}(n,M_2)$ as two stages of the same random graph process $\{\mathbb{G}(n,M)\}_M$. In particular, we may view $\mathbb{G}(n,M_1)$ as a graph obtained from $\mathbb{G}(n,M_2)$ by a random deletion of $M_2 - M_1$ edges. In order to show (8.3), we consider three cases which, together with the subsubsequence principle, yield the general result.

(i) $M_1\,\overline{\mathrm{ex}}(\mathbb{G}(n,M_2),G) \to \infty$.
Let H be a maximal G-free subgraph of $\mathbb{G}(n,M_2)$, and let $H' = H \cap \mathbb{G}(n,M_1)$. Then, clearly, $\mathrm{ex}(\mathbb{G}(n,M_1),G) \geq e(H')$ and we need to estimate $e(H')$. Note that $e(H')$ has a hypergeometric distribution with mean

$$\frac{M_1}{M_2} e(H) = M_1\,\overline{\mathrm{ex}}(\mathbb{G}(n,M_2),G),$$

which, by assumption, tends to infinity. Thus, a standard application of Chebyshev's inequality yields

$$e(H') = (1 + o_p(1)) M_1\,\overline{\mathrm{ex}}(\mathbb{G}(n,M_2),G),$$

as required.

(ii) $M_1\,\overline{\mathrm{ex}}(\mathbb{G}(n,M_2),G) \leq C$ *for some* $C > 0$, *but* $M_1 \to \infty$.
As $M_1 \to \infty$, *a.a.s.* one can find more than C disjoint edges in $\mathbb{G}(n,M_1)$, and thus $\mathrm{ex}(\mathbb{G}(n,M_1),G) > C \geq M_1\,\overline{\mathrm{ex}}(\mathbb{G}(n,M_2),G)$.

(iii) M_1 *is bounded.*
In this case *a.a.s.* $\overline{\mathrm{ex}}(\mathbb{G}(n,M_1),G) = 1$ (Exercise!) and (8.3) holds trivially. ■

Remark 8.7. If $\Delta(G) = 1$, in which case G is a disjoint union of edges and, possibly, isolated vertices, the result remains true provided $p_1 \geq n^{-1-o(1)}$ (Exercise!). However, there are counterexamples with smaller p_1 and p_2. For instance, the assertion is false if $G = 2K_2$, $p_1 = an^{-3/2}$ with $1 < a < 2$ and $p_2 = 2n^{-3/2}$ (Exercise!).

Remark 8.8. By almost the same argument it follows that if $F^{(1)}, F^{(2)}, \ldots$ is a sequence of graphs with $v(F^{(n)}) \to \infty$, $\Delta(G) \geq 2$, $p_1 = p_1(n) \leq p_2(n) = p_2$, and, furthermore, $\text{ex}(F^{(n)}_{p_1}, G) \xrightarrow{p} \infty$, then

$$\overline{\text{ex}}(F^{(n)}_{p_1}, G) \geq (1 + o_p(1))\, \overline{\text{ex}}(F^{(n)}_{p_2}, G),$$

where $F^{(n)}_p$ is the reliability network created by randomly destroying edges of $F^{(n)}$, independently, with probability $1 - p$ (Exercise!).

Proposition 8.6 suggests that, for a fixed a satisfying $\lim_{n\to\infty} \overline{\text{ex}}(K_n, G) < a < 1$, the property that

$$\overline{\text{ex}}(\mathbb{G}(n,p), G) \leq a, \tag{8.4}$$

although not monotone, behaves very much like a monotone one, and we may hope to find a threshold function for it.

A necessary condition for (8.4) is not hard to determine. Note that for any two graphs F and G we have

$$\text{ex}(F, G) \geq e_F - \#\{G \subseteq F\},$$

where $\#\{G \subseteq F\}$ is the number of copies of G contained in F, and that Property (8.4) is hereditary with respect to taking subgraphs of G. Thus, if for some $p = p(n)$ *a.a.s.* $\overline{\text{ex}}(\mathbb{G}(n,p), G) \leq a$, where $a < 1$, then, just as for the partition problem, the number of copies of each subgraph G' of G in $\mathbb{G}(n,p)$ should be comparable with the number of edges of $\mathbb{G}(n,p)$. Hence, $p = n^{-1/m^{(2)}(G)}$ is, again, our guess for the threshold. The reader is invited to compute the expectation and variance of appropriate random variables, making the above argument rigorous, and thus proving the following fact (Exercise!).

Proposition 8.9. *For every graph G with $\Delta(G) \geq 2$ and for every $0 < a < 1$, there exists a constant $c = c(G, a)$ such that* a.a.s.

$$\overline{\text{ex}}(\mathbb{G}(n,p), G) \geq a,$$

provided $p = p(n) \leq cn^{-1/m^{(2)}(G)}$. ■

Remark 8.10. In the rather uninteresting case in which $\Delta(G) = 1$, and thus $m^{(2)}(G) = 1/2$, one should instead assume that $pn^2 \to 0$ or, otherwise, slightly modify the assertion (Exercise!).

It is natural to conjecture that if the number of copies of G (or, more precisely, the number of copies of the subgraph of G most infrequent in $\mathbb{G}(n,p)$) is much larger than the number of edges of $\mathbb{G}(n,p)$, the value of $\overline{\text{ex}}(\mathbb{G}(n,p), G)$

approaches $\overline{\text{ex}}(K_n, G)$ (because of Proposition 8.4 it cannot drop any further). More formally, one may expect that the following is true.

Conjecture 8.11. For every graph G with $\Delta(G) \geq 2$ and for every $\eta > 0$, there exists $C = C(G, \eta)$ such that *a.a.s.*

$$\overline{\text{ex}}(\mathbb{G}(n,p), G) \leq (1+\eta)\,\overline{\text{ex}}(K_n, G)$$

whenever $p = p(n) \geq Cn^{-1/m^{(2)}(G)}$.

Unfortunately, at this moment we are able to verify this conjecture only for some special cases of G. In Section 8.2 we show its truth for $G = K_3$ and present an elementary proof of Frankl and Rödl (1986). An alternative proof for triangles is given in Section 8.5. It relies on a modified (sparse) version of the Szemerédi Regularity Lemma and on the observation that Conjecture 8.11 follows from a stronger one, Conjecture 8.35, stating a better-than-exponential upper bound on the probability of nonexistence of copies of G in a special model of a random graph. Conjecture 8.35 (or its weaker version, sufficient for the proof of Conjecture 8.11) is settled in the affirmative for G being an arbitrary cycle (Haxell, Kohayakawa and Łuczak 1995, Kreuter 1997, Kohayakawa, Kreuter and Steger 1998), or the complete graph K_4 (Kohayakawa, Łuczak and Rödl 1997). Thus, at least in these cases, Conjecture 8.11 holds. Yet another approach, which verifies Conjecture 8.11 for odd cycles, is presented by Haxell, Kohayakawa and Łuczak (1996).

Let us make a few comments on the connections between Conjecture 8.11 and Theorem 8.1. Note first that if G is bipartite, then, according to the Erdős–Stone–Simonovits result (8.1), we have $\overline{\text{ex}}(K_n, G) = o(1)$ and so Conjecture 8.11, if true, would imply the 1-statement of Theorem 8.1. However, if $\chi(G) \geq 3$, neither statement can be deduced from the other one, although, in a way, they both reflect the fact that in dense random graphs copies of G are distributed nearly as uniformly as they are in the complete graphs. The only case in which the partition and extremal properties get close to each other is that of $G = K_3$ and $r = 2$. This fact is utilized in the proof presented in Section 8.2.

Note also that Conjecture 8.11 and Proposition 8.9 would imply that the threshold function for the property $\overline{\text{ex}}(F, G) \leq a$ is not very much affected by the choice of a as long as $\lim_{n\to\infty} \overline{\text{ex}}(K_n, G) < a < 1$.

We close this introductory section with a brief account of further developments in the theory of partition and extremal properties of random structures. Some partial results for nonsymmetric Ramsey properties of random graphs were obtained by Kohayakawa and Kreuter (1997). A first step into the unexplored area of partition properties of random hypergraphs was taken in Rödl and Ruciński (1998). A few results on partition and extremal properties of random subsets of integers have appeared in Rödl and Ruciński (1995, 1997), Graham, Rödl and Ruciński (1996), and Kohayakawa, Łuczak and Rödl (1996).

8.2 TRIANGLES: THE FIRST APPROACH

Counting monochromatic triangles

The goal of this section is to present an elementary proof of Conjecture 8.11 when $G = K_3$. It will follow from a strengthening of the 1-statement of Theorem 8.1 for $G = K_3$ and $r = 2$. The strengthening claims not one but many monochromatic triangles in every two-coloring of $\mathbb{G}(n,p)$. There are so many that there is not enough room for them in any color class with fewer than $(\frac{1}{2} - \eta)\binom{n}{2}p$ edges. Only slightly weaker versions of this particular result were proved by Frankl and Rödl (1986) and Łuczak, Ruciński and Voigt (1992). The proof is based on a beautiful and simple idea of Goodman (1959), who applied it to bound the total number of triangles in a graph and its complement.

To fully appreciate this idea we begin by considering the deterministic case of two-colorings of the edges of the complete graph K_n, or equivalently, the random graph $\mathbb{G}(n,p)$ with $p = 1$. Thus, let us first ask, how many monochromatic triangles are guaranteed if one colors the edges of K_n with two colors, blue and red.

A trivial upper bound of $\frac{1}{4}\binom{n}{3}$ is provided by the probabilistic method (Exercise!). Surprisingly, this is asymptotically the right answer. For a given coloring and each $v = 1, 2, \ldots, n$, let b_v $[r_v]$ be the number of blue [red] edges incident with the vertex v. Then the total number of two-colored triangles is

$$\frac{1}{2}\sum_{v=1}^{n} b_v r_v = \frac{1}{2}\sum_{v=1}^{n} b_v(n-1-b_v) \le n(n-1)^2/8.$$

This implies Mantel's theorem (i.e., Turán's theorem for triangles – see, e.g., Bollobás (1998)). Indeed, let H be a graph with n vertices and $e_H > n^2/4$. Let us suppose that there is no triangle in H and count the ordered pairs (e,t), where e is an edge of H and t is a triple of vertices containing e. On the one hand, the number of such pairs is precisely $e_H(n-2)$. On the other hand, denoting by t_i, $i = 1, 2$, the number of triples of vertices containing i edges of H, there are $t_1 + 2t_2 \le 2(t_1 + t_2)$ such pairs. By the above upper bound on the number of two-colored triangles, treating the edges of H as those colored blue, we have

$$(n^2/4 + 3/4)(n-2) \le e_H(n-2) \le 2(t_1 + t_2) \le 2n(n-1)^2/8,$$

which is a contradiction for $n \ge 4$. Hence H contains a triangle.

After this deterministic rehearsal we should be ready to repeat the same argument for the random graph $\mathbb{G}(n,p)$.

Theorem 8.12. *For every $\varepsilon > 0$ there exists a constant $C = C(\varepsilon)$ such that if $np^2 \ge C$, then* a.a.s. *each two-coloring of the edges of $\mathbb{G}(n,p)$ results in at least $(1/4 - \varepsilon)\binom{n}{3}p^3$ monochromatic triangles.*

The probabilistic part of the proof is contained in the following technical lemma, the proof of which is left to the reader. (Exercise! – The only tools needed are Chebyshev's and Chernoff's inequalities.) Let T denote the number of triangles in $\mathbb{G}(n,p)$, $d_v = \deg(v)$ stand for the degree of vertex v, and N_v (also known as $N(v)$) be the neighborhood of v. Finally, for a vertex set A, let $e(A)$ be the number of edges induced in $\mathbb{G}(n,p)$ by A.

Lemma 8.13. *Suppose that $np^2 = \Omega(1)$. Then*

(i) $|T - \binom{n}{3}p^3| = o_p(n^3p^3)$;

(ii) $\max_{1\le v\le n} |d_v - (n-1)p| = o_p(np)$;

(iii) $\max_{1\le v\le n} |e(N_v) - \binom{d_v}{2}p| = o_p(n^2p^3)$.

(iv) *Moreover, for every $\varepsilon > 0$ there exists $C' = C'(\varepsilon)$ such that if $np^2 \ge C'$, then* a.a.s. *for every $v = 1,2,\dots,n$ and for each $A \subseteq N_v$, $e(A) > \binom{|A|}{2}p - \varepsilon n^2p^3$.* ∎

Proof of Theorem 8.12. For a given blue–red coloring of the edges of $\mathbb{G}(n,p)$, let B_v $[R_v]$ be the set of vertices adjacent to v by edges colored blue [red], and let z_v be the number of edges joining those pairs of neighbors of v which are adjacent to v by edges of different colors, that is,

$$z_v = e(B_v, R_v) = e(N_v) - e(B_v) - e(R_v).$$

Then, similarly to the deterministic case, there are precisely $\frac{1}{2}\sum_v z_v$ two-colored triangles. Recalling our previous notation $b_v = |B_v|$ and $r_v = |R_v|$, and using Lemma 8.13(iii, iv) with $\varepsilon/7$, we find that the number of two-colored triangles is bounded from above by

$$\frac{1}{2}\sum_v\left[\binom{d_v}{2}p - \binom{b_v}{2}p - \binom{r_v}{2}p + \frac{2}{7}\varepsilon n^2p^3 + o_p(n^2p^3)\right]$$
$$= \frac{p}{2}\sum_v b_v r_v + \frac{1}{7}\varepsilon n^3p^3 + o_p(n^3p^3),$$

provided $np^2 \ge C'(\varepsilon/7)$, where $C'(\varepsilon/7)$ comes from Lemma 8.13(iv).

Since by Lemma 8.13(ii), $b_v + r_v = d_v = np + o_p(np)$ for each v, uniformly, we find $\sum_v b_v r_v \le \frac{1}{4}n^3p^2 + o_p(n^3p^2)$. This, together with Lemma 8.13(i) yields Theorem 8.12 with $C(\varepsilon) = C'(\varepsilon/7)$. ∎

A Turán-type theorem for random graphs

As a consequence of Theorem 8.12, we will now derive a special case of Conjecture 8.11. It can be viewed as Mantel's theorem for random graphs.

Theorem 8.14. *For every $\eta > 0$ there exists $C = C(\eta) > 0$ such that if $p \geq Cn^{-1/2}$, then* a.a.s. *every subgraph of* $\mathbb{G}(n,p)$ *with at least* $(1/2+\eta)\binom{n}{2}p$ *edges contains a triangle.*

Proof. In order to repeat almost literally the argument used in the deterministic case, we need to show first that most edges belong to nearly the expected number of triangles. Let X_{ij} be the number of vertices joined to both vertices i and j. Then X_{ij} has the binomial distribution $\mathrm{Bi}(n-2,p^2)$ and the expectation around np^2. Let Z count the edges $\{i,j\}$ of $\mathbb{G}(n,p)$ for which $X_{ij} < (1-\varepsilon)np^2$, where $\varepsilon > 0$ satisfies

$$3(1/2+\eta)(1-\varepsilon) > 3/2 + 2\varepsilon. \tag{8.5}$$

We have

$$\mathbb{E}(Z) = \binom{n}{2} p\, \mathbb{P}(X_{12} < (1-\varepsilon)np^2) \leq \binom{n}{2} p e^{-\varepsilon^2 np^2/3}$$

via the Chernoff inequality (2.6). If $np^2 \to \infty$, then $\mathbb{E}(Z) = o(n^2 p)$ and by Markov's inequality (1.3), $Z = o_p(n^2 p)$. Otherwise, it is quite straightforward to show that $\mathbb{E}(Z(Z-1)) \sim (\mathbb{E}(Z))^2$ and thus, by Chebyshev's inequality *a.a.s.* $Z \leq 2\binom{n}{2}pe^{-(\varepsilon C)^2/3}$. Thus, in either case, *a.a.s.*

$$Z \leq 2\binom{n}{2} pe^{-(\varepsilon C)^2/3}. \tag{8.6}$$

Let us take $C = C(\varepsilon)$ as in Theorem 8.12, and so large that the addition of the term $-2e^{-(\varepsilon C)^2/3}$ to the leftmost brackets of (8.5) is negligible, in the sense that the inequality

$$3(1/2+\eta-2e^{-(\varepsilon C)^2/3})(1-\varepsilon) > 3/2+2\varepsilon \tag{8.7}$$

holds. Note that C depends on η through ε.

Suppose there exists a subgraph H of $\mathbb{G}(n,p)$ with $e_H \geq (1/2+\eta)\binom{n}{2}p$ and containing no triangle. We will show that this event implies either a violation of (8.6) or a violation of the conclusion of Theorem 8.12, both events having probability converging to 0. Indeed, assuming they are not violated, let us count in two ways the ordered pairs (e,t) where e is an edge of H and t is a triangle of $\mathbb{G}(n,p)$ containing e. On the one hand, there are at least $(1/2+\eta-2e^{-(\varepsilon C)^2/3})\binom{n}{2}p(1-\varepsilon)np^2$ such pairs. On the other hand, viewing the edges of H as the blue edges and the remaining edges of $\mathbb{G}(n,p)$ as the red ones, we conclude by Theorem 8.12 that there are no more than $2(\frac{3}{4}+\varepsilon)\binom{n}{3}p^3$ such pairs. This yields a contradiction to (8.7). ■

Remark 8.15. Let us note that the dependence of C on the parameter η in Theorem 8.14 is genuine. In other words, it is not true that there exists an absolute constant C such that for every $\eta > 0$ *a.a.s.* every subgraph H of

$\mathbb{G}(n, Cn^{-1/2})$ with at least $(1/2+\eta)\binom{n}{2}p$ edges contains a triangle. It follows that, for $G = K_3$, the equation (8.2) holds if and only if $p \gg n^{-1/2}$.

To find a counterexample, set $V_1 = \{1, 2, \ldots, n/2\}$, n even, and $V_2 = \{n/2+1, \ldots, n\}$, and fix $C > 0$. It can be proved routinely by the second moment method that, for sufficiently small $\eta = \eta(C)$, *a.a.s.* there are at least $3\eta\binom{n}{2}p$ edges with both endpoints in V_1 and with no common neighbor in V_2. Moreover, it is well known that at least one half of these edges form a bipartite graph. The union of this bipartite graph and the bipartite graph spanned in $\mathbb{G}(n,p)$ between V_1 and V_2 contains no triangle and has at least $(1/2+\eta)\binom{n}{2}p$ edges (Exercise!).

8.3 THE SZEMERÉDI REGULARITY LEMMA

The Szemerédi Regularity Lemma states that, roughly speaking, for every large graph there exists a partition of its vertex set into a small number of almost equal subsets, such that in most of the bipartite graphs induced by pairs of these subsets, the edges are, in a way, "uniformly" distributed. The Lemma was introduced as an important step in the proof of Szemerédi's celebrated density theorem (Szemerédi 1975) and soon after the graph theorists realized that the presence of such "uniform", or, as we will call them later, "regular" partitions could greatly simplify many existing proofs and lead to solutions of many open problems in graph theory. Nowadays the Szemerédi Regularity Lemma is one of the most powerful tools of modern graph theory (see, e.g., Bollobás (1998), Diestel (1996), Komlós and Simonovits (1996)), as well as of the theory of random structures. That is why we devote a whole section to various versions of this important (but purely deterministic) result.

Regular pairs and partitions

In order to state the Szemerédi Regularity Lemma in a mathematically rigorous way we need a few definitions. Throughout, H is a graph with vertex set $V(H)$ and edge set $E(H)$ and $0 < s \leq 1$ is a real number which will be called the *scaling factor*. The role played by s will soon become clear; here we only mention that the two most prominent cases are $s = 1$ and $s = \rho(H) = e_H/\binom{v_H}{2}$. For two disjoint subsets $U, W \subset V(H)$ the $(s; H)$-*density* $d_{s,H}(U, W)$ is defined as

$$d_{s,H}(U, W) = \frac{e_H(U, W)}{s|U||W|},$$

where $e_H(U, W)$ counts the edges of H joining U and W. For $0 < \varepsilon < 1$, we say that two disjoint subsets $U, W \subset V(H)$ form an $(s; H, \varepsilon)$-*regular pair* if for every pair of their subsets $U' \subseteq U$, $W' \subseteq W$, such that $|U'| \geq \varepsilon|U|$ and

$|W'| \geq \varepsilon |W|$, we have

$$|d_{s,H}(U', W') - d_{s,H}(U, W)| \leq \varepsilon,$$

that is, the $(s;H)$-density of any pair of large subsets of the pair (U, W) does not deviate much from the $(s;H)$-density of (U, W).

Furthermore, let $\Pi = (V_0, V_1, \ldots, V_k)$ be a partition of $V(H)$. We say that this partition is *$(s; H, \varepsilon, k)$-regular* if $|V_1| = |V_2| = \cdots = |V_k|$, $|V_0| \leq \varepsilon v_H$, and for all except at most $\varepsilon\binom{k}{2}$ choices of the indices $1 \leq i < j \leq k$, the pairs (V_i, V_j) are $(s; H, \varepsilon)$-regular. Note the special status of V_0, which, for that reason, will be called the *exceptional class* of the $(s; H, \varepsilon, k)$-regular partition $(V_0, V_1, \ldots, V_k)$. We say that a partition $(W_0, W_1, \ldots, W_{k'})$ is a *subpartition* of a partition $(V_0, V_1, \ldots, V_k)$ if for every $1 \leq i \leq k'$ there exists $1 \leq j_i \leq k$ such that $W_i \subseteq V_{j_i}$.

Clearly, every graph H on n vertices admits an $(s; H, \varepsilon, 1)$-regular partition and an $(s; H, \varepsilon, n)$-regular partition. In applications, however, one rather needs an $(s; H, \varepsilon, k)$-regular partition for some k which is bounded from below and from above by some constants m and M. The Szemerédi Regularity Lemma provides the existence of such a partition, with M depending on ε and m only.

The classic case

Let us start with the important case $s = 1$. It was considered by Szemerédi (1978) and later in most applications of his result. To simplify the notation, we drop the index s, when $s = 1$, using terms like pair density $d_H(U, W)$, (H, ε, k)-regular partitions, and so forth. Then, the *Szemerédi Regularity Lemma* can be stated as follows.

Lemma 8.16. *For every $\varepsilon > 0$ and a natural number m there exists $M = M(\varepsilon, m)$ such that every graph H on at least m vertices admits an (H, ε, k)-regular partition for some k, where $m \leq k \leq M$.*

The idea of the proof. It is remarkable that the proof of so deep and insightful a result is based on a very simple idea. Let $\Pi_k = (V_1, \ldots, V_k)$ be any partition of the set of vertices of a graph H, where $|V_1| = \cdots = |V_k|$. Associate with Π_k a real number $\operatorname{ind}\Pi_k$, called the *index* of Π_k, setting

$$\operatorname{ind}\Pi_k = \frac{1}{k^2}\sum_{i=1}^{k}\sum_{j=i+1}^{k}(d_H(V_i, V_j))^2.$$

Note that since the density of a pair of sets is not greater than one, the index of any partition is bounded by a half.

Now, let $\Pi'_{k'} = (W_1, \ldots, W_{k'})$ be a subpartition of Π_k into $k' \geq k$ equal parts. Then, clearly, for every $1 \leq i < j \leq k$, we have

$$d_H(V_i, V_j) = \frac{k^2}{(k')^2}\sum_{W_t \subset V_i}\sum_{W_u \subset V_j} d_H(W_t, W_u).$$

Thus, the Cauchy–Schwarz inequality gives

$$(d_H(V_i, V_j))^2 \leq \frac{(k')^2}{k^2} \sum_{W_t \subset V_i} \sum_{W_u \subset V_j} (d_H(W_t, W_u))^2 \tag{8.8}$$

and, consequently,

$$\operatorname{ind} \Pi'_{k'} - \operatorname{ind} \Pi_k \geq 0, \tag{8.9}$$

where equality holds in (8.8) (and thus in (8.9)) if and only if all terms on the right-hand side of (8.8) are equal.

Suppose that the partition Π_k is not (H, ε, k)-regular. Then a substantial fraction of the pairs (V_i, V_j) are not (H, ε)-regular, and one can pick a subpartition $\Pi'_{k'}$ so that for each such pair the densities on the right-hand side of (8.8) differ significantly from each other. The key observation in the proof of Lemma 8.16 is that these differences force the index to increase. More precisely, the following holds. If k is large enough and Π_k is not (H, ε, k)-regular, then there exists a subpartition $\Pi'_{k'}$ of Π_k such that k' is bounded from above by a function of k and the difference $\operatorname{ind} \Pi'_{k'} - \operatorname{ind} \Pi_k$ is bounded from below by some positive constant which depends only on ε but *not* on k.

From this statement the lemma follows almost immediately. We begin with any partition Π_{k_0} of $V(H)$ into $k_0 \geq m$ equal parts. If Π_{k_0} is not (H, ε, k_0)-regular, a new partition Π_{k_1} is constructed in such a way that its index is substantially larger than $\operatorname{ind} \Pi_{k_0}$. We continue this way until an (H, ε, k)-regular partition is found for some k. Since for every $r \geq 1$ the difference $\operatorname{ind} \Pi_{k_r} - \operatorname{ind} \Pi_{k_{r-1}}$ is bounded from below by a constant which depends only on ε and, on the other hand, the index of every partition is bounded from above by a half, the procedure is guaranteed to end after at most a number of steps, which depends on ε and m only.

Although it is quite easy to believe that this argument works, its detailed proof is rather tedious and must take care of several technicalities. For example, one needs to use a special form of the Cauchy–Schwarz inequality to control the growth of $\operatorname{ind} \Pi_k - \operatorname{ind} \Pi'_{k'}$. Furthermore, in the description above we tacitly assumed that k always divides both k' and n, whereas in the original proof of Lemma 8.16 this problem is solved by introducing the exceptional class which, for the sake of simplicity of presentation, does not appear in our outline. We omit the details, referring the reader to Szemerédi (1978), Diestel (1996), or Bollobás (1998). ■

The argument above can be easily modified to obtain stronger versions of Lemma 8.16. In particular, if $H_1, \dots, H_r$ are graphs on the same vertex set V, then one can mimic the proof with $\operatorname{ind} \Pi_k$ replaced by

$$\operatorname{ind} \Pi_k^r = \frac{1}{k^2 r} \sum_{\ell=1}^{r} \sum_{i=1}^{k} \sum_{j=i+1}^{k} (d_{H_\ell}(V_i, V_j))^2,$$

and find a partition which is (H_ℓ, ε, k)-regular simultaneously for each $\ell = 1, \ldots, r$. Thus, we arrive at the following strengthening of Lemma 8.16.

Lemma 8.17. *For all $\varepsilon > 0$ and natural numbers m and r there exists $M = M(\varepsilon, m, r)$ such that the following holds: For all graphs $H_1, \ldots, H_r$ on the same set V of at least m vertices, there exists a partition of V which is (H_ℓ, ε, k)-regular for some k, $m \le k \le M$, and every $\ell = 1, 2, \ldots, r$.* ■

Sparse regularity lemma

The Szemerédi Regularity Lemma has proved to be extremely useful in many combinatorial investigations. We must point out, however, two drawbacks of its applications. First, Szemerédi's argument gives a very poor upper bound on the value of $M(\varepsilon, m, \ell)$, which grows so quickly with $\varepsilon \to 0$ that it is almost useless for any quantitative estimates. The other problem is that the Szemerédi Regularity Lemma, as stated above, is meaningful only when one deals with graphs of large density. For a graph H with n vertices and, say, maximum degree at most $\sqrt{n}$, each partition of $V(H)$ into k equal parts is (H, ε, k)-regular, provided n is large enough. This is because the density of the bipartite subgraph induced by any two sets of size $\Omega(n)$ is $O(1/\sqrt{n}) = o(1)$ and, therefore, does not measure effectively the distribution of the edges. Nonetheless, as noticed independently by Kohayakawa (1997) and Rödl (personal communication), a simple generalization of the Szemerédi Regularity Lemma will efficiently work for sparse graphs too. The key observation is that the proof of Lemma 8.16 still works for the scaled densities $d_{s,H}$ instead of the ordinary densities d_H, provided there exists a constant b such that $d_{s,H}(X,Y) \le b$ for all pairs of large sets X, Y. The constant b is an upper bound for an appropriately scaled index function and guarantees that the procedure of taking subpartitions terminates after a bounded number of steps.

More precisely, for $b \ge 1$ and $\beta > 0$, call a graph H *$(s; b, \beta)$-bounded* if for every pair of disjoint subsets $U, W \subset V(H)$ with $|U|, |W| \ge \beta v_H$ we have $d_s(U, W) \le b$. Then, a "scaled" version of the Szemerédi Regularity Lemma can be stated as follows.

Lemma 8.18. *For all $\varepsilon > 0$, $b \ge 1$ and natural numbers m and r there exist $\beta = \beta(\varepsilon, b, m, r) > 0$ and $M = M(\varepsilon, b, m, r)$ such that the following holds: For every choice of scaling factors $s_1, \ldots, s_r$, and $(s_\ell; b, \beta)$-bounded graphs H_ℓ, $\ell = 1, \ldots, r$, on the same set V of at least m vertices, there exists a partition of V which is $(s_\ell; H_\ell, \varepsilon, k)$-regular for some k, $m \le k \le M$, and every $\ell = 1, 2, \ldots, r$.* ■

Besides the classic case of $s_\ell \equiv 1$, the other instance of Lemma 8.18 which will be used in the forthcoming sections is the case in which $s_\ell = \rho(H_\ell) = e_{H_\ell} / \binom{v_{H_\ell}}{2}$. To simplify the notation in the latter case, we say that a graph H is *(b, β)-bounded* if it is $(\rho(H); b, \beta)$-bounded, a pair is *sparsely (H, ε)-regular* if it is $(\rho(H); H, \varepsilon)$-regular, and a partition is *sparsely (H, ε, k)-regular* if it is $(\rho(H); H, \varepsilon, k)$-regular. Below we state this special version of Lemma 8.18.

Lemma 8.19. *For all $\varepsilon > 0$, $b \geq 1$ and natural numbers m and r there exist $\beta = \beta(\varepsilon, b, m, r) > 0$ and $M = M(\varepsilon, b, m, r)$ such that the following holds: For every choice of (b, β)-bounded graphs $H_1, \ldots, H_r$ on the same set V of at least m vertices, there exists a partition of V which is sparsely (H_ℓ, ε, k)-regular for some k, $m \leq k \leq M$, and every $\ell = 1, 2, \ldots, r$.* ■

8.4 A PARTITION THEOREM FOR RANDOM GRAPHS

This section is entirely devoted to presenting an outline of the proof of the 1-statement of Theorem 8.1. For more details, see Rödl and Ruciński (1995).

Uniformly dense graphs

For $0 < d \leq 1$ and $\xi > 0$, we say that a graph F is *(ξ, d)-dense*, if for every $V \subseteq V(F)$ with $|V| \geq \xi v_F$ the induced subgraph $F[V]$ has density at least d. Note that it suffices to demand this for subsets V with $|V| = \lceil \xi v_F \rceil$; the property then holds for larger subsets by averaging over all their subsets with exactly $\lceil \xi v_F \rceil$ elements. Observe further that the complete graph K_n is (ξ, d)-dense for all choices of $\xi > 0$ and $0 < d \leq 1$. It can be easily verified (Exercise!) that the following is true (Rödl and Ruciński 1995, Lemma 2).

Proposition 8.20. *For each $0 < d \leq 1$ and each graph G, there exists $\xi > 0$ such that every (ξ, d)-dense graph on n vertices contains $\Theta(n^{v(G)})$ copies of G.* ■

Thus, in a sense, (ξ, d)-dense graphs imitate complete graphs.

The heart of the proof of Theorem 8.1 is the following deterministic lemma which utilizes the Szemerédi Regularity Lemma in a "canonical" way.

Lemma 8.21. *For all $0 < \xi' \leq 1$ and $0 < d \leq 1$, and for every natural number r there exist $\xi > 0$, $\nu > 0$, and n_0, such that if F is a (ξ, d)-dense graph on $n \geq n_0$ vertices and $E(F) = E_1 \cup \cdots \cup E_r$, then there exist $\ell_0 \in [r]$ and $V \subseteq V(F)$, $|V| \geq \nu n$, for which the subgraph of F consisting of the vertices of V and of the edges of $E_{\ell_0} \cap F[V]$ is (ξ', d')-dense, where $d' = d/20r$.*

Remark 8.22. This lemma, together with Proposition 8.20 and inequality (2.6), implies Theorem 8.1 in the case in which p is a constant. Indeed, by Chernoff's inequality $\mathbb{G}(n, p)$ is then *a.a.s.* $(\xi, p/2)$-dense for any fixed ξ (Exercise!). Thus, by Lemma 8.21, for every r-coloring at least one of the color classes is $(\xi', p/40r)$-dense on a large subset V of $[n]$, and contains, by Proposition 8.20, many (monochromatic) copies of G. We will see later in this section how Lemma 8.21 can be applied in the sparse range of p.

Proof. Set

$$q = \left\lceil \frac{2}{d} \right\rceil, \qquad t = \left\lceil \frac{2.05}{\xi'} \right\rceil, \tag{8.10}$$

$$m = R(q, \underbrace{t, \ldots, t}_{r}), \tag{8.11}$$

where $R(q, \underbrace{t, \ldots, t}_{r})$ is the Ramsey number (see, e.g., Graham, Rothschild and Spencer (1992) for definition). Furthermore, set

$$\varepsilon = \min\left\{ \frac{1}{20t}, \frac{d}{100r}, \frac{1}{2rm} \right\}, \tag{8.12}$$

$$\xi = \frac{q(1-\varepsilon)}{M}, \qquad \nu = \frac{t(1-\varepsilon)}{M}, \qquad \text{and} \qquad n_0 = \frac{20M}{1-\varepsilon}, \tag{8.13}$$

where $M = M(\varepsilon, 2m, r)$ is the constant provided by the Szemerédi Regularity Lemma 8.17.

Let F be a (ξ, d)-dense graph on $n \geq n_0$ vertices, and let $E(F) = E_1 \cup \cdots \cup E_r$ be a partition of the edge set of F. Denote by $H_\ell = F[E_\ell]$ the (spanning) subgraph of F consisting of the edges of E_ℓ, $\ell = 1, \ldots, r$. By Lemma 8.17 there exists a partition $V(F) = C_0 \cup C_1 \cup \cdots \cup C_k$, $2m \leq k \leq M$, which is (H_ℓ, ε, k)-regular for each $\ell = 1, \ldots, r$.

As, by (8.12), at least $(1 - r\varepsilon)\binom{k}{2} > (1 - \frac{1}{m})\frac{k^2}{2}$ pairs (C_i, C_j) are (H_ℓ, ε)-regular for each $\ell = 1, 2, \ldots, r$, it follows from Turán's theorem that there are m sets, $C_1, \ldots, C_m$ say, such that all $\binom{m}{2}$ pairs of them are (H_ℓ, ε)-regular for each ℓ.

Consider the $(r+1)$-coloring of $[m]^2$, $[m]^2 = D_0 \cup D_1 \cup \cdots \cup D_r$, where

$$\{i, j\} \in D_0 \qquad \text{if} \qquad d_F(C_i, C_j) < \frac{d}{2} \tag{8.14}$$

and, for $\ell \in [r]$,

$$\{i, j\} \in D_\ell \qquad \text{if} \qquad d_{H_\ell}(C_i, C_j) \geq \frac{d}{2r}.$$

(Note that a pair may belong to more than one set D_ℓ.) By (8.11), there exists either a subset $K \subset [m]$, $|K| = q$, $[K]^2 \subset D_0$, or a subset $L \subset [m]$, $|L| = t$, $[L]^2 \subset D_\ell$, for some $\ell \in [r]$. The first option is impossible, since then, putting $x = |C_i|$, the set $C = \bigcup_{j \in K} C_j$ would have, by (8.14) and (8.10), density

$$\rho(F[C]) = \frac{e(F[C])}{\binom{|C|}{2}} < \frac{\binom{q}{2}\frac{d}{2}x^2 + q\binom{x}{2}}{\binom{qx}{2}} < \frac{d}{2} + \frac{1}{q} \leq d\,.$$

This contradicts the fact that F is (ξ, d)-dense, because, by (8.13), $|C| = qx \geq q\frac{(1-\varepsilon)n}{M} = \xi n$. Thus, for some $\ell_0 \in [r]$, there are t sets, $C_1, \ldots, C_t$ say, which satisfy

$$d_{H_{\ell_0}}(C_i, C_j) \geq \frac{d}{2r} \qquad \text{for all} \qquad \{i,j\} \in [t]^2.$$

We will prove that the graph $H = H_{\ell_0}[V]$, where $V = C_1 \cup \cdots \cup C_t$, is $(\xi', \frac{d}{20r})$-dense. (Note that, by (8.13) again, $|V| = tx \geq t\frac{(1-\varepsilon)n}{M} = \nu n$ as required.)

Consider $V' \subset V$ of size

$$|V'| = \lceil \xi' |V| \rceil = \lceil \xi' tx \rceil.$$

Then, by (8.10),

$$|V'| \leq \left\lceil \left(\frac{2.05}{\xi'} + 1 \right) \xi' x \right\rceil.$$

Since by (8.13) we have $x \geq \frac{(1-\varepsilon)n}{M} \geq 20$ and also $\xi' \leq 1$, the right-hand side above can further be bounded from above by $3.1x$, leading to

$$|V'| \leq 3.1x. \tag{8.15}$$

Set $x_i = |V' \cap C_i|$, $i = 1, \ldots, t$. Then, owing to (8.15), we infer that

$$\sum_{i=1}^{t} \binom{x_i}{2} \leq 3\binom{x}{2} + \binom{0.1x}{2} < 3.01\binom{x}{2}. \tag{8.16}$$

Let $V' = V'' \cup V'''$ be a partition, where $V'' = \bigcup_{i:x_i \leq \varepsilon x} V' \cap C_i$. Clearly, $|V''| \leq t\varepsilon x$, and thus, by (8.10) and (8.12),

$$|V'''| \geq (\xi' - \varepsilon)tx \geq 2x. \tag{8.17}$$

Let us bound from below the number of edges in the graph $H[V''']$. By the definition of V''' and by the choice of $C_1, \ldots, C_t$,

$$e(H[V''']) \geq \sum\sum \left(\frac{d}{2r} - \varepsilon \right) x_i x_j,$$

where the double summation is taken over all pairs i, j, $1 \leq i < j \leq t$, such that $x_i \geq \varepsilon x$ and $x_j \geq \varepsilon x$. But, by (8.16) and (8.17),

$$\sum\sum x_i x_j = \binom{\sum x_i}{2} - \sum \binom{x_i}{2} \geq \binom{2x}{2} - 3.01\binom{x}{2} > 0.99\binom{x+1}{2},$$

where the single summations are taken over all i satisfying $x_i \geq \varepsilon x$. Hence

$$e(H[V''']) \geq 0.99 \left(\frac{d}{2r} - \varepsilon \right) \binom{x+1}{2}. \tag{8.18}$$

Finally, by (8.15), (8.18) and (8.12),

$$\rho(H[V']) = \frac{e(H[V'])}{\binom{|V'|}{2}} \geq \frac{e(H[V'''])}{\binom{|V'|}{2}} \geq \frac{0.99(\frac{d}{2r} - \varepsilon)\binom{x+1}{2}}{\binom{3.1x}{2}}$$
$$\geq \frac{0.99}{9.7}\left(\frac{d}{2r} - \varepsilon\right) \geq \frac{1}{9.8}\left(\frac{d}{2r} - \varepsilon\right) \geq \frac{d}{20r},$$

proving that the graph H is $(\xi', \frac{d}{20r})$-dense. ∎

Sketch of proof of the 1-statement of Theorem 8.1

The proof proceeds by double induction on the number of colors r and the number of edges e_G. As often happens with induction, it helps to generalize the statement a little. Our strengthening touches all three aspects of Theorem 8.1: the random graph space, the property in question, and the probability with which this property is held by the random graph.

(1) We replace the random graph $\mathbb{G}(n,p)$ with the reliability network Γ_p, where Γ is a (ξ, d)-dense graph on n vertices.

(2) We replace the partition property $\Gamma_p \to (G)_r$ with the property that every r-coloring results in $\Omega(n^{v_G} p^{e_G})$ monochromatic copies of G. Moreover, only copies contained in complete subgraphs of Γ count. We call such copies *nested*. Note that when $\Gamma = K_n$, every copy of G is nested.

(3) We replace the convergence of probability to 1 with the condition that the probability of the opposite event is $2^{-\Omega(n^2 p)}$. (In what follows we will often use the phrase *high probability*, meaning precisely this.)

More formally, we prove the following general result which implies the 1-statement of Theorem 8.1.

Theorem 8.23. *For every graph G with at least one edge, for all integers $r \geq 1$ and all real numbers $0 < d \leq 1$, there exist positive numbers ξ, a, b, C, and n_0 such that if*

(i) $n \geq n_0$,

(ii) *Γ is a (ξ, d)-dense graph with n vertices, and*

(iii) $p \geq Cn^{-1/m^{(2)}(G)}$,

then, with probability at least $1 - 2^{-bn^2 p}$, every r-coloring of the edges of Γ_p results in at least $an^{v_G} p^{e_G}$ monochromatic, nested copies of G.

Sketch of proof. For both initial cases, $e_G = 1$, r arbitrary, and $r = 1$, e_G arbitrary, every copy of G is automatically monochromatic, and all we need in order to validate Theorem 8.23 is to show that Γ_p contains sufficiently many nested copies of G with sufficiently high probability. Proposition 8.20 shows that Γ contains $\Theta(n^{v_G})$ complete subgraphs K_{v_G}, and a standard application of Theorem 2.14 gives the required result (Exercise!).

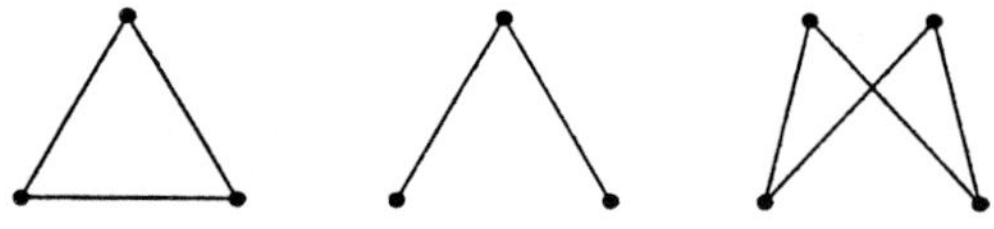

Fig. 8.2 Graphs G, J, and JJ for $G = K_3$.

Assume now that $e_G \geq 2$ and $r \geq 2$, and that Theorem 8.23 is true for all instances with either fewer than r colors or fewer than e_G edges. Our strategy is to apply the two-round exposure technique (*cf.* Section 1.1), that is, to represent Γ_p as a union of two independent random graphs Γ_{p_1} and Γ_{p_2}, where $p_1 + p_2 - p_1 p_2 = p$, and p_1 and p_2 are suitably chosen. It is planned that p_2 will be sufficiently bigger than p_1, but both of the same order of magnitude.

Let $J = J(G, e)$ be the graph obtained from G by the removal of one fixed edge e, and let $JJ = JJ(G, e)$ be obtained from the union of two copies of G sharing e by the removal of e (see Figure 8.2, where $G = K_3$, $J = P_2$ and $JJ = C_4$). By the induction assumption, with high probability, there are many monochromatic, nested copies of J in every r-coloring of Γ_{p_1}. Thus, there are many edges of Γ, which, when added to a monochromatic, nested copy of J, form a copy of G, provided there are not too many copies of JJ. It is here where we use Lemmas 2.51 and 2.52. Indeed, we are satisfied with an upper bound on the number of copies of JJ on a subset of edges of Γ_{p_1}. By a standard application of the Cauchy–Schwarz inequality, there are $\Theta(n^2)$ edges in Γ such that each of them "closes" as many as $\Omega(n^{v_G - 2} p^{e_G - 1})$ monochromatic, nested copies of J. Most of these edges are not in Γ_p. Let us denote the subgraph of Γ consisting of all such edges by F, and associate with each edge u of F the color most frequently appearing in the monochromatic copies of J in Γ_p, which, together with u, form a copy of G. The colors associated in the above way with the edges of F vary from edge to edge, which naturally imposes a partition $E(F) = E_1 \cup \cdots \cup E_r$. Write H_ℓ for the spanning subgraph of F with the edge set E_i, $i = 1, \ldots, r$.

In the second round we would like to apply the induction assumption with $r - 1$ colors to one of the graphs H_ℓ, $\ell = 1, 2, \ldots, r$. For this, however, we have to show that one of these graphs is (ξ', d')-dense for some $\xi' > 0$ and $d' > 0$. It turns out that it is easier to show first that with sufficiently high probability the graph F is (ξ_0, d_0)-dense for some ξ_0 and d_0, and then apply Lemma 8.21. Indeed, the proof that F is (ξ_0, d_0)-dense is similar to showing that F has many edges, which we have just described. Then, by Lemma 8.21, there is a color $\ell_0 \in [r]$ and a large subset $V \subseteq V(F)$ such that the graph $H = H_{\ell_0}[V]$ satisfies the assumptions of Theorem 8.23.

We are now justified in applying the induction assumption with $r - 1$ colors to H. Thus, provided that color ℓ_0 has not been used for the random graph

H_{p_2}, we might conclude that the second round produces with high probability plenty of monochromatic, nested copies of G in H_{p_2}. On the other hand, every time color ℓ_0 is used on an edge of H_{p_2}, it produces $\Omega(n^{v_G-2}p_2^{e_G-1})$ nested copies of G of color ℓ_0. Hence, if color ℓ_0 is used for H_{p_2} at least $\Omega(n^2p_2)$ times, we are done. It remains to clear the case in which the selected color is used only a few times on the edges of H_{p_2}. As all of the above holds also for $(1-\delta)p_2$ instead of p_2, this last case follows from Lemma 2.52 alone (Exercise!).

No matter how the adversary colored the edges emerging from the first round, the outcome of the second round should be successful. Therefore, the probability of failure in the second round must be much smaller than the reciprocal of the number of all possible r-colorings h of the edges of Γ_{p_1}. The number of edges of Γ_{p_1} is, by Chernoff's inequality, with high probability, fewer than n^2p_1, and thus the number of such colorings does not exceed $r^{n^2p_1}$. The probability of failure in the second round is forced to be sufficiently small by choosing p_2 sufficiently bigger than p_1.

Let us now organize the whole proof a little bit more rigorously. Let $\mathcal{A}$ be the event that there is an r-coloring $\bar{h}: E(\Gamma_p) \to [r]$ with fewer than $an^{v_G}p^{e_G}$ monochromatic, nested copies of G.

For a copy J' of J in Γ, let $\text{cl}(J')$ be the set of all edges $u \in \Gamma$ such that $J' \cup \{u\}$ is isomorphic to G. Given an r-coloring $h: E(\Gamma_{p_1}) \to [r]$, define the edge sets

$$E_\ell(h) = \{u \in E(\Gamma) \setminus E(\Gamma_{p_1}) : |\{J' \subseteq \Gamma_{p_1} : u \in \text{cl}(J') \text{ and } h(J') = \ell\}| \geq z\},$$

where $z = cn^{v_G-2}p_1^{e_G-1}$ for some c, and set $H_\ell(h) = (V(\Gamma), E_\ell(h))$, $\ell = 1, \ldots, r$.

Let $\mathcal{B}$ be the event that for every $h: E(\Gamma_{p_1}) \to [r]$ there exist an $\ell_0 \in [r]$ and a set $V \subset V(\Gamma) = [n]$, $|V| \geq \nu n$, such that the graph $H_{\ell_0}(h)[V]$ is (ξ', d')-dense and that $|E(\Gamma_{p_1})| \leq n^2p_1$. Conditioning on Γ_{p_1} and fixing h, let $\mathcal{A}_h$ be the event that there is an extension of h, $\bar{h}: E(\Gamma_p) \to [r]$, that is, $\bar{h} = h$ when restricted to $E(\Gamma_{p_1})$, such that there are fewer than $an^{v_G}p^{e_G}$ monochromatic, nested copies of G. Then,

$$\mathbb{P}(\mathcal{A}) \leq \mathbb{P}(\neg\mathcal{B}) + \sum_{K \in \mathcal{B}} \mathbb{P}(\mathcal{A} \mid \Gamma_{p_1} = K)\,\mathbb{P}(\Gamma_{p_1} = K)$$

and

$$\mathbb{P}(\mathcal{A} \mid \Gamma_{p_1} = K) = \mathbb{P}\left(\bigcup_h \mathcal{A}_h \mid \Gamma_{p_1} = K\right) \leq r^{n^2p_1}\,\mathbb{P}(\mathcal{A}_{h_0} \mid \Gamma_{p_1} = K),$$

where the summation is taken over all r-colorings h of the edges of $\Gamma_{p_1} = K$ and h_0 maximizes the conditional probability.

We have just outlined the proofs of the inequalities

$$\mathbb{P}(\mathcal{B}) > 1 - 2^{-\Omega(n^2p_1)}$$

and

$$\mathbb{P}(\mathcal{A}_h \mid \Gamma_{p_1} = K) \leq 2^{-\Omega(n^2 p_2)},$$

the latter for every $K \in \mathcal{B}$ and for every r-coloring h of the edges of K. These two facts imply that $\mathbb{P}(\mathcal{A}) = 2^{-\Omega(n^2 p)}$ for some small a. ∎

8.5 TRIANGLES: AN APPROACH WITH PERSPECTIVE

In Section 8.2 we used Goodman's elegant idea to verify Conjecture 8.11 for triangles and showed that the 1-statement of Theorem 8.1 is valid for $G = K_3$ and $r = 2$. Now we present an entirely new approach based on the sparse version of the Szemerédi Regularity Lemma (Lemma 8.19) and a better-than-exponential estimate of the probability that a sufficiently "dense" and "regular" random graph contains a copy of a given graph G. So far, the method verifies Conjecture 8.11 and yields Theorem 8.1 (for an arbitrary number r of colors) only in a few small cases of G. But we hope that the fundamental Conjecture 8.35 stated below will soon be proved, paving the road to a complete solution of Conjecture 8.11. In this section we will restrict ourselves to the simplest case $G = K_3$.

We switch now to the uniform random graph $\mathbb{G}(n, M)$. It has the advantage over the binomial model $\mathbb{G}(n,p)$ that the relative density $\rho(\mathbb{G}(n, M))$ is fixed and equal to $M/\binom{n}{2}$. We set

$$\rho_M = \rho(\mathbb{G}(n, M)) = M \bigg/ \binom{n}{2},$$

for convenience. Thus, in this section we will prove the following result.

Theorem 8.24. *For every $\eta > 0$ there exists $C = C(\eta) > 0$ such that if $M \geq Cn^{3/2}$ then* a.a.s. *every subgraph of $\mathbb{G}(n, M)$ with at least $(1/2 + \eta)M$ edges contains a triangle.*

By Proposition 1.12, Theorem 8.24 implies Theorem 8.14.

The idea of proof

The proof we give contains probabilistic as well as deterministic ingredients. To some extent, the general framework is analogous to that of the proof of Theorem 8.23. The notion of a (ξ, d)-dense graph is replaced by ξ-uniformity, and Lemma 8.26 below has the flavor of Lemma 8.21. Both are consequences of the Szemerédi Regularity Lemma. However, in Section 8.4, owing to the chosen method of proof, we were able to use its dense version, Lemma 8.17, despite the fact that Theorem 8.23 deals with sparse random graphs. Here we do not have this option. We will directly apply the sparse version of the Szemerédi Regularity Lemma in the form of Lemma 8.19.

The straightforward approach, so successful in the vertex-coloring case (see Section 7.6), would be to show that the expected number of the triangle-free subgraphs of $\mathbb{G}(n, M)$ with $M' = (1/2 + \eta)M$ edges tends to 0. There are, roughly, 2^M subgraphs of $\mathbb{G}(n, M)$ with M' edges and each such subgraph can be viewed as a random graph $\mathbb{G}(n, M')$ on its own (formally, turn to the random graph process $\{\mathbb{G}(n, M)\}_M$ and consider its subprocess of M' specified steps). Unfortunately, Theorem 3.11 implies that if, say, $n^{3/2} \leq M' \leq \frac{1}{5}n^2$, then $\mathbb{P}(\mathbb{G}(n, M') \not\supseteq K_3) = e^{-\Theta(M)}$, which may not be sufficient. However, the lower bound on $\mathbb{P}(\mathbb{G}(n, M) \not\supseteq K_3)$ was obtained via a bound on the probability that $\mathbb{G}(n, M)$ is bipartite. The main idea of this proof is that $\mathbb{G}(n, M)$ is so far from being bipartite that each subgraph with M' edges contains a highly regular tripartite subgraph which is then extremely likely to contain a triangle.

Before making the above argument rigorous, we must decide how to define the tripartite structure in precise, mathematical terms. For our purposes, given $n, \rho, \varepsilon > 0$, an *(n, ρ, ε)-triplet* is a tripartite graph T with a specified tripartition $V(T) = V_1 \cup V_2 \cup V_3$ such that $|V_1| = |V_2| = |V_3| \geq n$, each of the pairs (V_1, V_2), (V_2, V_3) and (V_1, V_3) is sparsely (T, ε)-regular and the number of edges in each of the three bipartite graphs induced by these pairs satisfies

$$e_T(V_i, V_j) \geq \rho |V_i||V_j|.$$

If, moreover, $e_T(V_i, V_j) = \lceil \rho |V_i||V_j| \rceil$ for each pair, then the triplet is said to be *exact.*

Hence, in order to prove Theorem 8.24, and thus Conjecture 8.11 for triangles, we will show first that *a.a.s.* every subgraph of $\mathbb{G}(n, M)$ with substantially more than half of its edges contains a large triplet (Corollary 8.27), and then that *a.a.s.* every such triplet contains a triangle (Lemma 8.32). By conditioning on the number of edges in the triplet, we will reduce our considerations to tripartite random graphs with a fixed number of edges, and show that, if highly regular, they contain no triangle with probability $(o(1))^M$ (Lemma 8.30).

Similarly one can conduct the proof of the 1-statement of Theorem 8.1 for $G = K_3$ and an arbitrary number r of colors. We state without proof an appropriate fact as part (ii) of Lemma 8.26.

Uniformly sparse graphs

We say that a graph F on n vertices is *ξ-uniform* if for every pair of disjoint subsets X and Y of vertices of F such that $|X|, |Y| \geq \xi n$, its $(\rho(F); F)$-density is close to 1, or, more precisely,

$$1 - \xi < d_{\rho(F),F}(X, Y) = \frac{e_F(X, Y)}{\rho(F)|X||Y|} < 1 + \xi,$$

and the relative density of the subgraph induced by X in F is also close to 1, that is,

$$1-\xi<\frac{\rho(F[X])}{\rho(F)}=\frac{e(F[X])}{\rho(F)\binom{|X|}{2}}<1+\xi,$$

where $\rho(F)=e_F/\binom{n}{2}$.

Lemma 8.25. *If $M/n\to\infty$ and $\xi>0$, then the random graph $\mathbb{G}(n,M)$ is a.a.s. ξ-uniform.* ∎

The above lemma follows (Exercise!) by an easy application of Chernoff's bound for the hypergeometric distribution (see Theorem 2.10 and the inequality (2.9)).

Next we show, by a nontrivial application of the sparse version of the Szemerédi Regularity Lemma, that every sufficiently dense subgraph of a ξ-uniform graph contains a large triplet.

Lemma 8.26.

(i) *For every $0<\eta\le 1/2$ and $\varepsilon>0$ there exist $\xi=\xi(\eta,\varepsilon)>0$ and n_0 such that every spanning subgraph H of a ξ-uniform graph F on $n\ge n_0$ vertices satisfying $e_H\ge(1/2+\eta)e_F$ contains a $(\xi n,0.1\eta\rho(F),\varepsilon)$-triplet.*

(ii) *For every natural number r and $\varepsilon>0$ there exist $\xi>0$ and n_0 such that for every partition $E(F)=E_1\cup\cdots\cup E_r$ of the edges of a ξ-uniform graph F with $n\ge n_0$ vertices, there exists an $\ell_0\in[r]$ such that the graph $H_{\ell_0}=F[E_{\ell_0}]$ contains a $(\xi n,\rho(F)/2r,\varepsilon)$-triplet.*

Proof. We will only prove part (i). Before plunging into this detailed and slightly tedious proof we strongly encourage the reader to formulate and prove the special case of part (i) when $F=K_n$ (Exercise!).

For the general case, we may assume $\varepsilon\le 1$. Set $\varepsilon'=\varepsilon\eta/15$ and $m=\lceil 20/\eta\rceil$ and apply Lemma 8.19 with ε', $b=2$ and $r=1$. Let $\xi=\min(\beta,1/2M,\varepsilon')$, where $\beta=\beta(\varepsilon',2,m,1)$ and $M=M(\varepsilon',2,m,1)$ are as in Lemma 8.19. Let F be a ξ-uniform graph on n vertices and let H be a spanning subgraph of F satisfying $e_H\ge(1/2+\eta)e_F$. Since ξ is smaller than η, the graph H is $(2,\xi)$-bounded (Exercise!) and thus, by Lemma 8.19, there is a sparsely (H,ε',k)-regular partition $\Pi=(V_0,V_1,\dots,V_k)$ of vertices of H with $m\le k\le 1/2\xi$. Note that $|V_i|\ge\frac{1-\varepsilon'}{k}n\ge\xi n$ for $i\ge 1$.

Call a pair (V_i,V_j), $1\le i<j\le k$, *good* if it is sparsely (H,ε')-regular and satisfies

$$e_H(V_i,V_j)\ge 0.1\eta\rho(F)|V_i||V_j|. \tag{8.19}$$

Our goal is to show that the auxiliary graph, the vertices of which are the sets $V_1,\dots,V_k$ and the edges represent good pairs, has more than $k^2/4$ edges. Then, by Mantel's theorem, there is a triangle in this graph. This triangle consists of three sets V_{i_1},V_{i_2},V_{i_3}, which induce a tripartite subgraph T of H, with every pair being sparsely (H, ε')-regular and satisfying (8.19). Since, by (8.19),

$$\rho(T) \geq 0.3\eta\rho(F)|V_1|^2 / \binom{3|V_1|}{2} \geq \frac{1}{15}\eta\rho(F) \geq \frac{1}{15}\eta\rho(H),$$

these pairs are also sparsely (T, ε)-regular, and it follows that T is a $(\xi n, 0.1\eta\rho(F), \varepsilon)$-triplet.

To obtain the required lower bound on the number of good pairs, we will first bound from above the total number of edges of H which are not within the good pairs. It will turn out that the majority of the edges of H is indeed between the good pairs and, as there cannot be too many edges between any fixed pair, the lower bound on the number of good pairs will follow.

The edges not within the good pairs can be classified into four groups:

(a) *Edges with at least one endpoint in the exceptional class V_0*. For each $i = 1, \ldots, k$, let W_i be a subset of vertices such that $W_i \supseteq V_0$, $W_i \cap V_i = \emptyset$ and $|W_i| = \lceil \varepsilon' n \rceil \geq \xi n$. Then,

$$e(H[V_0]) \leq e(F[W_1]) < (1+\xi)\binom{|W_1|}{2}\rho(F) < 0.01\eta e_F$$

and, similarly, for each $i = 1, \ldots, k$,

$$e_H(V_0, V_i) \leq e_F(W_i, V_i) < (1+\xi)|W_i||V_i|\rho(F) < 0.20\eta e_F / k,$$

yielding a total upper bound of $0.21\eta e_F$ on the number of these edges.

(b) *Edges with both endpoints in the same class V_i, for any $i = 1, \ldots, k$*. The number of edges of H contained in the set V_i is

$$e(H[V_i]) \leq e(F[V_i]) < (1+\xi)\binom{|V_i|}{2}\rho(F) < (1+\xi)e_F/mk < 0.06\eta e_F/k.$$

Hence the total number of edges in this category is fewer than $0.06\eta e_F$.

(c) *Edges between the pairs (V_i, V_j), $1 \leq i < j \leq k$, which are not sparsely (H, ε')-regular.* Note that, since F is ξ-uniform, for all $1 \leq i, j \leq k$

$$e_H(V_i, V_j) \leq (1+\xi)e_F \frac{|V_i||V_j|}{\binom{n}{2}}.$$

Thus, the number of edges in this category is bounded from above by

$$\varepsilon' \binom{k}{2}(1+\xi)e_F \frac{|V_i||V_j|}{\binom{n}{2}} < 0.07\eta e_F.$$

(d) *Edges between the pairs (V_i, V_j) which violate (8.19)*. There are no more than

$$\binom{k}{2} 0.1\eta e_F \frac{|V_i||V_j|}{\binom{n}{2}} < 0.1\eta e_F$$

such edges.

Consequently, at least

$$e_H - 0.44\eta e_F > (1+\eta)e_F/2$$

edges of H join good pairs (V_i, V_j).

On the other hand, we have just noticed in point (c) above that no bipartite graph spanned in H by a pair (V_i, V_j), $1 \le i < j \le k$, has more than

$$(1+\xi)e_F \frac{|V_i||V_j|}{\binom{n}{2}} \le (1+\eta)2e_F/k^2$$

edges. Thus, among the pairs (V_i, V_j), $1 \le i, j \le k$, there must be more than

$$\frac{(1+\eta)e_F/2}{2(1+\eta)e_F/k^2} = \frac{1}{4}k^2$$

good ones, and the assertion (i) follows.

The second part can be proved in an analogous way, but, instead of Mantel's theorem, one must use Turán's and Ramsey's theorems (in this order) as in the proof of Lemma 8.21. Since this result is needed only for an alternative proof of Theorem 8.1 in the very special case $G = K_3$, we leave the proof to the reader (Exercise!). ■

Lemma 8.26(i) together with Lemma 8.25 have the following consequence.

Corollary 8.27. *For every $0 < \eta \le 1/2$ and $\varepsilon > 0$ there exists $\xi = \xi(\eta, \varepsilon) > 0$ such that if $M/n \to \infty$, then* a.a.s. *every subgraph H of $\mathbb{G}(n, M)$ with $e(H) \ge (1/2 + \eta)M$ contains a $(\xi n, 0.1\eta\rho_M, \varepsilon)$-triplet.* ■

The conclusions of Lemma 8.26 and Corollary 8.27 may be strengthened to the existence of an exact triplet by the following simple fact, the proof of which is left to the reader. (Exercise! – Use Theorem 2.10.)

Lemma 8.28. *For every $\varepsilon > 0$ there exists $C = C(\varepsilon) > 0$ such that if B is a bipartite graph with bipartition (V_1, V_2), $|V_1| = |V_2| = n$, with L edges, and the pair (V_1, V_2) is sparsely (B, ε)-regular, then for every K with $Cn \le K \le L$ there is a subgraph B' of B with K edges and such that (V_1, V_2) is sparsely $(B', 2\varepsilon)$-regular.* ■

Corollary 8.29. *Every (n, ρ, ε)-triplet with $\rho n \ge C(\varepsilon)$ contains an exact $(n, \rho, 2\varepsilon)$-triplet.* ■

Tripartite random graphs

In order to state the main probabilistic ingredient of our argument we need to introduce one more model of a random graph. Let $\mathbb{G}_3(n, M)$ be a graph chosen uniformly at random from the family of all tripartite graphs F with vertex set $V = V_1 \cup V_2 \cup V_3$, where $|V_1| = |V_2| = |V_3| = n$, such that for each $1 \le i < j \le 3$, $e_F(V_i, V_j) = M$. For this random graph, although "genuinely" tripartite, the probability of containing no triangls does not drop down to $(o(1))^M$. Indeed, splitting $V_1 = V' \cup V''$, $|V'| = |V''| = n/2$, if $M \le n^2/4$, then with probability at least 16^{-M} there is no edge between V_2 and V' and no edge between V_3 and V'' (Exercise!). Hence, with at least this probability, there is no triangle in $\mathbb{G}_3(n, M)$. To make the appearence of triangles more likely, we condition on the event that each of the three bipartite subgraphs of $\mathbb{G}_3(n, M)$ has a highly regular structure, that is, that $\mathbb{G}_3(n, M)$ is a triplet.

Lemma 8.30. *For every* $0 < \varepsilon \le 0.01$ *and natural numbers* n *and* $M \ge (8/\varepsilon)^{1/2} n^{3/2}$,

$$\mathbb{P}(\mathbb{G}_3(n, M) \text{ is a triangle-free } (n, M/n^2, \varepsilon)\text{-triplet}) \le \varepsilon^{M/16}. \tag{8.20}$$

We deduce Lemma 8.30 from the following result on random subsets of a regular pair of sets.

Lemma 8.31. *Let* $0 < \varepsilon \le 0.01$, $t \ge 1$, $M \ge 4n^2/(\varepsilon t)$ *and let* H *be a bipartite graph with bipartition* (V', V'') *such that* $|V'| = |V''| = n$, $e(H) = M$ *and the pair* (V', V'') *is sparsely* (H, ε)*-regular. Furthermore, let* S'_t *and* S''_t *be two random sets of size* t *picked independently and uniformly from all* t*-element subsets of, respectively,* V' *and* V''*. Then, the probability that there is no edge between* S'_t *and* S''_t *is smaller than* $\varepsilon^{t/4}$.

Proof. We may assume that $t < \varepsilon n$, since otherwise the assumption that (V', V'') is sparsely (H, ε)-regular implies that there is always an edge between S'_t and S''_t. We will show that with probability at least $1 - \varepsilon^{t/3}$ the neighborhood of the set S'_t contains all but at most εn vertices of V''. This is all we need, as the probability that S''_t is contained in a fixed set of size at most εn is not greater than

$$\binom{\varepsilon n}{t} \Big/ \binom{n}{t} \le \varepsilon^t$$

and thus

$$\mathbb{P}(e_H(S'_t, S''_t) = 0) \le \varepsilon^{t/3} + \varepsilon^t < \varepsilon^{t/4}.$$

Let us generate S'_t sequentially, picking its elements $s_1, \ldots, s_t$ one by one, uniformly at random, from all currently available vertices of V'. For $i = 1, \ldots, t$, let $S_i = \{s_1, \ldots, s_i\}$ and let W_i denote the set of all vertices of V'' which are not adjacent to any vertices of S_i. Set for convenience $W_0 = V''$. Furthermore, let B_i, $i = 0, \ldots, t$, be the set of all vertices of $V' \setminus S_i$ with fewer than $\rho(H)|W_i|$ neighbors in W_i, where, recall, $\rho(H) = \rho = M/\binom{2n}{2}$. Note that $d_{\rho,H}(V', V'') = (2n-1)/n \ge 1.5$ for $n \ge 2$.

Suppose that $|W_t| > \varepsilon n$, and thus $|W_i| > \varepsilon n$ for each $0 \le i \le t$. Then, for each $0 \le i \le t$, $|B_i| < \varepsilon n$, since otherwise the pair (B_i, W_i) with its density

$d_{\rho,H}(B_i, W_i)$ smaller than 1 would yield a contradiction with the assumption that the pair (V', V'') is sparsely (H, ε)-regular. Hence, for each $0 \le i \le t-1$,

$$\mathbb{P}(s_{i+1} \in B_i) = \frac{|B_i|}{n-i} < \frac{\varepsilon n}{n-t} \le 1.02\varepsilon.$$

The supposition that $|W_t| > \varepsilon n$ has also another consequence. Let us consider how many times the event $\{s_{i+1} \in B_i\}$ holds. Each time we choose s_{i+1} outside B_i, the size of W_i is decreased by at least $\rho(H)|W_i| > \rho\varepsilon n$, that is, $|W_{i+1}| - |W_i| \le -\rho(H)|W_i| < -\rho\varepsilon n$. This means, however, that the event $\{|W_t| > \varepsilon n\}$ implies that $\{s_{i+1} \in B_i\}$ holds at least $t/2$ times, since otherwise $|W_t| - |W_0| < -\frac{t}{2}\rho\varepsilon n < -\varepsilon tM/4n \le -n$, yielding a contradiction. The probability that $\{s_{i+1} \in B_i\}$ holds for at least $t/2$ indices $i = 1, \ldots, t$ can be bounded from above by

$$\binom{t}{\lceil t/2 \rceil}(1.02\varepsilon)^{\lceil t/2 \rceil} < (4.08\varepsilon)^{t/2} \le \varepsilon^{t/3},$$

because $\varepsilon < (4.08)^{-3}$. Consequently, $\mathbb{P}(|W_t| > \varepsilon n) \le \varepsilon^{t/3}$. ∎

Proof of Lemma 8.30. For given $\varepsilon > 0$, n and M let $\mathcal{A}_{i,j}$ be the event that the pair (V_i, V_j) is sparsely $(\mathbb{G}_3(n, M), \varepsilon)$-regular and let $\mathcal{A} = \mathcal{A}_{1,2} \cap \mathcal{A}_{1,3} \cap \mathcal{A}_{2,3}$. Further, let $\mathcal{K}$ be the event that $\mathbb{G}_3(n, M) \not\supseteq K_3$. Hence the event that $\mathbb{G}_3(n, M)$ is a triangle-free $(n, M/n^2, \varepsilon)$-triplet is $\mathcal{K} \cap \mathcal{A}$.

Let $\mathcal{D}$ denote the event that at least $n/2$ vertices of V_1 each have more than $t = \lceil M/2n \rceil$ neighbors in both V_2 and V_3. It can be easily verified (Exercise!) that the conjunction $\mathcal{A}_{1,2} \cap \mathcal{A}_{1,3}$ implies $\mathcal{D}$ with plenty of room to spare. Thus

$$\mathbb{P}(\mathcal{K} \cap \mathcal{A}) \le \mathbb{P}(\mathcal{K} \cap \mathcal{D} \cap \mathcal{A}_{2,3}).$$

Denote by $N_i(v)$ the set of all neighbors of a vertex $v \in V_1$ which belong to V_i, $i = 2, 3$, and consider the random vectors $\mathbf{D}_i = (|N_i(v)| : v \in V_1)$, $i = 2, 3$. Let Λ be the set of all pairs of integer vectors of length n and sum M, such that at some $\lceil n/2 \rceil$ coordinates the entries of both vectors are greater than t. Furthermore, let $\mathbb{G}_3(n, M)[V_2, V_3]$ be the subgraph of $\mathbb{G}_3(n, M)$ induced by V_2 and V_3, and let $\mathcal{H}$ be the set of all bipartite graphs with vertex set (V_2, V_3) which satisfy property $\mathcal{A}_{2,3}$. Then, by the law of total probability,

$$\begin{aligned}\mathbb{P}(&\mathcal{K} \cap \mathcal{D} \cap \mathcal{A}_{2,3}) \\ &= \sum_{(\mathbf{s}_2, \mathbf{s}_3) \in \Lambda} \sum_{H \in \mathcal{H}} \mathbb{P}(\mathcal{K} \mid \{\mathbb{G}_3(n, M)[V_2, V_3] = H\} \cap \{\mathbf{D}_2 = \mathbf{s}_2\} \cap \{\mathbf{D}_3 = \mathbf{s}_3\}) \\ &\qquad \times \mathbb{P}(\{\mathbf{D}_2 = \mathbf{s}_2\} \cap \{\mathbf{D}_3 = \mathbf{s}_3\})\, \mathbb{P}(\mathbb{G}_3(n, M)[V_2, V_3] = H).\end{aligned}$$

Clearly, to complete the proof it is enough to show that

$$\mathbb{P}(\mathcal{K} \mid \{\mathbb{G}_3(n, M)[V_2, V_3] = H\} \cap \{\mathbf{D}_2 = \mathbf{s}_2\} \cap \{\mathbf{D}_3 = \mathbf{s}_3\}) \le \varepsilon^{M/16}$$

for every triple $\mathbf{s}_2, \mathbf{s}_3, H$. Fix one such triple. For a vertex $v \in V_1$, let $\mathcal{M}_v$ be the event that $e_H(N_2(v), N_3(v)) = 0$. The event $\mathcal{K}$ implies $\bigcap_{v \in V_1} \mathcal{M}_v$, which, in turn, implies $\bigcap_{v \in \bar{V}_1} \mathcal{M}_v$, where $\bar{V}_1$ is the subset of those vertices of V_1 for which $s_2(v), s_3(v) \geq t$.

Now observe that with all the degrees fixed the choices of neighborhoods $N_2(v)$ and $N_3(v)$ are independent of each other for all $v \in V_1$. Moreover, for any $t_1, t_2 \geq t$,

$$\mathbb{P}(e_H(S'_{t_1}, S''_{t_2}) = 0) \leq \mathbb{P}(e_H(S'_t, S''_t) = 0),$$

where S'_{t_1} and S''_{t_2} are random sets defined as in Lemma 8.31 but of sizes t_1 and t_2, respectively. Hence, by Lemma 8.31,

$$\begin{aligned}
&\mathbb{P}\big(\mathcal{K} \mid \{\mathbb{G}_3(n, M)[V_2, V_3] = H\} \cap \{\mathbf{D}_2 = \mathbf{s}_2\} \cap \{\mathbf{D}_3 = \mathbf{s}_3\}\big) \\
&\quad \leq \mathbb{P}\Big(\bigcap_{v \in \bar{V}_1} \mathcal{M}_v \mid \{\mathbb{G}_3(n, M)[V_2, V_3] = H\} \cap \{\mathbf{D}_2 = \mathbf{s}_2\} \cap \{\mathbf{D}_3 = \mathbf{s}_3\}\Big) \\
&\quad = \prod_{v \in \bar{V}_1} \mathbb{P}(e_H(S'_{s_2(v)}, S''_{s_3(v)}) = 0) \leq \mathbb{P}(e_H(S'_t, S''_t) = 0)^{|\bar{V}_1|} \leq \left(\varepsilon^{t/4}\right)^{n/2} \\
&\quad \leq \varepsilon^{M/16}.
\end{aligned}$$

■

Proof of Theorem 8.24

The last ingredient of the proof of Theorem 8.24 is the following lemma.

Lemma 8.32. *For every $\eta > 0$ there exists $\varepsilon = \varepsilon(\eta) > 0$ such that for every $\xi > 0$ there exists $C = C(\eta, \varepsilon, \xi) < \infty$ such that if $M \geq Cn^{3/2}$, then* a.a.s. *every $(\xi n, \eta p_M, \varepsilon)$-triplet contained in $\mathbb{G}(n, M)$ contains a triangle.*

Proof. In view of Corollary 8.29, it suffices to prove that *a.a.s.* every exact $(\xi n, \eta p_M, \varepsilon)$-triplet contained in $\mathbb{G}(n, M)$ contains a triangle. We will do this with $\varepsilon < 0.01$ so small that $\delta = 64\varepsilon^{\eta/16} < 1$, and with $C = (\eta\sqrt{\xi\varepsilon/2})^{-1}$.

For any given ℓ with $\xi n \leq \ell \leq n/3$, let V_1, V_2, V_3 be three disjoint subsets of $[n]$ such that $|V_1| = |V_2| = |V_3| = \ell$. Let $\mathbb{G}(n, M)[V_1, V_2, V_3]$ be the tripartite subgraph of $\mathbb{G}(n, M)$ induced by V_1, V_2 and V_3 and let $\mathbb{G}_3(\ell; K_{12}, K_{13}, K_{23})$ be the random tripartite graph with vertex set $V_1 \cup V_2 \cup V_3$ and K_{ij} edges between the sets V_i and V_j, $1 \leq i < j \leq 3$. Furthermore, let $M' = M'(\ell) = \lceil \eta p_M \ell^2 \rceil$.

As a guideline for the forthcoming estimates, note that there are at most 8^n choices of V_1, V_2, V_3. Observe also that for any property $\mathcal{P}$ the probability that $\mathbb{G}_3(\ell; K_{12}, K_{13}, K_{23})$ contains a spanning subgraph satisfying $\mathcal{P}$ which has M' edges across each of the three pairs (V_1, V_2), (V_1, V_3) and (V_2, V_3), can be bounded from above by the probability that $\mathbb{G}_3(\ell, M') \in \mathcal{P}$ multiplied by $\binom{K_{12}}{M'}\binom{K_{13}}{M'}\binom{K_{23}}{M'}$ (Exercise!). Due to our choice of C, $M'/\ell^{3/2} \geq \eta p_M \ell^{1/2} > 2\eta C(\ell/n)^{1/2} = (8/\varepsilon)^{1/2}$, and thus we may apply Lemma 8.30 to $\mathbb{G}_3(\ell, M')$.

Consequently, for any $K_{12}, K_{13}, K_{23} \leq 2\rho_M \ell^2$ and sufficiently large n,

$$\begin{aligned}
&\mathbb{P}(\mathbb{G}(n,M)[V_1,V_2,V_3] \text{ contains a triangle-free exact } (\ell,\eta\rho_M,\varepsilon)\text{-triplet} \\
&\qquad\qquad\qquad\qquad \mid e_{\mathbb{G}(n,M)}(V_i,V_j) = K_{ij},\ 1 \leq i < j \leq 3) \\
&= \mathbb{P}(\mathbb{G}_3(n; K_{12}, K_{13}, K_{23}) \text{ contains a triangle-free exact } (\ell,\eta\rho_M,\varepsilon)\text{-triplet}) \\
&\leq \binom{K_{12}}{M'}\binom{K_{13}}{M'}\binom{K_{23}}{M'} \mathbb{P}(\mathbb{G}_3(\ell, M') \text{ is a triangle-free } (\ell,\eta\rho_M,\varepsilon)\text{-triplet}) \\
&\leq 2^{K_{12}+K_{13}+K_{23}} \varepsilon^{M'/16} \leq 2^{6\rho_M \ell^2} \varepsilon^{\eta\rho_M \ell^2/16} = \delta^{\rho_M \ell^2}.
\end{aligned}$$

Moreover, owing to Theorem 2.10, the probability that the random tripartite graph $\mathbb{G}(n,M)[V_1,V_2,V_3]$ has more than $2\rho_M\ell^2$ edges in any of the three bipartite graphs it forms, can be bounded from above by $3e^{-3\rho_M\ell^2/8}$. Hence, by the law of total probability,

$$\begin{aligned}
&\mathbb{P}(\mathbb{G}(n,M)[V_1,V_2,V_3] \text{ contains a triangle-free exact } (\ell,\eta\rho_M,\varepsilon)\text{-triplet}) \\
&\qquad\qquad\qquad \leq \delta^{\rho_M \ell^2} + 3e^{-3\rho_M \ell^2/8}.
\end{aligned}$$

Finally, summing over all $\ell \geq \xi n$ and subsets V_1, V_2, V_3,

$$\begin{aligned}
&\mathbb{P}(\mathbb{G}(n,p) \text{ contains a triangle-free exact } (\xi n,\eta\rho_M,\varepsilon)\text{-triplet}) \\
&\leq \sum_{\ell=\xi n}^{n/3} 8^n \left(\delta^{\rho_M \ell^2} + 3e^{-3\rho_M \ell^2/8}\right) \leq n 8^n \left(\delta^{2M\xi^2} + 3e^{-3M\xi^2/4}\right) = o(1).
\end{aligned}$$

■

Proof of Theorem 8.24. The theorem now follows from Corollary 8.27 and Lemma 8.32, by first choosing $\varepsilon = \varepsilon(0.1\eta)$ as in Lemma 8.32, then $\xi = \xi(\eta,\varepsilon)$ as in Corollary 8.27, and finally $C = C(0.1\eta,\varepsilon,\xi)$ as in Lemma 8.32 again. ■

The argument we have just presented is more involved than the method based on Goodman's idea described in Section 8.2, but it has two big advantages. First, it has more potential for generalizations; for example, one can easily observe that it can be modified to give a new proof of the 1-statement of Theorem 8.1 in the case $G = K_3$ (Exercise!). More importantly, there is some hope that it can be generalized to show extremal results for graphs other than triangles – see the last subsection of this chapter. Second, the same method can give some information on the structure of maximal triangle-free subgraphs of a random graph.

Triangle-free subgraphs

A structural strengthening of Mantel's result is the Stability Theorem (see, e.g., Bollobás (1978, p. 340) and Simonovits (1983)), which states not only that the Turán graph (i.e., the balanced complete bipartite graph) maximizes the number of edges in a triangle-free graph, but that every triangle-free graph with the number of edges close to $n^2/4$ looks very much like the Turán graph. A precise statement of the Stability Theorem goes as follows.

Proposition 8.33. *For every $\alpha > 0$ there exists $\beta > 0$ such that every triangle-free graph with n vertices and at least $n^2/4 - \beta n^2$ edges can be turned into the Turán graph by adding and/or deleting at most αn^2 edges.* ∎

Now, let us assume that H is a triangle-free subgraph of $\mathbb{G}(n, M)$ with $M \geq Cn^{3/2}$ and $e_H \geq M/2$, say. As in Lemma 8.26 we apply to H the sparse version of the Szemerédi Regularity Lemma (Lemma 8.19) for some small $\varepsilon > 0$ and very large m. In such a way we obtain a partition $\Pi = (V_0, V_1, \ldots, V_k)$, in which at least $k^2/4 - \beta k^2$ out of the $\binom{k}{2}$ pairs of sets $V_1, \ldots, V_k$ are "good", that is, are sparsely (H, ε)-regular and contain a fair number of edges. On the other hand in such a partition we *a.a.s.* do not find a triplet since, as we have shown in Lemma 8.32, *a.a.s.* each triplet which is contained in $\mathbb{G}(n, M)$ contains a triangle, provided the random graph is dense enough. Thus, the auxiliary "partition" graph, with vertices representing the sets $V_1, \ldots, V_k$ and edges between the good pairs, is triangle-free and, owing to Proposition 8.33, has a "bipartite-like" structure. This, in turn, implies that H itself must be bipartite-like as well. Although the technical details behind the above idea are not very appealing, we believe that the reader can convert it into a rigorous argument and show the following result (Advanced Exercise!).

Theorem 8.34. *For every constant $\eta > 0$ there exists $C = C(\eta)$ such that for every $M \geq Cn^{3/2}$* a.a.s. *each triangle-free subgraph of $\mathbb{G}(n, M)$ with at least $M/2$ edges can be made bipartite by omitting at most ηM edges.* ∎

Note also that *a.a.s.* the random graph $\mathbb{G}(n, M)$ contains no bipartite subgraph with more than, say, $M/2 + n \log n$ edges (Exercise!), and thus Theorem 8.34 immediately implies Theorem 8.24.

It is, maybe, worthwhile to mention that the threshold function for the property that the largest triangle-free subgraph of a graph is bipartite has not yet been found. The remark made at the end of Section 8.2 suggests that the threshold function $\widehat{M}$ for this property (if it exists) satisfies $\widehat{M} \gg Cn^{3/2}$. On the other hand, Babai, Simonovits and Spencer (1990) proved that for $M \sim n^2/4$, the largest triangle-free subgraph of $\mathbb{G}(n, M)$ is *a.a.s.* bipartite.

A stronger conjecture

We conclude this section with some comments on Conjecture 8.11 for graphs G other than triangles. As we have already observed, unlike the proof from Section 8.2, the argument presented in this section can be easily used for any graph G, provided one could show a result similar to Lemma 8.30. More specifically, let G be any graph with vertex set $\{1, 2, \ldots, k\}$ and $\mathbb{F}_G(n, M)$ be a graph chosen uniformly at random from all k-partite graphs F with vertex set $V_1 \cup \cdots \cup V_k$ such that $|V_1| = \cdots = |V_k| = n$, and $e_F(V_i, V_j) = M$ for all

$1 \le i < j \le k$ for which $\{i, j\}$ is an edge of G and $e_F(V_i, V_j) = 0$ otherwise. Let $\mathcal{A}$ be the event that each pair (V_i, V_j) is sparsely $(\mathbb{F}_G(n, M), \varepsilon)$-regular and let $\mathcal{G}$ be the event that $\mathbb{F}_G(n, M)$ contains no copy of G. Then the main probabilistic problem concerning extremal properties of random graphs (and, we believe, one of the most important open questions in the theory of random graphs) is to verify the following conjecture of Kohayakawa, Łuczak and Rödl (1997).

Conjecture 8.35. For every graph G and every $\alpha > 0$ there exist constants $\varepsilon = \varepsilon(G, \alpha)$ and $C = C(G, \alpha)$ such that

$$\mathbb{P}(\mathcal{G} \cap \mathcal{A}) \le \alpha^M,$$

provided $M \ge Cn^{2-1/m^{(2)}(G)}$.

It is not hard to see that this conjecture would imply Theorem 8.1 and, coupled with the Erdős–Stone–Simonovits Theorem, would also settle in the affirmative Conjecture 8.11, even in its stronger, extremal version analogous to Theorem 8.34. Unfortunately, at this moment we are unable to prove it for general G. One can easily see that the statement holds trivially when G is a tree. Füredi (1994), following ideas of Kleitman and Winston (1982), proved that a slightly weaker statement holds for cycles of length four, even if we replace $\mathbb{F}_G(n, M)$ by $\mathbb{G}(n, M)$. Haxell, Kohayakawa and Łuczak (1995) and Kohayakawa, Kreuter and Steger (1998) extended this result to every cycle of even length. A somewhat weaker version of Conjecture 8.35, which, however, is sufficient for showing Conjecture 8.11, was proved by Kreuter (1997) for odd cycles, and by Kohayakawa, Łuczak and Rödl (1997) for $G = K_4$. It seems that with a substantial amount of work, methods from Haxell, Kohayakawa and Łuczak (1995) and Kohayakawa, Łuczak and Rödl (1997) can be used to prove Conjecture 8.35 for some other graphs, for example, the complete bipartite graphs $K_{2,\ell}$. The two Kuratowski graphs, $K_{3,3}$ and K_5, are the smallest instances of G for which Conjecture 8.35 remains open.

9

Random Regular Graphs

A *regular* graph is a graph with the same degree (i.e., number of edges) at each vertex; we say that the graph is *r-regular* if the common degree is r. Traditionally, a 3-regular graph is also called *cubic*.

Note that an r-regular graph with n vertices has $rn/2$ edges; hence rn has to be even. Moreover, $r \leq n-1$. Conversely, it is easily seen that there exist r-regular graphs with n vertices whenever rn is even and $n > r \geq 0$.

We define the *random r-regular graph* $\mathbb{G}(n,r)$ to be a random graph with the uniform distribution over all r-regular graphs on n given vertices, say $[n]$. We assume, whenever talking about $\mathbb{G}(n,r)$, that rn is even and $n > r$.

In this chapter we will study random r-regular graphs for a fixed $r \geq 1$, letting the number of vertices n tend to infinity. (If r is odd, we tacitly assume n to be even.) The reader may think of r as quite small, for example, 3 or 4. (Smaller values of r are too simple, and are exceptions to many of the results below.) We will here not consider the case $r \to \infty$ (as some function of n), which is rather different.

As we will see below, random r-regular graphs turn out to have properties quite different from the two basic models $\mathbb{G}(n,p)$ (the binomial random graph) and $\mathbb{G}(n,M)$ (the uniform random graph). For example, they are sparse but typically connected.

This chapter is to a large extent based on Janson (1995b), which also contains similar results for random regular directed graphs; see further, for example, Cooper, Frieze and Molloy (1994). For further results, including several topics not covered here, see Frieze and Łuczak (1992) (chromatic number) and the recent survey by Wormald (1999b).

Example 9.1. The cases $r = 0$ and $r = 1$ are trivial. A 0-regular graph is an empty graph, and a 1-regular graph is the same as a perfect matching, that is, a set of disjoint edges covering all the vertices. The random graph $\mathbb{G}(n, 1)$ is thus obtained by randomly choosing one of the $(n-1)!! = n!/2^{n/2}(n/2)!$ partitions of the vertex set into $n/2$ pairs.

Example 9.2. Also the case $r = 2$ is rather simple. It is obvious that every component in a 2-regular graph is a cycle, so $\mathbb{G}(n, 2)$ is obtained by a random partition of the vertex set into cycles of length at least 3. This is similar to the cycle decomposition of a permutation, although there the cycles are oriented. Hence, the study of $\mathbb{G}(n, 2)$ becomes similar to the study of random permutations.

The definition of random regular graphs that we have given is conceptually simple, but it is not so easy to use. (For example, there is no simple formula for the total number of r-regular graphs on n vertices, so we do not even know the probability of obtaining a given r-regular graph. An asymptotic formula was given by Bender and Canfield (1978); see Corollary 9.8 below.) Of course, we may obtain $\mathbb{G}(n, r)$ from $\mathbb{G}(n, p)$ (for any $p \in (0, 1)$, but $p = r/n$ seems natural) or $\mathbb{G}(n, M)$ (with $M = rn/2$) by conditioning on the graph being r-regular, but the probability of this event is exponentially small and this procedure is not very useful.

Fortunately, there is an efficient way to generate $\mathbb{G}(n, r)$ that is useful both for theoretical studies as here and (for small r) for the practical problem of constructing random regular graphs in simulations. This is the *configuration model*, which will be described in the next section.

Note also that there are other natural ways to generate regular graphs at random. For example, we may construct r independent random perfect matchings of the n given vertices (n even) and take the union of them, conditioning on the event that there are no multiple edges. Another possibility is to take $\lfloor r/2 \rfloor$ independent Hamilton cycles on the vertices, together with a perfect matching in the case in which r is odd.

A different approach is to construct the graph sequentially by adding edges one by one, at each step randomly (uniformly) choosing between all remaining edges that do not increase any vertex degree above r. (It was shown by Ruciński and Wormald (1992) that this process *a.a.s.* leads to an r-regular graph.)

It should be emphasized that these constructions do *not* give uniformly distributed r-regular graphs. However, the resulting distributions are interesting in their own right. Moreover, it has been proved in several cases (and conjectured in others) that the distribution is not too far from the uniform one, in the sense that properties holding *a.a.s.* for one of the distributions also hold *a.a.s.* for the other; this, which is expressed by saying that the distributions are *contiguous*, will be studied in Sections 9.5 and 9.6.

9.1 THE CONFIGURATION MODEL

Most work on random regular graphs is based on the following construction, due in different versions to Bender and Canfield (1978) and Bollobás (1980, 1985). We will use Bollobás's version, which has become standard.

Given a set V, which is to be the vertex set of the graph, we associate disjoint r-element sets to the elements in V. In order to be specific, let n and r be positive integers (with rn even), take $V = [n]$, and consider the set $W = [n] \times [r]$; A *configuration* is a partition of W into $rn/2$ pairs; these pairs are called *edges* of the configuration and the points in W are called *half-edges.* The natural projection of the set W onto $V = [n]$ (ignoring the second coordinate) projects each configuration F to a multigraph $\pi(F)$ on V. Note that $\pi(F)$ may contain loops (arising from edges in F between two half-edges corresponding to the same vertex in V) and multiple edges (arising from sets of two or more edges in F whose endpoints correspond to the same pair of vertices in V). Thus $\pi(F)$ is, in general, not a simple graph. Note, however, that $\pi(F)$ is an r-regular multigraph (with multiple edges and loops counted in the natural way). In particular, if $\pi(F)$ lacks loops and multiple edges, it is an r-regular graph.

There are $(rn-1)!! = (rn)!/2^{rn/2}(rn/2)!$ different configurations on W (Exercise!). When we talk about a *random configuration*, we will always (unless we explicitly specify a different distribution) mean a configuration chosen at random, uniformly among all possibilities. Each r-regular graph on V is the projection of the same number of configurations (*viz.* $r!^n$), and it follows that if we take the projection $\pi(F)$ of a random configuration and condition on it being a simple graph, we obtain a random r-regular graph on V with the uniform distribution over all such graphs (Exercise!). This is thus a construction of the random graph $\mathbb{G}(n, r)$.

It turns out that it is often advantageous to allow loops and multiple edges and work with r-regular multigraphs, if necessary afterwards conditioning to simple graphs. We thus define the *random r-regular multigraph* $\mathbb{G}^*(n, r)$ to be the multigraph $\pi(F)$ obtained from a random configuration F. Note that $\mathbb{G}^*(n, r)$ does *not* have the uniform distribution over all r-regular multigraphs on V, because different multigraphs arise from different numbers of configurations. In fact, it is easily seen that the probability of obtaining a given multigraph is proportional to a weight consisting of the product of a factor $1/2$ for each loop and a factor $1/j!$ for each multiple edge of multiplicity j. (In particular, the weight for any simple graph is 1, in accordance with the fact stated above that the conditional distribution over the r-regular simple graphs is uniform.)

Remark 9.3. It may be shown that the distribution of $\mathbb{G}^*(n, r)$ is contiguous to the uniform distribution, in the sense of Section 9.5; see Janson (1995b).

An important feature of the configuration model is that the probability of obtaining a simple graph, $\mathbb{P}(\mathbb{G}^*(n, r)$ is simple) is bounded below by some positive number (depending on r) for all $n > r$; more precisely, as will be shown in Section 9.2, the probability converges to $\exp(-(r^2-1)/4)$ as $n \to \infty$. Hence, the probability that an event occurs for the random r-regular graph $\mathbb{G}(n,r)$ is bounded by a constant times the probability that the event occurs for $\mathbb{G}^*(n,r)$, which equals the probability that a corresponding event occurs for a random configuration. In particular, any event holding *a.a.s.* for $\mathbb{G}^*(n,r)$ also holds *a.a.s.* for $\mathbb{G}(n,r)$ (Exercise!).

It follows directly from the definitions that the probability that any given set of k disjoint edges on W is contained in a random configuration is given by

$$p_k = \frac{(rn-2k-1)!!}{(rn-1)!!} = \frac{1}{(rn-1)(rn-3)\cdots(rn-2k+1)}. \tag{9.1}$$

This will be used repeatedly in the sequel. In particular, we will use the following asymptotical results.

Lemma 9.4.

(i) *If $m \geq 2$ is even, then*

$$(m-1)!! = \sqrt{2}m^{m/2}e^{-m/2}(1+O(1/m)).$$

(ii) *If k is fixed and $n \to \infty$, then*

$$p_k \sim (rn)^{-k}.$$

(iii) *If $rn - 2k \to \infty$, then*

$$p_k \sim n^{-k}e^k(r-2k/n)^{rn/2-k}r^{-rn/2}.$$

Proof. (i) follows from $(m-1)!! = m!/2^{m/2}(m/2)!$ and Stirling's formula, (ii) follows directly from (9.1), and (iii) follows from (9.1) and (i) (Exercise!). ∎

9.2 SMALL CYCLES

The foundation of our results for random r-regular graphs is the study of the numbers of small cycles, due to Bollobás (1980, 1985) and Wormald (1981b).

Given a (multi)graph G, we let $Z_k = Z_k(G)$ denote the number of cycles of length k in G. Here for simple graphs we let $k = 3, 4, \dots$, but for multigraphs we let $k = 1, 2, \dots$, where Z_1 is the number of loops and Z_2 is the number of pairs of parallel edges. Note that a multigraph is simple if and only if $Z_1 = Z_2 = 0$.

Taking the graph G to be our random graph $\mathbb{G}(n,r)$ or $\mathbb{G}^*(n,r)$, Z_k becomes a random variable. We then have the following theorem. By joint convergence of an infinite number of variables we mean joint convergence of every finite subset, which is equivalent to convergence in $\mathbb{R}^\infty$.

Theorem 9.5. *Let $\lambda_k = \frac{1}{2k}(r-1)^k$ and let $Z_{k\infty} \in \text{Po}(\lambda_k)$ be independent Poisson distributed random variables, $k = 1, 2, 3, \ldots$ Then the random variables $Z_k(\mathbb{G}^*(n,r))$ converge in distribution to $Z_{k\infty}$, $Z_k(\mathbb{G}^*(n,r)) \xrightarrow{d} Z_{k\infty}$ as $n \to \infty$, jointly for all k.*

Proof. We use the method of moments, more precisely Theorem 6.10. We begin by computing the expectation of $Z_k = Z_k(\mathbb{G}^*(n,r))$. Each k-cycle in $\mathbb{G}^*(n,r)$ arises from a set of k edges in the corresponding configuration, such that the endpoints of the edges match properly when they are projected to V; with a slight abuse of language, we call such a set of k edges a k-cycle on W. Let a_k be the number of possible k-cycles on W. The probability that a given one of them is contained in a random configuration is p_k given in (9.1), and thus $\mathbb{E}\, Z_k = a_k p_k$.

In order to calculate a_k, we consider oriented cycles, with a specified initial vertex and a specified direction, and note that each (unoriented) k-cycle corresponds to $2k$ oriented ones. Hence the number of oriented k-cycles on W is $2ka_k$. Moreover, an oriented k-cycle on W consists of k edges that can be written $\big((v_i, x_i), (v_{i+1}, y_{i+1})\big)$ (with indices taken modulo k), and it is thus described by a sequence of k distinct vertices $v_1, \ldots, v_k \in V$ and, for each $i = 1, \ldots, k$, two distinct indices $p_i, q_i \in [r]$. This description is unique, and thus

$$2ka_k = (n)_k\big(r(r-1)\big)^k. \tag{9.2}$$

For fixed k and r we thus have, as $n \to \infty$, $a_k \sim \frac{1}{2k} n^k r^k (r-1)^k$ and, by Lemma 9.4(ii), $p_k \sim (rn)^{-k}$. Consequently,

$$\mathbb{E}\, Z_k = a_k p_k \sim \frac{1}{2k}(r-1)^k = \lambda_k.$$

In other words, $\mathbb{E}\, Z_k \to \lambda_k$ as $n \to \infty$, for each $k \geq 1$.

Before proceeding, let us note that the same argument shows that if H is any (multi)graph with v vertices and e edges, then the expected number of copies of H in $\mathbb{G}^*(n,r)$ is $O(n^v) \cdot O(n^{-e}) = O(n^{v-e})$ (Exercise!). In particular, if H is a graph with more edges than vertices, that is, $v < e$, the expected number of copies of H is $O(n^{-1})$.

Next, we compute factorial moments. We begin with $\mathbb{E}(Z_k)_2$ and indicate the small modifications needed in the general case later.

Note that $(Z_k)_2$ is the number of ordered pairs of two distinct k-cycles in $\mathbb{G}^*(n,r)$. The two k-cycles may or may not intersect, and we write $(Z_k)_2 = Y' + Y''$ where Y' is the number of ordered pairs of (vertex) disjoint k-cycles, while Y'' is the number of ordered pairs of distinct k-cycles having at least one common vertex.

The number Y'' may be further decomposed according to the number of common vertices and edges, and their relative positions. This expresses Y'' as the sum of a number of terms Y_j'', where the number of terms depends on k but not on n, and each term counts the number of copies of some (multi)graph H_j in $\mathbb{G}^*(n,r)$, where H_j (being the union of two distinct cycles with at least one common vertex) is connected and has more than one cycle. Each such H_j has more edges than vertices and thus, by the result just shown, $\mathbb{E}\,Y_j'' = O(n^{-1})$ for each j. Summing over all j, we finally obtain

$$\mathbb{E}\,Y'' = O(n^{-1}).$$

Hence the main term is $\mathbb{E}\,Y'$, which we compute in the same way as $\mathbb{E}\,Z_k$. If a_{kk} denotes the number of ordered pairs of possible k-cycles on W that project to disjoint cycles on V, we obtain as above, orienting both cycles,

$$(2k)^2 a_{kk} = (n)_{2k}\bigl(r(r-1)\bigr)^{2k} \sim (2ka_k)^2.$$

The probability that a given pair of disjoint k-cycles on W is contained in a random configuration is $p_{2k} \sim (rn)^{-2k} \sim p_k^2$, and consequently,

$$\mathbb{E}\,Y' = a_{kk}p_{2k} \sim (a_k p_k)^2 \sim \lambda_k^2.$$

$$\mathbb{E}(Z_k)_2 = \mathbb{E}\,Y' + \mathbb{E}\,Y'' = \lambda_k^2 + o(1).$$

The same argument applies to any factorial moment $\mathbb{E}(Z_k)_m$, and more generally to any joint factorial moment $\mathbb{E}(Z_1)_{m_1}(Z_2)_{m_2}\cdots(Z_l)_{m_l}$; we now consider the number of sequences of $m_1 + m_2 + \cdots + m_l$ distinct cycles such that the first m_1 have length 1, the next m_2 have length 2, etc. (Here $l \geq 1$ and $m_1, \ldots, m_l \geq 0$ are any fixed integers.) As before we write this number as $Y' + Y''$, where Y' counts the sequences of disjoint cycles, and split Y'' further according to the pattern of intersection of the cycles into a sum of terms Y_j'', each counting the number of copies of some graph H_j in $\mathbb{G}^*(n,r)$. The graphs H_j that appear here are unions of cycles, each having at least one component with more than one cycle, and it is easy to see that each H_j has more edges than vertices. Hence, we again obtain $\mathbb{E}\,Y_j'' = O(n^{-1})$ and $\mathbb{E}\,Y'' = O(n^{-1})$ (Exercise!).

For Y', the same argument as above yields $\mathbb{E}\,Y' \sim \lambda_1^{m_1}\lambda_2^{m_2}\cdots\lambda_l^{m_l}$, and summing we obtain

$$\mathbb{E}(Z_1)_{m_1}(Z_2)_{m_2}\cdots(Z_l)_{m_l} \to \lambda_1^{m_1}\lambda_2^{m_2}\cdots\lambda_l^{m_l}. \tag{9.3}$$

By Theorem 6.10, this implies that the joint distribution of $Z_1, \ldots, Z_l$ converges to the joint distribution of $Z_{1\infty}, \ldots, Z_{l\infty}$, which completes the proof. (Note that the right-hand side of (9.3) equals $\mathbb{E}(Z_{1\infty})_{m_1}(Z_{2\infty})_{m_2}\cdots(Z_{l\infty})_{m_l}$.) ∎

Recall that $\mathbb{G}^*(n,r)$ is simple if and only if $Z_1 = Z_2 = 0$, and that $\mathbb{G}^*(n,r)$ conditioned on $Z_1 = Z_2 = 0$ yields $\mathbb{G}(n,r)$.

Corollary 9.6. *Let λ_k and $Z_{k\infty}$ be as in Theorem 9.5. Then the random variables $Z_k(\mathbb{G}(n,r))$ converge in distribution to $Z_{k\infty}$, $Z_k(\mathbb{G}(n,r)) \xrightarrow{d} Z_{k\infty}$ as $n \to \infty$, jointly for all $k \geq 3$.*

Proof. Directly from Theorem 9.5, conditioning on $Z_1 = Z_2 = 0$ (Exercise!). ∎

Corollary 9.7. *If $n \to \infty$, then*

$$\mathbb{P}(\mathbb{G}^*(n,r) \textit{ is simple}) \to e^{-(r^2-1)/4} > 0.$$

Proof. Theorem 9.5 yields

$$\mathbb{P}(Z_0 = Z_1 = 0) \to \mathbb{P}(Z_{0\infty} = Z_{1\infty} = 0) = e^{-\lambda_1 - \lambda_2},$$

where $\lambda_1 + \lambda_2 = \frac{1}{2}(r-1) + \frac{1}{4}(r-1)^2 = \frac{1}{4}(r^2-1)$. ∎

Corollary 9.8. *The number L_n of labelled r-regular graphs on n nodes satisfies, as $n \to \infty$ for fixed r,*

$$L_n \sim \sqrt{2}e^{-(r^2-1)/4}(r^{r/2}e^{-r/2}/r!)^n n^{rn/2}.$$

Proof. The number of configurations is $(rn-1)!! \sim \sqrt{2}(rn/e)^{rn/2}$, the proportion of them that yield simple graphs is $e^{-(r^2-1)/4} + o(1)$ by Corollary 9.7, and there are $r!^n$ such configurations corresponding to each r-regular graph. ∎

Theorem 9.9. *Any property that holds* a.a.s. *for $\mathbb{G}^*(n,r)$ holds* a.a.s. *for $\mathbb{G}(n,r)$ too.*

Proof. Suppose that the property $\mathcal{P}$ holds *a.a.s.* for $\mathbb{G}^*(n,r)$. Then Corollary 9.7 yields

$$\begin{aligned}
\mathbb{P}(&\mathbb{G}(n,r) \text{ does not have } \mathcal{P}) \\
&= \mathbb{P}(\mathbb{G}^*(n,r) \text{ does not have } \mathcal{P} \mid \mathbb{G}^*(n,r) \text{ is simple}) \\
&= \frac{\mathbb{P}(\mathbb{G}^*(n,r) \text{ does not have } \mathcal{P} \text{ and is simple})}{\mathbb{P}(\mathbb{G}^*(n,r) \text{ is simple})} \\
&\leq \frac{\mathbb{P}(\mathbb{G}^*(n,r) \text{ does not have } \mathcal{P})}{\mathbb{P}(\mathbb{G}^*(n,r) \text{ is simple})} \to 0.
\end{aligned}$$

∎

The converse does not hold, as the trivial example of not containing a loop shows.

9.3 HAMILTON CYCLES

In the preceding section we studied the numbers of cycles of a given fixed length in a random regular graph. Let us now instead study Hamilton cycles, that is, cycles of length n (where n as usual is the number of vertices). We let $H(G)$ denote the number of Hamilton cycles in a (multi)graph G, and let $H_n = H(\mathbb{G}(n,r))$, $H_n^* = H(\mathbb{G}^*(n,r))$.

In the notation of Section 9.2, thus $H_n^* = Z_n$ and by (9.2) and (9.1),

$$\mathbb{E}\, H_n^* = a_n p_n = \frac{n!\,(r(r-1))^n}{2n}\frac{(rn-2n-1)!!}{(rn-1)!!}. \tag{9.4}$$

If $r=0$ or 1, then there is clearly never any Hamilton cycle at all in $\mathbb{G}(n,r)$ or $\mathbb{G}^*(n,r)$. If $r=2$, then (9.4) yields, using Stirling's formula and Lemma 9.4(i),

$$\mathbb{E}\, H_n^* = \frac{n!\,2^n}{2n(2n-1)!!} \sim \sqrt{\frac{\pi}{4n}}$$

and thus $\mathbb{E}\, H_n^* \to 0$ as $n \to \infty$. Hence, by the first moment method, there is *a.a.s.* no Hamilton cycle in $\mathbb{G}^*(n,2)$. By Theorem 9.9, there is also *a.a.s.* no Hamilton cycle in $\mathbb{G}(n,2)$.

Let us thus assume $r \geq 3$. Then (9.4), Stirling's formula and Lemma 9.4 yield

$$\begin{aligned}\mathbb{E}\, H_n^* &\sim \Big(\frac{\pi}{2n}\Big)^{1/2}\Big(\frac{r(r-1)n}{e}\Big)^n (r-2)^{rn/2-n} r^{-rn/2} e^n n^{-n}\\ &= \sqrt{\tfrac{\pi}{2}}\, n^{-1/2}\Big(\frac{(r-1)(r-2)^{r/2-1}}{r^{r/2-1}}\Big)^n.\end{aligned} \tag{9.5}$$

Moreover, it is easily verified that

$$A(r) = \frac{(r-1)(r-2)^{r/2-1}}{r^{r/2-1}} > 1, \qquad r \geq 3,$$

for example because $A(3) = 2/\sqrt{3}$, while for $r \geq 4$,

$$A(r) = (r-1)\Big(1+\frac{2}{r-2}\Big)^{-(r-2)/2} \geq (r-1)e^{-1} > 1.$$

Consequently, $\mathbb{E}\, H_n^* \to \infty$ as $n \to \infty$, which suggests (but does not prove) that $\mathbb{G}^*(n,r)$ typically has lots of Hamilton cycles when $r \geq 3$.

In order to study the existence of Hamilton cycles further, we calculate the variance of H_n^*. This is considerably more involved than the calculation just given for the expectation, and we give for the moment only the result (Frieze, Jerrum, Molloy, Robinson and Wormald 1996), postponing the proof to the next section.

Lemma 9.10. *If $r \geq 3$, then*

$$\mathbb{E}(H_n^*)^2 \sim \frac{r}{r-2}\frac{\pi}{2n}A(r)^{2n} \sim \frac{r}{r-2}(\mathbb{E}\, H_n^*)^2 \tag{9.6}$$

and thus

$$\operatorname{Var}(H_n^*)/(\mathbb{E}\, H_n^*)^2 \to \frac{2}{r-2}.$$

Recall that if X_n is any sequence of random variables with $\mathbb{E}\,X_n > 0$ and $\operatorname{Var}(X_n)/(\mathbb{E}\,X_n)^2 \to 0$, or equivalently $\mathbb{E}\,X_n^2/(\mathbb{E}\,X_n)^2 \to 1$, then Chebyshev's inequality yields $X_n/\mathbb{E}\,X_n \xrightarrow{p} 1$ and, in particular, $\mathbb{P}(X_n > 0) \to 1$; see (1.2) and Section 3.1. For H_n^* this fails, but just barely; $\operatorname{Var}(H_n^*)/(\mathbb{E}\,H_n^*)^2$ converges to a positive number. This suggests that $H_n^*/\mathbb{E}\,H_n^*$ converges in distribution to a nondegenerate distribution, and we will soon show that this indeed is the case.

We have so far considered H_n^*, but the situation is the same for H_n; we will later show the corresponding results (Robinson and Wormald 1994):

$$\mathbb{E}\,H_n \sim e\,\mathbb{E}\,H_n^* \sim e\sqrt{\tfrac{\pi}{2}}n^{-1/2}A(r)^n, \tag{9.7}$$

$$\mathbb{E}\,H_n^2/(\mathbb{E}\,H_n)^2 \sim e^{-2/(r-1)}\frac{r}{r-2} \tag{9.8}$$

and thus

$$\operatorname{Var}(H_n)/(\mathbb{E}\,H_n)^2 \to e^{-2/(r-1)}\frac{r}{r-2} - 1. \tag{9.9}$$

Again, the second moment method fails. For example, for $r = 3$, the inequality (3.2) yields $\limsup \mathbb{P}(H_n = 0) \le \lim \operatorname{Var}(H_n)/(\mathbb{E}\,H_n)^2 = \frac{3}{e} - 1 \approx 0.104$, while the sharper inequality (3.4), using $\mathbb{P}(H_n > 0) \ge (\mathbb{E}\,H_n)^2/\mathbb{E}\,H_n^2 \to e/3$, yields $\limsup \mathbb{P}(H_n = 0) \le 1 - \lim(\mathbb{E}\,H_n)^2/\mathbb{E}\,H_n^2 = 1 - \frac{e}{3} \approx 0.094$. This shows that Hamilton cycles appear in at least 90% of the realizations of $\mathbb{G}(n,3)$ when n is large, but this estimate is not sharp. Indeed, as we will show below, $\mathbb{G}(n,r)$ *a.a.s.* has a Hamilton cycle for any $r \ge 3$. This was generally believed for a long time, and explicitly conjectured by Bollobás (1981b); partial results were given by various methods by Bollobás (1983) ($r \ge 10^7$), Fenner and Frieze (1984) ($r \ge 796$) and Frieze (1988) ($r \ge 85$), but the general result was not proved until the papers by Robinson and Wormald (1992, 1994).

The main idea in the proof by Robinson and Wormald is to condition on the number of small cycles, compute the conditional variance which turns out to be rather small, and use Chebyshev's inequality for the conditioned variables. This is thus a conditioned version of the second moment method. The argument can be regarded as an *analysis of variance*; the main point is that most of the total variance is explained by the variance between the groups, whence the variance within the groups is relatively small; see Cooper, Frieze, Molloy and Reed (1996) for further comments (and results).

Remark 9.11. One can similarly study random *bipartite* regular graphs. It turns out that in this case $\mathbb{E}\,H_n^2/(\mathbb{E}\,H_n)^2 \to 1$ and thus the second moment method applies, at least provided $r = 3$ (Robinson and Wormald 1984). This probably holds also for $r \ge 4$, but as far as we know no one has yet verified this. The fact that *a.a.s.* there exists a Hamilton cycle in a random bipartite regular graph with $r \ge 4$ was proved in Robinson and Wormald (1994) by the method indicated in Remark 9.39, below.)

We state a general theorem (Janson 1995b) that will be applicable also in several similar situations below. In our applications we let $X_{in} = Z_{in}$, the number of cycles of length i in a random regular (multi)graph. The proof, which is based on the argument in Robinson and Wormald (1992) and related to the projection methods discussed in Section 6.4, will be given in the next section. We define $0^0 = 1$.

Theorem 9.12. *Let $\lambda_i > 0$ and $\mu_i \geq 0$, $i = 1, 2, \ldots$, be constants, let $\delta_i = \mu_i/\lambda_i - 1 \geq -1$, and suppose that for each n there are random variables X_{in}, $i = 1, 2, \ldots$, and Y_n (defined on the same probability space), such that X_{in} is non-negative integer valued and $\mathbb{E}\, Y_n \neq 0$ (at least for large n) and, furthermore, the following conditions are satisfied:*

(A1) *$X_{in} \xrightarrow{d} X_{i\infty}$ as $n \to \infty$, jointly for all i, where $X_{i\infty} \in \mathrm{Po}(\lambda_i)$ are independent Poisson random variables;*

(A2) *For any finite sequence $x_1, \ldots, x_m$ of non-negative integers,*

$$\frac{\mathbb{E}(Y_n \mathbf{1}[X_{1n} = x_1, \ldots, X_{mn} = x_m])}{\mathbb{E}\, Y_n} \to \prod_{i=1}^{m} \frac{\mu_i^{x_i}}{x_i!} e^{-\mu_i} \qquad \text{as } n \to \infty;$$

(A3) $\sum_i \lambda_i \delta_i^2 < \infty$;

(A4) $$\frac{\mathbb{E}\, Y_n^2}{(\mathbb{E}\, Y_n)^2} \to \exp\Big(\sum_i \lambda_i \delta_i^2\Big) \qquad \text{as } n \to \infty.$$

Then

$$\frac{Y_n}{\mathbb{E}\, Y_n} \xrightarrow{d} W = \prod_{i=1}^{\infty} (1 + \delta_i)^{X_{i\infty}} e^{-\lambda_i \delta_i} \qquad \text{as } n \to \infty; \tag{9.10}$$

moreover, this and the convergence in (A1) *hold jointly. Furthermore, the normalized variables $Y_n / \mathbb{E}\, Y_n$ are uniformly square integrable, and $\mathbb{E}\, W^2 = \lim_{n\to\infty} \mathbb{E}\, Y_n^2 / (\mathbb{E}\, Y_n)^2$.*

The infinite product defining W converges, so W is well defined. This is not obvious, so we state it together with some useful properties as the next theorem. The proof is postponed to the next section.

Theorem 9.13. *Suppose that $\lambda_i > 0$ and $\delta_i \geq -1$ and $X_{i\infty}$ are as in Theorem 9.12 and that* (A3) *holds. Then the infinite product defining W in* (9.10) *converges a.s. and in L^2, $\mathbb{E}\, W = 1$ and $\mathbb{E}\, W^2 = \exp(\sum_i \lambda_i \delta_i^2)$. Furthermore, the event $W > 0$ equals, up to a set of probability zero, the event that $X_{i\infty} > 0$ for some i with $\delta_i = -1$. In particular, $W > 0$ a.s. if and only if $\delta_i > -1$ for every i.*

Note that a.s. here and below is used in the standard sense "with probability 1"; see Remark 1.2.

Remark 9.14. The values of λ_i, μ_i and δ_i are prescribed by (A1) and (A2). If $\sup_n \mathbb{E} X_{in}^2 < \infty$ for each i, as is the case in all our applications, then $\{X_{in}\}_n$ and $\{X_{in}Y_n/\mathbb{E} Y_n\}_n$ are uniformly integrable (for fixed i) and we obtain (by Theorem 9.12 or directly from (A1) and (A2))

$$\mathbb{E} X_{in} \to \mathbb{E} X_{i\infty} = \lambda_i$$
$$\mathbb{E}(X_{in}Y_n/\mathbb{E} Y_n) \to \mathbb{E} X_{i\infty}W = \mathbb{E} X_{i\infty}(1+\delta_i)^{X_{i\infty}}e^{-\lambda_i\delta_i} = \lambda_i(1+\delta_i) = \mu_i$$

and thus

$$\lambda_i = \lim_{n\to\infty} \mathbb{E} X_{in}$$
$$\mu_i = \lim_{n\to\infty} \mathbb{E}(X_{in}Y_n)/\mathbb{E} Y_n$$
$$\delta_i = \lim_{n\to\infty} \mathbb{E}(X_{in}Y_n)/(\mathbb{E} X_{in}\,\mathbb{E} Y_n) - 1 = \lim_{n\to\infty} \operatorname{Cov}(X_{in}, Y_n)/(\mathbb{E} X_{in}\,\mathbb{E} Y_n).$$

Remark 9.15. It follows from the proof that if we have several sequences of variables $Y_n^{(j)}$, each satisfying the conditions of Theorem 9.12 (possibly with different $\delta_i^{(j)}$) with the same variables X_{in}, then the $Y_n^{(j)}$ converge jointly. It follows, for example, by the uniform square integrability, that $\mathbb{E} Y_n^{(1)}Y_n^{(2)}/(\mathbb{E} Y_n^{(1)}\,\mathbb{E} Y_n^{(2)}) \to \mathbb{E} W^{(1)}W^{(2)} = \exp\bigl(\sum_i \lambda_i\delta_i^{(1)}\delta_i^{(2)}\bigr)$.

Remark 9.16. If (A1) holds, then (A2) is easily seen to be equivalent to

(A2′) For any finite sequence $x_1, \ldots, x_m$ of non-negative integers,

$$\frac{\mathbb{E}(Y_n \mid X_{1n} = x_1, \ldots, X_{mn} = x_m)}{\mathbb{E} Y_n} \to \prod_{i=1}^{m}(1+\delta_i)^{x_i}e^{-\lambda_i\delta_i} \qquad \text{as } n \to \infty.$$

We will in all our applications of Theorem 9.12 verify condition (A2) by the method of moments argument used by Robinson and Wormald (1992). We state this step as a separate lemma.

Lemma 9.17. *Suppose that $Y_n \geq 0$ and define (when $\mathbb{E} Y_n > 0$) a new probability measure Q_n by $Q_n(A) = \mathbb{E}(Y_nI_A)/\mathbb{E} Y_n$ for every event A in the probability space Ω_n where X_{in} and Y_n are defined. Then* (A2) *is equivalent to*

(A2″) *Under the measures Q_n, $X_{in} \xrightarrow{d} \operatorname{Po}(\mu_i)$, jointly for all i with independent limits.*

In particular, (A2) *follows if*

(A2‴) $\mathbb{E}_{Q_n}\bigl((X_{in})_{j_1}\cdots(X_{mn})_{j_n}\bigr) \to \prod_1^m \mu_i^{j_i}$ *as $n \to \infty$, for every finite sequence $j_1, \ldots, j_m$ of non-negative integers.*

Proof. The first statement is clear by the definition of Q_n, and the second follows by the method of moments (Theorem 6.10). ■

Remark 9.18. (A2‴) can also be written

$$\mathbb{E}\big(Y_n(X_{in})_{j_1}\cdots(X_{mn})_{j_m}\big)/\,\mathbb{E}\,Y_n \to \prod_1^m \mu_i^{j_i},$$

avoiding explicit mention of Q_n (Exercise!).

For Hamilton cycles, the general theorem yields the following result; we postpone the proof.

Theorem 9.19. *Let $r \geq 3$ be fixed and let $H_n = H(\mathbb{G}(n,r))$ be the number of Hamilton cycles in a random r-regular graph on n vertices. Then*

$$\frac{H_n}{\mathbb{E}\,H_n} \xrightarrow{d} W = \prod_{\substack{i\geq 3\\ i\ odd}} \left(1 - \frac{2}{(r-1)^i}\right)^{Z_{i\infty}} e^{1/i}, \tag{9.11}$$

where $Z_{i\infty} \in \operatorname{Po}\Big(\frac{(r-1)^i}{2i}\Big)$ are independent Poisson random variables.

As a consequence, we obtain the existence result by Robinson and Wormald (1992, 1994).

Theorem 9.20. *If $r \geq 3$ is fixed, then $\mathbb{G}(n,r)$* a.a.s. *has a Hamilton cycle.*

Proof. The limiting random variable W in Theorem 9.19 is positive a.s. by Theorem 9.13, since $\delta_i = -2/(r-1)^i > -1$. Hence Theorem 9.19 implies

$$\mathbb{P}(H_n \geq 1) = \mathbb{P}\big(H_n/\,\mathbb{E}\,H_n > 0\big) \to \mathbb{P}(W > 0) = 1. \qquad \blacksquare$$

An immediate consequence of the existence of a Hamilton cycle is that the graph contains a perfect matching (provided n is even).

Corollary 9.21. *If $r \geq 3$ is fixed and n is even, then $\mathbb{G}(n,r)$* a.a.s. *contains a perfect matching.* ■

Remark 9.22. The existence of a Hamilton cycle also proves 2-connectivity *a.a.s.* of $\mathbb{G}(n,r)$ for $r \geq 3$. (Again, this is false if $r \leq 2$.) However, this is not the historical path. Indeed, connectivity was proved much earlier and by much simpler methods, which also show the stronger result that $\mathbb{G}(n,r)$ *a.a.s.* is r-connected (Bollobás 1981b, 1985, Wormald 1981a); see also Łuczak (1992).

The limiting distribution in Theorem 9.19 is more complicated than the ones that usually appear in the theory of random graphs. The Poisson variables $Z_{i\infty}$ that appear have, however, a simple interpretation as the numbers of small cycles in the random regular graph. More precisely, see Theorem 9.12, if we let Z_{in} be the number of cycles of length i in the graph, then $Z_{in} \xrightarrow{d} Z_{i\infty}$ as $n \to \infty$, and this convergence in distribution holds jointly, for all i, together with $H_n/\mathbb{E}\,H_n \xrightarrow{d} W$. Hence the theorem can be interpreted as saying that $H_n/\mathbb{E}\,H_n$ is, with large probability, well approximated by the infinite product with $Z_{i\infty}$ replaced by the small cycle count Z_{in}, or rather by a finite product $\prod_{k=1}^{K}\left(1-2/(r-1)^{2k+1}\right)^{Z_{2k+1,n}}e^{1/(2k+1)}$. (We have to let $n\to\infty$ first and then $K\to\infty$ in order to get convergence, because the factors $e^{1/(2k+1)}$, which are necessary convergence factors in (9.11), act as divergence factors for any finite n.)

It should perhaps not be surprising that H_n thus essentially is determined by the numbers of small cycles. Similar results hold for the standard models $\mathbb{G}(n,p)$ and $\mathbb{G}(n,M)$, where for certain ranges of p and M, $\log H_n$ is well approximated by a linear function of the number of edges, and of the number of paths of length 2, respectively; see Janson (1994b). Those results are simpler, however, since $\log H_n$ then is asymptotically normal, while in the present case $\log H_n - \log\mathbb{E}\,H_n$ converges to $\log W$ which has a rather complicated infinitely divisible distribution. Moreover, for $\mathbb{G}(n,p)$ and $\mathbb{G}(n,M)$ it suffices to use one small subgraph count in the approximation, whereas here we need an infinite sequence. (There are parallels to this in $\mathbb{G}(n,p)$ for other functionals; see Section 6.4, Barbour, Janson, Karoński and Ruciński (1990) and Janson (1994a).)

Note that every small subgraph count for random regular graphs can be essentially expressed in terms of small cycle counts, since a random regular graph *a.a.s.* has no small multicyclic subgraphs. This explains why only cycle counts appear in Theorem 9.19 (as $Z_{i\infty}$), but it seems mysterious that only the odd cycle counts appear, while the even cycle counts are asymptotically independent of H_n. (Recall in this connection that for random bipartite cubic graphs, where there are no odd cycles at all, $\mathbb{E}\,H_n^2/(\mathbb{E}\,H_n)^2\to 1$; see Remark 9.11.)

We may extend Theorem 9.19 to multigraphs.

Theorem 9.23. *Let $r\geq 3$ be fixed and let $H_n^*=H(\mathbb{G}^*(n,r))$ be the number of Hamilton cycles in a random r-regular multigraph on n vertices. Then*

$$\frac{H_n^*}{\mathbb{E}\,H_n^*}\xrightarrow{d} W=\prod_{\substack{i\geq 1\\ i\ odd}}\left(1-\frac{2}{(r-1)^i}\right)^{Z_{i\infty}}e^{1/i}, \tag{9.12}$$

where $Z_{i\infty}\in\operatorname{Po}\left(\frac{(r-1)^i}{2i}\right)$ are independent Poisson random variables.

As above, the variables $Z_{i\infty}$ are the limits of the cycle counts Z_{in}.

Note that the limit distribution for multigraphs in Theorem 9.23 differs from the one for graphs in Theorem 9.19 only by the additional factor $\left(1-\frac{2}{r-1}\right)^{Z_{1\infty}}e$, where $Z_{1\infty}\in\operatorname{Po}\left(\frac{r-1}{2}\right)$ should be interpreted as (the limit of) the number of 1-cycles, or loops, in the multigraph.

If $r\geq 4$, so that $1-\frac{2}{r-1}>0$, then the limit random variable W is a.s. strictly positive also in Theorem 9.23, and thus the theorem implies that a random r-regular multigraph *a.a.s.* contains a Hamilton cycle. On the other hand, if $r = 3$, then $1 - \frac{2}{r-1} = 0$, and thus $W = 0$ when $Z_{1\infty} > 0$ but, as before, $W > 0$ a.s. when $Z_{1\infty} = 0$. Consequently, it follows from the joint convergence of (H_n^*, Z_{1n}) to $(W, Z_{1\infty})$ that a random cubic multigraph *a.a.s.* is Hamiltonian if and only if it lacks loops, in the sense that as $n \to \infty$, the probability tends to 0 that a random cubic multigraph has one of these properties but not the other. (One implication is obvious for all cubic multigraphs; the other holds only asymptotically.)

Remark 9.24. The joint convergence of (H_n^*, Z_{1n}, Z_{2n}) and the fact that $\mathbb{P}(Z_{1\infty} = Z_{2\infty} = 0) > 0$ imply that the limit in Theorem 9.23 remains true if we condition on $Z_{1n} = Z_{2n} = 0$, that is, if we let $(X \mid A)$ denote the random variable X conditioned on the event A,

$$\left(\frac{H_n^*}{\mathbb{E}\,H_n^*} \,\middle|\, Z_{1n} = Z_{2n} = 0\right) \xrightarrow{d} (W \mid Z_{1\infty} = Z_{2\infty} = 0)$$

$$\stackrel{\mathrm{d}}{=} e \prod_{\substack{i \geq 3 \\ i \text{ odd}}} \left(1 - \frac{2}{(r-1)^i}\right)^{Z_{i\infty}} e^{1/i}.$$

The left-hand side is the distribution of the number $H_n = H(\mathbb{G}(n,r))$ of Hamilton cycles in a random regular graph, normalized by the expectation $\mathbb{E}\,H_n^*$ for the random multigraph. Since the uniform integrability of $H_n^*/\mathbb{E}\,H_n^*$ (see Theorem 9.12) survives the conditioning, taking the expectation yields

$$\mathbb{E}\,H_n/\mathbb{E}\,H_n^* = \mathbb{E}(H_n^* \mid Z_{1n} = Z_{2n} = 0)/\mathbb{E}\,H_n^* \to \mathbb{E}(W \mid Z_{1\infty} = Z_{2\infty} = 0) = e,$$

and we recover Theorem 9.19 as a corollary to Theorem 9.23.

Remark 9.25. We can use the conditioning argument in Remark 9.24 in all cases where Theorem 9.12 applies. We obtain, if $Y_n' = (Y_n \mid X_{1n} = X_{2n} = 0)$,

$$\mathbb{E}\,Y_n'/\mathbb{E}\,Y_n \to \mathbb{E}(W \mid X_{1\infty} = X_{2\infty} = 0) = \exp(-\lambda_1\delta_1 - \lambda_2\delta_2) \tag{9.13}$$

and similarly

$$\mathbb{E}(Y_n')^2/(\mathbb{E}\,Y_n)^2 \to \mathbb{E}(W^2 \mid X_{1\infty} = X_{2\infty} = 0)$$

$$= \exp(-2\lambda_1\delta_1 - 2\lambda_2\delta_2 + \sum_{i=3}^{\infty} \lambda_i\delta_i^2);$$

combining these we obtain

$$\mathbb{E}(Y_n')^2/(\mathbb{E}\,Y_n')^2 \to \exp\Big(\sum_{i=3}^{\infty} \lambda_i\delta_i^2\Big) = \exp(-\lambda_1\delta_1^2 - \lambda_2\delta_2^2)\,\mathbb{E}\,W^2. \tag{9.14}$$

For the Hamilton cycles in Theorem 9.23 we have $\delta_1 = -2/(r-1)$, $\delta_2 = 0$ and $\lambda_1 = (r-1)/2$, and we see that (9.13) and (9.14) yield (9.7) and (9.8) from (9.5) and (9.6).

9.4 PROOFS

We prove the assertions in the previous section.

Proof of Theorem 9.13. We begin by observing that, since $X_{i\infty} \in \mathrm{Po}(\lambda_i)$, $\mathbb{E}(1+\delta_i)^{X_{i\infty}} = e^{\lambda_i(1+\delta_i)-\lambda_i} = e^{\lambda_i\delta_i}$ and $\mathbb{E}\big((1+\delta_i)^{X_{i\infty}}\big)^2 = e^{\lambda_i(1+\delta_i)^2-\lambda_i} = e^{\lambda_i(2\delta_i+\delta_i^2)}$. Thus, defining

$$W^{(m)} = \prod_1^m (1+\delta_i)^{X_{i\infty}} e^{-\lambda_i\delta_i},$$

$W^{(m)}$ is a product of independent variables with mean 1, and hence the sequence $(W^{(m)})_{m=1}^{\infty}$ is a martingale; *cf.* Section 2.4 and, for example, Chung (1974, Chapter 9). The martingale is L^2-bounded because

$$\mathbb{E}(W^{(m)})^2 = \prod_1^m e^{\lambda_i\delta_i^2} = \exp\Big(\sum_1^m \lambda_i\delta_i^2\Big) \to \exp\Big(\sum_1^\infty \lambda_i\delta_i^2\Big) < \infty \quad \text{as } m\to\infty.$$

Thus, by the martingale convergence theorem (see, e.g., Chung (1974, Section 9.4)), the limit $W = \lim_{m\to\infty} W^{(m)}$ exists a.s. and in L^2, with $\mathbb{E}\,W = \lim_{m\to\infty}\mathbb{E}\,W^{(m)} = 1$ and $\mathbb{E}\,W^2 = \exp\big(\sum_1^\infty \lambda_i\delta_i^2\big)$. We observe for future use that

$$\mathbb{E}\,|W - W^{(m)}|^2 = \mathbb{E}\,W^2 - \mathbb{E}(W^m)^2 = \exp\Big(\sum_1^\infty \lambda_i\delta_i^2\Big) - \exp\Big(\sum_1^m \lambda_i\delta_i^2\Big). \tag{9.15}$$

In order to show that $W \neq 0$ a.s., except when $X_{i\infty} > 0$ for some i with $\delta_i = -1$, let us break the product defining W into two parts $W_1 = \prod_{\delta_i<-1/2}$ and $W_2 = \prod_{\delta_i\ge -1/2}$. For W_1 we observe that although there may conceivably be infinitely many indices in $I_1 = \{i : \delta_i < -1/2\}$, Condition (A3) implies that $\mathbb{E}\sum_{I_1} X_{i\infty} = \sum_{I_1}\lambda_i \le 4\sum_{I_1}\lambda_i\delta_i^2 < \infty$, and similarly $\sum_{I_1}|\lambda_i\delta_i| < \infty$. Hence there are a.s. only finitely many non-zero $X_{i\infty}$, $i \in I_1$, and $W_1 = e^{-\sum_{I_1}\lambda_i\delta_i}\prod_{I_1}(1+\delta_i)^{X_{i\infty}}$, where the product really is a finite product which is positive unless some factor vanishes, that is, unless $X_{i\infty} > 0$ for some i with $\delta_i = -1$.

For W_2 we define $\widetilde{\delta}_i = -\delta_i/(1+\delta_i)$, $i \in I_2 = \{i : \delta_i \ge -\frac{1}{2}\}$, and note that $\sum_{I_2}\lambda_i\widetilde{\delta}_i^2 < \infty$ and thus the argument above shows that

$$\widetilde{W}_2 = \prod_{I_2}(1+\widetilde{\delta}_i)^{X_{i\infty}} e^{-\lambda_i\widetilde{\delta}_i}$$

converges a.s. with $\widetilde{W}_2 < \infty$. However, since $(1+\delta_i)(1+\tilde{\delta}_i) = 1$,

$$W_2\widetilde{W}_2 = \prod_{I_2} e^{-\lambda_i(\delta_i+\widetilde{\delta}_i)} = \exp\Big(-\sum_{I_2}\lambda_i\delta_i^2/(1+\delta_i)\Big) > 0,$$

so $W_2 > 0$ a.s., which completes the proof that $W > 0$ a.s. except when some factor vanishes. ■

Proof of Theorem 9.12. For this proof we make, without loss of generality, two simplifying assumptions. First we may assume $\mathbb{E}\, Y_n = 1$. Secondly, we invoke a theorem by Skorokhod (1956) which implies that although the variables X_{in} and Y_n originally may be defined on different probability spaces for different n, and $X_{i\infty}$ on yet another probability space, we may replace them by other variables having the same distributions, such that they become defined on a single probability space and the convergence in (A1) holds a.s., i.e. $X_{in} \to X_{i\infty}$ a.s. as $n \to \infty$ for each i.

Fix a large integer m and define the functions

$$f_n(x_1, \dots, x_m) = \mathbb{E}(Y_n \mid X_{1n} = x_1, \dots, X_{mn} = x_m)$$
$$f_\infty(x_1, \dots, x_m) = \lim_{n\to\infty} f_n(x_1, \dots, x_m) = \prod_1^m (1+\delta_i)^{x_i} e^{-\lambda_i \delta_i},$$

where we used Assumption (A2) (in the form (A2′)). Consider the random variable

$$Y_n^{(m)} = \mathbb{E}(Y_n \mid X_{1n}, \dots, X_{mn}) = f_n(X_{1n}, \dots, X_{mn}).$$

By our simplifying assumption $X_{in} \to X_{i\infty}$ a.s., we have a.s. $X_{in} = X_{i\infty}$ for all $i \le m$ and sufficiently large n; thus

$$\lim_{n\to\infty} Y_n^{(m)} = \lim_{n\to\infty} f_n(X_{1\infty}, \dots, X_{m\infty}) = f_\infty(X_{1\infty}, \dots, X_{m\infty}) = W^{(m)} \quad \text{a.s.,} \tag{9.16}$$

with $W^{(m)}$ as above. Hence, by Fatou's lemma

$$\liminf_{n\to\infty} \mathbb{E}(Y_n^{(m)})^2 \ge \mathbb{E}(W^{(m)})^2 = \exp\Big(\sum_1^m \lambda_i \delta_i^2\Big).$$

Consequently, since $Y_n^{(m)}$ is a conditional expectation of Y_n,

$$\begin{aligned}\limsup_{n\to\infty} \mathbb{E}\,|Y_n - Y_n^{(m)}|^2 &= \limsup_{n\to\infty} \big(\mathbb{E}\, Y_n^2 - \mathbb{E}(Y_n^{(m)})^2\big) \\ &\le \exp\Big(\sum_1^\infty \lambda_i \delta_i^2\Big) - \exp\Big(\sum_1^m \lambda_i \delta_i^2\Big).\end{aligned} \tag{9.17}$$

Using Chebyshev's inequality (1.2), (9.16), (9.17) and (9.15), we now obtain, for every $\varepsilon > 0$,

$$\begin{aligned}
&\limsup_{n\to\infty} \mathbb{P}(|Y_n - W| > 3\varepsilon) \\
&\quad \le \limsup_{n\to\infty} \mathbb{P}(|Y_n - Y_n^{(m)}| > \varepsilon) + \limsup_{n\to\infty} \mathbb{P}(|Y_n^{(m)} - W^{(m)}| > \varepsilon) \\
&\qquad\quad + \mathbb{P}(|W^{(m)} - W| > \varepsilon) \\
&\quad \le \varepsilon^{-2} \limsup_{n\to\infty} \mathbb{E}\,|Y_n - Y_n^{(m)}|^2 + 0 + \varepsilon^{-2}\,\mathbb{E}\,|W - W^{(m)}|^2 \\
&\quad \le 2\varepsilon^{-2}\left[\exp\Big(\sum_1^\infty \lambda_i \delta_i^2\Big) - \exp\Big(\sum_1^m \lambda_i \delta_i^2\Big)\right]. \qquad (9.18)
\end{aligned}$$

We now let $m \to \infty$, keeping ε fixed. The right-hand side of (9.18) tends to 0, so the left-hand side, which does not depend on m, has to vanish for each $\varepsilon > 0$, which proves $Y_n \xrightarrow{p} W$.

Finally, by (A4) and Theorem 9.13, $\mathbb{E}\,Y_n^2 \to \exp(\sum_i \lambda_i \delta_i^2) = \mathbb{E}\,W^2$, and this, together with the already established convergence in distribution, implies that the random variables Y_n are uniformly square integrable. ∎

Proof of Theorems 9.19 and 9.23. Most steps are the same for the random graphs and multigraphs, so we do both theorems together. In the proof we let $i \ge 1$ in the multigraph case but $i \ge 3$ in the graph case.

The assertion (A1), with $\lambda_i = \frac{1}{2i}(r-1)^i$, is Theorem 9.5 and Corollary 9.6.

For (A2) we use Lemma 9.17. The computation of factorial moments is done in Robinson and Wormald (1992) (at least for $\mathbb{G}(n, 3)$), using generating functions to keep track of the different possibilities for intersections of cycles. Here we will give a slightly different argument, using matrices. We restrict ourselves for the time being to random multigraphs, which (sometimes) are easier to handle than graphs.

The measure Q_n in Lemma 9.17 has in this case the following interpretation. Consider the set $\overline{\Omega}_n$ of all pairs $(\widetilde{G}, \tilde{H})$ of a configuration $\widetilde{G}$ (on the nr half-edges $[n]\times[r]$) and a set $\tilde{H}$ of edges in $\widetilde{G}$ that projects to a Hamilton cycle in the multigraph G^* obtained by projecting $\widetilde{G}$. Pick one of these pairs $(\widetilde{G}, \tilde{H})$ at random, uniformly over $\overline{\Omega}_n$, and take the projection G^*. This defines a random r-regular multigraph which has distribution Q_n, since the probability of obtaining a specific multigraph G^* is proportional to the number of pairs $(\widetilde{G}, \tilde{H}) \in \overline{\Omega}$ projecting to G^*, which is proportional to $H(G^*)$ (Exercise!). Furthermore, by symmetry, we obtain the same distribution Q_n by picking the pair $(\widetilde{G}, \tilde{H})$ at random in the subset $\overline{\Omega}'_n$ of all such pairs $(\widetilde{G}, \tilde{H})$ such that the edges in $\tilde{H}$ join half-edges in $V \times \{1, 2\}$. But these pairs are just the pairs $(\tilde{H} \cup \tilde{H}', \tilde{H})$, where $\tilde{H}$ is a configuration on $[n] \times \{1, 2\}$ projecting to a Hamilton cycle and $\tilde{H}'$ is any configuration on $[n] \times \{3, \dots, r\}$. Picking an element in $\overline{\Omega}'_n$ at random is thus the same as picking such $\tilde{H}$ and $\tilde{H}'$ independently at random, and after projecting we see that Q_n equals the distribution of the union of a random Hamilton cycle and a random $(r-2)$-regular multigraph $\mathbb{G}^*(n, r-2)$ (both defined on the same vertex set $[n]$).

We will, for simplicity, only verify

$$\mathbb{E}_{Q_n} Z_{kn} \to \mu_k \qquad \text{as } n \to \infty, \tag{9.19}$$

for a suitable μ_k; the extension to mixed higher factorial moments is routine, following the argument in the proof of Theorem 9.5 (Exercise!).

Fix $k \geq 1$ and assume $n > k$. There are $\frac{1}{2k}(n)_k \sim \frac{1}{2k}n^k$ k-cycles in the complete graph on $[n]$. For each such cycle C we fix an orientation and label the edges by $1, \ldots, k$, starting at an arbitrary edge. Consider a configuration $\tilde{H} \cup \tilde{H}'$, where as above $\tilde{H}$ is a configuration on $W_1 = [n] \times \{1, 2\}$ projecting to a Hamilton cycle and $\tilde{H}'$ is any configuration on $W_2 = [n] \times \{3, \ldots, r\}$. If $\tilde{C}$ is a set of k edges in $\tilde{H} \cup \tilde{H}'$ that projects to C, we say that an edge in $\tilde{C}$ is of *type 1* if it belongs to $\tilde{H}$ and *type 2* if it belongs to $\tilde{H}'$. Let us calculate the expected number of such "cycles" $\tilde{C}$ in $\tilde{H} \cup \tilde{H}'$ that project to C and have edges of types $i_1, \ldots, i_k$, where $(i_j)_1^k \in \{1, 2\}^k$ is a given sequence and the edges in $\tilde{C}$ are ordered according to the ordering in the projection C.

Let us say that the vertex in C between edges l and $l+1$ has type $i_l i_{l+1}$. There are two choices of half-edges in $\tilde{C}$ at a vertex of type 11 (the two possible orderings of the two corresponding half-edges in W_1), at a vertex of type 12 or 21 there are $2(r-2)$ choices (one half-edge from W_1 and one from W_2), and at a vertex of type 22 there are $(r-2)(r-3)$ choices (two different half-edges from W_2). Hence, if b_{ij} is the number of choices of the two half-edges in $\tilde{C}$ at a vertex of type ij, we have $b_{11} = 2$, $b_{12} = b_{21} = 2(r-2)$ and $b_{22} = (r-2)(r-3)$.

Each choice of half-edges over all k vertices specifies $\tilde{C}$ completely, and it remains to compute the probability that this $\tilde{C}$ actually is a subset of $\tilde{H} \cup \tilde{H}'$. Note that if there are k_1 indices l with $i_l = 1$, and thus $k_2 = k - k_1$ with $i_l = 2$, then each choice of half-edges specifies k_1 edges in $\tilde{H}$ and k_2 in $\tilde{H}'$. The probability that the latter k_2 occur is, by Lemma 9.4(ii), $\sim ((r-2)n)^{-k_2}$. For $\tilde{H}$ we observe that there are $C(n) = 2^n n!/2n = 2^{n-1}(n-1)!$ configurations that project to a Hamilton cycle. If we specify k_1 edges such that their projection is a union of paths, then the number of configurations containing these edges is $C(n-k_1) = 2^{n-k_1-1}(n-k_1-1)!$, since we may contract these edges and obtain a one-to-one correspondence with Hamilton cycles on a set of $n-k_1$ vertices; hence the probability that $\tilde{H}$ contains the given set of edges is asymptotically equal to $(2n)^{-k_1}$. On the other hand, a set of edges on W_1 whose projection contains a cycle of length smaller than n is never a subset of $\tilde{H}$.

Since $\tilde{H}$ and $\tilde{H}'$ are independent, it follows that the expected number of "cycles" $\tilde{C}$ in $\tilde{H} \cup \tilde{H}'$ that project to C and have edges of types $i_1, \ldots, i_k$ is asymptotically given by, letting $i_{k+1} = i_1$, and defining $r_1 = 2$, $r_2 = r-2$,

$$\prod_{l=1}^{k} b_{i_l i_{l+1}} (2n)^{-k_1} ((r-2)n)^{-k_2} = n^{-k} \prod_{l=1}^{k} b_{i_l i_{l+1}} r_{i_l}^{-1},$$

provided $k_2 > 0$; in the remaining case $i_1 = \cdots = i_k = 1$, the expected number is 0. If we set $a_{ij} = b_{ij}/r_i$, sum over all sequences $(i_j)_1^k$ and multiply by the number $\frac{1}{2k}(n)_k$ of k-cycles C, we obtain

$$\mathbb{E}_{Q_n} Z_{kn} \to \frac{1}{2k} \sum_{(i_l)\neq(1,\ldots,1)} \prod_{l=1}^{k} a_{i_l i_{l+1}} = \frac{1}{2k}\Big(\sum_{(i_l)} \prod_{l=1}^{k} a_{i_l i_{l+1}} - a_{11}^k\Big). \tag{9.20}$$

This equation is of the form (9.19), and we proceed to evaluate the right-hand side. We introduce the matrix

$$A = (a_{ij}) = \begin{pmatrix} 1 & r-2 \\ 2 & r-3 \end{pmatrix} \tag{9.21}$$

and can rewrite (Exercise!) (9.20) as

$$\mathbb{E}_{Q_n} Z_{kn} \to \frac{1}{2k}(\operatorname{Tr} A^k - 1). \tag{9.22}$$

It is easily verified (Exercise!) that the matrix A has eigenvalues $r-1$ and -1, for example by solving the characteristic equation $(1-\lambda)(r-3-\lambda)-2(r-2) = 0$; hence $\operatorname{Tr} A^k = (r-1)^k + (-1)^k$, and we have verified

$$\mathbb{E}_{Q_n} Z_{kn} \to \mu_k$$

with

$$\mu_k = \frac{1}{2k}\big((r-1)^k + (-1)^k - 1\big). \tag{9.23}$$

As claimed above, (A2‴) follows similarly (Exercise!), and (A2) follows by Lemma 9.17.

With this value of μ_k, we obtain

$$\delta_k = \frac{\mu_k}{\lambda_k} - 1 = \frac{(-1)^k - 1}{(r-1)^k} = \begin{cases} -2/(r-1)^k, & k \text{ odd}, \\ 0 & k \text{ even}. \end{cases} \tag{9.24}$$

(It is the fact that δ_k vanishes for even k that makes the number of Hamilton cycles asymptotically independent of the even cycle counts. The calculation above gives no intuitive explanation for this phenomenon.)

Having verified (A2) and thus (A2′) for the multigraph case, we note that the graph case follows immediately by specializing to $x_1 = x_2 = 0$, using, in particular,

$$\frac{\mathbb{E}\, H_n}{\mathbb{E}\, H_n^*} = \frac{\mathbb{E}(H_n^* \mid Z_{1n} = Z_{2n} = 0)}{\mathbb{E}\, H_n^*} \to e^{-\lambda_1\delta_1 - \lambda_2\delta_2} = e,$$

which we obtained in a different way in Remark 9.24.

The values of λ_i and δ_i yield

$$\sum_1^\infty \lambda_i\delta_i^2 = \sum_{i \text{ odd}} \frac{2}{i}(r-1)^{-i} = \log\Big(1+\frac{1}{r-1}\Big) - \log\Big(1-\frac{1}{r-1}\Big)$$
$$= \log\frac{r}{r-2},$$

and thus (for the graph case) $\sum_3^\infty \lambda_i\delta_i^2 = \log\frac{r}{r-2} - \frac{2}{r-1}$. Hence (A3) holds, and the variance condition (A4) says

$$\mathbb{E}\, H_n^2/(\mathbb{E}\, H_n)^2 \to \frac{r}{r-2}e^{-2/(r-1)} \tag{9.25}$$

in the graph case (this is (9.8)), and in the multigraph case

$$\mathbb{E}\, H_n^{*2}/(\mathbb{E}\, H_n^*)^2 \to \frac{r}{r-2}. \tag{9.26}$$

The latter result was stated in Lemma 9.10, which is proved below; this lemma thus completes the verification of the conditions in Theorem 9.12 for the multigraph case, and Theorem 9.23 follows.

The graph case, Theorem 9.19, follows from Theorem 9.23 by conditioning as we observed in Remark 9.24. Alternatively, Theorem 9.23 implies (9.25) by Remark 9.25, which verifies (A4) for the graph case; thus Theorem 9.19 follows by Theorem 9.12 (Exercise!). ■

Remark 9.26. For $r = 3$, (9.25) is proved directly in Robinson and Wormald (1984). Robinson and Wormald (1992) use this case and they state the result for $r \geq 4$ (when the proof is more difficult) in Robinson and Wormald (1994).

It remains to prove the variance estimate Lemma 9.10; we follow essentially Frieze, Jerrum, Molloy, Robinson and Wormald (1996). (The estimate follows by a long, and not very illuminating, calculation. It would be desirable to find another simpler, or more conceptual proof.)

Proof of Lemma 9.10. We first obtain an exact formula for $\mathbb{E}(H_n^*)^2$ by a combinatorial argument, similar to the argument used above in the proof of (9.20). This formula ((9.28) below) is rather complicated, however, and a non-trivial analytic argument then is needed to complete the proof.

Consider the family of all pairs $(\tilde{H}, \tilde{H}')$ of two n-cycles on $W = [n] \times [r]$, that is, two sets of edges on W that project to Hamilton cycles on $[n]$; if $\tilde{G}$ is a random configuration, then

$$\mathbb{E}(H_n^*)^2 = \sum_{\tilde{H},\tilde{H}'} \mathbb{P}(\tilde{H} \cup \tilde{H}' \subseteq \tilde{G}).$$

If the intersection $\tilde{H} \cap \tilde{H}'$ consists of b edges on W, then, by (9.1),

$$\mathbb{P}(\tilde{H} \cup \tilde{H}' \subseteq \tilde{G}) = p_{2n-b} = (rn-4n+2b-1)!!/(rn-1)!!.$$

Hence, if a_n is the number of n-cycles $\tilde{H}$ on W, given by (9.2), and $N(b)$ is the number of n-cycles $\tilde{H}'$ that intersect a given $\tilde{H}$ in b edges on W (this number is clearly independent of $\tilde{H}$), then

$$\mathbb{E}(H_n^*)^2 = \sum_{b=0}^{n} a_n N(b) p_{2n-b}. \tag{9.27}$$

If $b = n$, then $N(b) = 1$ ($\tilde{H}' = \tilde{H}$).

If $1 \le b < n$, we decompose $N(b) = \sum_{a=1}^{b} N(b,a)$, where $N(b,a)$ is the number of n-cycles $\tilde{H}'$ such that the b edges on W in $\tilde{H} \cap \tilde{H}'$, when projected to $[n]$, form a disjoint paths. (These paths are subpaths of the Hamilton cycle H on $[n]$ obtained by projecting $\tilde{H}$.)

To compute $N(b,a)$, consider first the number of ways to choose the a paths in H. Fix an orientation of H. Choosing an initial vertex which begins one of the paths, we then can choose the successive lengths of the paths in $\binom{b-1}{a-1}$ ways, and the lengths of the gaps between them in $\binom{n-b-1}{a-1}$ ways. Further, there are n possible initial vertices, and a of them can be used for each set of a paths. Consequently, the number of sets of a paths with total length b is

$$\frac{n}{a}\binom{b-1}{a-1}\binom{n-b-1}{a-1} = \frac{an}{b(n-b)}\binom{b}{a}\binom{n-b}{a}.$$

Next, by collapsing the a paths into single vertices, we see that each *oriented* Hamilton cycle H' on $[n]$ containing these paths defines an oriented Hamilton cycle on $n-b$ vertices and, furthermore, an orientation of each of the a paths; this yields a bijection, and thus the number of such oriented Hamilton cycles H' equals

$$2^a(n-b-1)!$$

Finally, each such H' corresponds to

$$(r-2)^{n+a-b}(r-3)^{n-a-b}$$

oriented n-cycles on W, since the half-edges may be chosen in $(r-2)(r-3)$ ways at each of the $n-a-b$ vertices disjoint from the given paths, and in $r-2$ ways at each of the $2a$ endpoints of the paths. Ignoring the orientation yields an additional factor $\frac{1}{2}$, and multiplying the factors together we obtain

$$N(b,a) = \frac{an}{b(n-b)} 2^{a-1}(r-2)^{n+a-b}(r-3)^{n-a-b}(n-b-1)!\binom{b}{a}\binom{n-b}{a}.$$

The same formula holds for the case $b = 0$ too, in which case we set $a = 0$ and $N(0) = N(0,0)$, provided we interpret a/b as 1.

Consequently, we obtain from (9.27) and (9.4) the formula

$$\frac{\mathbb{E}(H_n^*)^2}{(\mathbb{E}\,H_n^*)^2} = \frac{1}{\mathbb{E}\,H_n^*} + \sum_{0\le a\le b<n} \frac{N(b,a)p_{2n-b}}{a_n p_n^2}$$

$$= \frac{1}{\mathbb{E}\,H_n^*} + \sum_{0\le a\le b<n} \frac{an^2}{b(n-b)^2} 2^a r^{-n}(r-1)^{-n}(r-2)^{n+a-b}(r-3)^{n-a-b}$$

$$\times \binom{b}{a}\binom{n-b}{a} \frac{(n-b)!}{n!} \frac{(rn-4n+2b-1)!!(rn-1)!!}{(rn-2n-1)!!^2}. \tag{9.28}$$

The second step of the proof is to estimate this double sum.

First consider the case $r=3$. In this case, because of the factor $(r-3)^{n-a-b}$ in (9.28), we only have to consider terms with $a+b=n$, and (9.28) simplifies to

$$\frac{\mathbb{E}(H_n^{*2})}{(\mathbb{E}\,H_n^*)^2} = \frac{1}{\mathbb{E}\,H_n^*} + \sum_{a=1}^{n/2} \frac{n^2}{a(n-a)} 2^a 6^{-n} \frac{(n-a)!\,(n-2a-1)!!\,(3n-1)!!}{(n-2a)!\,n!\,(n-1)!!^2}. \tag{9.29}$$

Using Stirling's formula, the terms in the sum can be written

$$\frac{1}{\sqrt{2\pi n}} \frac{n^2}{a\sqrt{n-a}\sqrt{n-2a+1}} e^{ng(a/n)} \Big(1 + O\Big(\frac{1}{n-2a+1}\Big)\Big), \tag{9.30}$$

where

$$g(x) = x\log 2 - \log 6 + (1-x)\log(1-x) - \tfrac{1}{2}(1-2x)\log(1-2x) + \tfrac{3}{2}\log 3.$$

Hence the sum is dominated by the terms where $g(a/n)$ is close to the maximum of g on $[0,1/2]$. We differentiate and find

$$g'(x) = \log 2 - \log(1-x) + \log(1-2x),$$

and thus the equation $g'(x_0) = 0$ has the unique root $x_0 = 1/3$ in $(0,1/2)$. Moreover,

$$g''(x_0) = \frac{1}{1-x_0} - \frac{2}{1-2x_0} = -\frac{9}{2} < 0$$

and it follows that x_0 is the unique maximum point of g in the closed interval $[0,1/2]$; the maximum value is thus

$$g(x_0) = \tfrac{1}{3}\log 2 - \log 6 + \tfrac{2}{3}\log\tfrac{2}{3} - \tfrac{1}{6}\log\tfrac{1}{3} + \tfrac{3}{2}\log 3 = 0,$$

and, consequently, $g(x) < 0$ for $x \neq x_0$.

Since $g(x_0) = g'(x_0) = 0$ and $g''(x_0) < 0$, a Taylor expansion shows that if $\delta > 0$ is small enough, then for some $c_1 > 0$

$$g(x) \leq -c_1(x - x_0)^2, \qquad |x - x_0| \leq 2\delta. \tag{9.31}$$

Furthermore, the set $[0, 1/2] \setminus (x_0 - \delta, x_0 + \delta)$ is compact, and g is continuous and negative there; hence, for some $c_2 > 0$, $g(x) \leq -c_2$ when $|x - x_0| \geq \delta$. It follows that the terms in (9.29) with $|a/n - x_0| > \delta$ are exponentially small, and rewriting the sum of the remaining terms as an integral, we obtain from (9.29) and (9.30)

$$\frac{\mathbb{E}(H_n^{*2})}{(\mathbb{E}\, H_n^*)^2} = \frac{1}{\sqrt{2\pi n}} \int_{n(x_0-\delta)}^{n(x_0+\delta)} h_n(\lfloor t \rfloor/n) e^{ng(\lfloor t \rfloor/n)}\, dt + o(1)$$

where, for $|x - x_0| \leq 2\delta$,

$$h_n(x) = \frac{n^2}{nx\sqrt{n - nx}\sqrt{n - 2nx + 1}}\Bigl(1 + O\Bigl(\frac{1}{n}\Bigr)\Bigr) = h(x) + O\Bigl(\frac{1}{n}\Bigr),$$

with

$$h(x) = \frac{1}{x\sqrt{1-x}\sqrt{1-2x}}.$$

The variable substitution $t = nx_0 + y\sqrt{n}$ yields

$$\frac{\mathbb{E}(H_n^{*2})}{(\mathbb{E}\, H_n^*)^2} = \frac{1}{\sqrt{2\pi n}} \int_{-\delta\sqrt{n}}^{\delta\sqrt{n}} h_n\bigl(x_0 + y/\sqrt{n} + O(1/n)\bigr) e^{ng(x_0 + y/\sqrt{n} + O(1/n))} dt + o(1).$$

For fixed y, $h_n\bigl(x_0 + y/\sqrt{n} + O(1/n)\bigr) \to h(x_0) = 2^{-1/2}9$ and, by a Taylor expansion using $g(x_0) = g'(x_0) = 0$ again,

$$ng\bigl(x_0 + y/\sqrt{n} + O(1/n)\bigr) = \tfrac{1}{2} g''(x_0) y^2 + o(1).$$

Consequently, the dominated convergence theorem, justified by (9.31), yields

$$\begin{aligned}\frac{\mathbb{E}(H_n^{*2})}{(\mathbb{E}\, H_n^*)^2} &\to \frac{1}{\sqrt{2\pi}} \int_{-\infty}^{\infty} h(x_0) e^{-\frac{1}{2}|g''(x_0)| y^2}\, dy \\ &= h(x_0)|g''(x_0)|^{-1/2} = 3,\end{aligned}$$

as asserted by (9.26).

The proof for $r \geq 4$ is similar, but requires a more complicated two-variable analysis, and we omit some details. First, using Stirling's formula, the terms in the sum in (9.28) may be written (Exercise!)

$$\frac{1}{2\pi n} h(a/n, b/n) e^{ng(a/n, b/n)} \Bigl(1 + O\Bigl(\frac{1}{\min(a, b - a, n - a - b) + 1}\Bigr)\Bigr),$$

where

$$\begin{aligned} g(x,y) = &\, x\log 2 - \log r - \log(r-1) + (1+x-y)\log(r-2) \\ &+ (1-x-y)\log(r-3) + y\log y + 2(1-y)\log(1-y) \\ &- (y-x)\log(y-x) - 2x\log x - (1-x-y)\log(1-x-y) \\ &+ \left(\tfrac{r}{2} - 2 + y\right)\log(r-4+2y) + \tfrac{r}{2}\log r - (r-2)\log(r-2) \end{aligned}$$

and (with a minor modification when $x = 0$, $x = y$ or $x + y = 1$)

$$h(x,y) = y^{-1/2}(1-y)^{-1}(y-x)^{-1/2}(1-x-y)^{-1/2}.$$

The partial derivatives of g are

$$\begin{aligned} \frac{\partial g}{\partial x} &= \log 2 + \log(r-2) - \log(r-3) + \log(y-x) - 2\log x + \log(1-x-y), \\ \frac{\partial g}{\partial y} &= -\log(r-2) - \log(r-3) + \log y - 2\log(1-y) - \log(y-x) \\ &\quad + \log(1-x-y) + \log(r-4+2y). \end{aligned}$$

It is easily verified (Exercise!) that $(x_0, y_0) = \big(2(r-2)/r(r-1), 2/r\big)$ is a critical point of g, and routine calculations show that $g(x_0, y_0) = 0$ and that the Hessian matrix $D^2 g$ of second derivatives is negative definite at (x_0, y_0), so that g has a local maximum there, and $\operatorname{Det}\big(D^2 g(x_0, y_0)\big) = r^3(r-1)^2/4(r-2)(r-3)$. Moreover, with further effort (Frieze, Jerrum, Molloy, Robinson and Wormald 1996), it can be shown that (x_0, y_0) is the unique global maximum point of g in the domain $\{(x,y) : 0 \le x \le y \le 1 - x\}$.

The proof is now completed similarly to the case $r = 3$. Substituting first $a = \lfloor xn \rfloor$, $b = \lfloor yn \rfloor$ and then $x = x_0 + z_1/\sqrt{n}$, $y = y_0 + z_2/\sqrt{n}$, one finally obtains, arguing as above,

$$\begin{aligned} \frac{\mathbb{E}(H_n^{*2})}{(\mathbb{E}\,H_n^*)^2} &\to \frac{1}{2\pi}\int_{-\infty}^{\infty}\int_{-\infty}^{\infty} h(x_0,y_0)e^{-\frac{1}{2}(z_1,z_2)D^2 g(x_0,y_0)(z_1,z_2)^t}\,dz_1\,dz_2 \\ &= h(x_0,y_0)\operatorname{Det}\big(D^2 g(x_0,y_0)\big)^{-1/2} \\ &= \frac{r}{r-2}, \end{aligned}$$

which verifies (9.26). ∎

Remark 9.27. A similar, but even more complicated, variance calculation for the number of cycles of length $l(n)$, where $l(n)$ is a given sequence with $l(n) \to \infty$ but $l(n) < n$ (with some margin), is given by Garmo (1999).

9.5 CONTIGUITY OF RANDOM REGULAR GRAPHS

We will in this section study different models of random regular (multi)graphs. Our main objective is to show that in many cases, different constructions yield random (multi)graphs that have distributions that are contiguous in the sense of Section 9.6; thus, although the distributions are different, they are not too different. (More precisely, two contiguous random objects have the same *qualitative* properties; anything that holds *a.a.s.* for one of them holds *a.a.s.* for the other, but quantitative properties such as (asymptotical) probabilities may differ. A formal definition of contiguity is given in Section 9.6, together with some properties used below.)

We denote contiguity of two random (multi)graphs $\mathbb{G}_n^{(1)}$ and $\mathbb{G}_n^{(2)}$ by $\mathbb{G}_n^{(1)} \approx \mathbb{G}_n^{(2)}$. Recall that this is an asymptotic property as $n \to \infty$.

Besides $\mathbb{G}(n,r)$ and $\mathbb{G}^*(n,r)$ defined earlier, we will in this section also study the intermediate $\mathbb{G}'(n,r)$, which we define as a random regular multigraph $\mathbb{G}^*(n,r)$ conditioned to have no loops; thus $\mathbb{G}'(n,r)$ may have multiple edges but no loops.

Moreover, we let $\mathbb{H}(n)$ denote a random Hamilton cycle on $[n]$; this is a random 2-regular graph, but it obviously has a different distribution than $\mathbb{G}(n,2)$ or $\mathbb{G}^*(n,2)$ (except for some very small n).

We will also study multigraphs obtained by adding several random regular (multi)graphs; we use the notation $+$ for the union of independent random (multi)graphs on the same vertex set $[n]$. Note that the union of simple graphs, in general, is not a simple graph, because double edges may arise. Hence we also define the "simple sum" $\oplus$ by defining $\mathbb{G}_1 \oplus \mathbb{G}_2$ to be $\mathbb{G}_1 + \mathbb{G}_2$ conditioned on being simple.

Example 9.28. $\mathbb{G}(n,1) + \mathbb{G}(n,1) + \mathbb{G}(n,1)$ denotes the random 3-regular multigraph obtained by taking independently three perfect matchings at random, while $\mathbb{G}(n,1) \oplus \mathbb{G}(n,1) \oplus \mathbb{G}(n,1)$ denotes the random 3-regular graph obtained by taking three disjoint perfect matchings at random.

After these preliminaries, we begin by observing that we have already studied $\mathbb{H}(n) + \mathbb{G}^*(n,r-2)$ in the proof of Theorem 9.23, where the distribution of this multigraph was denoted by Q_n. If we let P_n denote the distribution of $\mathbb{G}^*(n,r)$, then $dQ_n/dP_n = H_n^*/\,\mathbb{E}\,H_n^*$, and Proposition 9.49 and Theorem 9.23 show that P_n and Q_n are contiguous, provided the limit W in Theorem 9.23 satisfies $\mathbb{E}\,W = 1$ and $W > 0$ a.s. By Theorem 9.13, $\mathbb{E}\,W = 1$ always holds, while $W > 0$ a.s. if and only if $\delta_i > -1$ for all i; since $\delta_i = -2(r-1)^{-i}$, this holds for $r \geq 4$ but not for $r = 3$, in which case $\delta_1 = -1$. We have proved the following.

Theorem 9.29. *If $r \geq 4$, then the random multigraphs $\mathbb{H}(n) + \mathbb{G}^*(n,r-2)$ and $\mathbb{G}^*(n,r)$ are contiguous.* ∎

For $r = 3$, Theorem 9.29 does not hold; this can be seen directly because $\mathbb{H}(n) + \mathbb{G}^*(n,1)$ never contains any loop, while $\mathbb{G}^*(n,3)$ contains loops with positive limit probability. Indeed, this is the only obstacle, and we have the following substitute when $r = 3$.

Theorem 9.30. *The random cubic multigraphs $\mathbb{H}(n) + \mathbb{G}^*(n,1)$ and $\mathbb{G}'(n,3)$ are contiguous.*

Proof. Let P_n be the distribution of $\mathbb{G}'(n,3)$ and Q_n the distribution of $\mathbb{H}(n)+\mathbb{G}^*(n,1)$. Then $dQ_n/dP_n = H'/\mathbb{E}\,H'$, where $H' = H(\mathbb{G}'(n,3))$, the number of Hamilton cycles in $\mathbb{G}'(n,3)$. It follows from Theorem 9.23 by conditioning on $Z_{1n} = 0$, using the same argument as in Remark 9.24, that

$$\frac{H'_n}{\mathbb{E}\,H'_n} \xrightarrow{d} W' = \prod_{\substack{i\geq 2\\ i \text{ odd}}} \left(1 - \frac{2}{2^i}\right)^{Z_{i\infty}} e^{1/i}.$$

Since this limit variable satisfies $W' > 0$ a.s. and $\mathbb{E}\,W' = 1$, the result follows by Proposition 9.49. ∎

If we condition on the multigraphs being simple in the last two theorems, using Proposition 9.50(i), we obtain the following corollary for graphs. (Note that $r = 3$ is no longer special.)

Corollary 9.31. *If $r \geq 3$, then $\mathbb{H}(n) \oplus \mathbb{G}(n, r-2)$ and $\mathbb{G}(n,r)$ are contiguous.* ∎

In particular, taking $r = 3$, we see that $\mathbb{H}(n) \oplus \mathbb{G}(n,1)$ and $\mathbb{G}(n,3)$ are contiguous. Since $\mathbb{G}(n,1)$ is a perfect matching, and thus $\mathbb{H}(n) \oplus \mathbb{G}(n,1)$ by definition always contains a perfect matching, it follows that $\mathbb{G}(n,3)$ *a.a.s.* contains one too. This gives another proof of Corollary 9.21 for $r = 3$. A similar argument and induction yields Corollary 9.21 for any odd $r \geq 3$.

We proceed to study more general unions, and begin with the small cycles in them.

Theorem 9.32. *Let $\mathbb{G} = \mathbb{G}_1 + \cdots + \mathbb{G}_m$, where $m \geq 1$ and $\mathbb{G}_1, \ldots, \mathbb{G}_m$ are independent random regular (multi)graphs such that $\mathbb{G}_i$ is a copy of either $\mathbb{G}^*(n, r_i)$ (with $r_i \geq 1$) or $\mathbb{H}(n)$; in the latter case we let $r_i = 2$. Let $r = r_1 + \cdots + r_m$; thus $\mathbb{G}$ is an r-regular multigraph. Let further m_H be the number of $\mathbb{H}(n)$ among the $\mathbb{G}_i$. Then*

$$Z_k(\mathbb{G}) \xrightarrow{d} \mathrm{Po}(\mu_k), \qquad \text{as } n \to \infty,$$

jointly for all $k \geq 1$ with independent limits, with

$$\mu_k = \frac{(r-1)^k + (m-1)(-1)^k - m_H}{2k}, \qquad k \geq 1.$$

Proof. We argue as we did for the special case $\mathbb{H}(n) + \mathbb{G}^*(n, r-2)$ in the proof of Theorem 9.23. We now have m types and let $b_{ij} = r_i r_j$ when $i \neq j$, $b_{ii} = r_i(r_i - 1)$; thus $a_{ij} = b_{ij}/r_i = r_j - \delta_{ij}$, where δ_{ij} is Kronecker's delta. We have to exclude the cases when all edges in the cycle are of the same type and this type corresponds to a graph $\mathbb{H}(n)$, and we obtain as before $Z(\mathbb{G}) \xrightarrow{d} \mathrm{Po}(\mu_k)$, jointly for all k, with

$$\mu_k = \frac{1}{2k}\left(\mathrm{Tr}(A^k) - m_H\right).$$

Here A is the $m \times m$ matrix $(a_{ij})_{i,j=1}^m = (r_j - \delta_{ij})_{i,j=1}^m$. Thus $A + I$ is the matrix $(r_j)_{i,j=1}^m$, which has rank 1 and therefore $m - 1$ vanishing eigenvalues; the last eigenvalue has to equal $\mathrm{Tr}(A + I) = r$. Hence, A has the eigenvalues $r - 1$ and -1 (with multiplicity $m - 1$), and

$$\mathrm{Tr}(A^k) = (r - 1)^k + (m - 1)(-1)^k. \qquad \blacksquare$$

Before proving further instances of contiguity, let us look at a counter-example.

Example 9.33. Consider the three different 2-regular multigraphs $\mathbb{G}^*(n, 2)$, $\mathbb{H}(n)$, and $\mathbb{G}(n, 1) + \mathbb{G}(n, 1)$. Theorem 9.32 shows that in all three cases, $Z_k \to \mathrm{Po}(\mu_k)$ (jointly for all k), with

$$\mu_k = \begin{cases} \frac{1}{2k}, & \text{for } \mathbb{G}^*(n,2), \\ \frac{1+(-1)^k}{2k}, & \text{for } \mathbb{G}(n,1) + \mathbb{G}(n,1), \\ 0, & \text{for } \mathbb{H}(n). \end{cases}$$

It follows that the conclusion of Corollary 9.54 is violated for any pair of these three random 2-regular multigraphs; hence no two of them are contiguous.

This remains true if we condition on the multigraphs being simple, since we then still have $Z_k \to \mathrm{Po}(\mu_k)$ for $k \geq 3$, which is just as good (or bad) for the application of Corollary 9.54. For example, $\mathbb{G}(n, 2)$ and $\mathbb{G}(n, 1) \oplus \mathbb{G}(n, 1)$ are not contiguous.

We will see below (Theorem 9.43) that such counter-examples occur only for 2-regular multi-graphs; in all instances of Theorem 9.32 with $r \geq 3$, $\mathbb{G}$ is contiguous to either $\mathbb{G}^*(n, r)$ or $\mathbb{G}'(n, r)$.

One conceivable method to prove this result is to follow the proofs of Theorems 9.29 and 9.30. We thus try to apply Theorem 9.12 with $X_{in} = Z_i(\mathbb{G}^*(n, r))$ and Y_n equal to the number of decompositions of the random multigraph $\mathbb{G}^*(n, r)$ as a union $G_1 + \cdots + G_m$, where G_i is any r_i-regular multigraph when $\mathbb{G}_i$ is of the type $\mathbb{G}^*(n, r_i)$, and G_i is a Hamilton cycle when $\mathbb{G}_i$ is of the type $\mathbb{H}(n)$. The measure Q_n in Lemma 9.17 then is the distribution of $\mathbb{G}$.

Condition (A1) holds by Theorem 9.5, with $\lambda_k = \frac{1}{2k}(r - 1)^k$, and (A2′) holds by Theorem 9.32, with $\mu_k = \frac{1}{2k}\big((r-1)^k + (m-1)(-1)^k - m_H\big)$. Hence

$$\delta_k = \frac{\mu_k}{\lambda_k} - 1 = \frac{(m - 1)(-1)^k - m_H}{(r - 1)^k} \tag{9.32}$$

and

$$\begin{aligned}\sum_{k=1}^{\infty}\lambda_k\delta_k^2 &= \sum_{k=1}^{\infty}\frac{\left((m-1)(-1)^k - m_H\right)^2}{2k(r-1)^k}\\
&= \sum_{k=1}^{\infty}\left((m-1)^2+m_H^2\right)^2\frac{1}{2k}\Big(\frac{1}{r-1}\Big)^k - \sum_{k=1}^{\infty}(m-1)m_H\frac{1}{k}\Big(-\frac{1}{r-1}\Big)^k\\
&= -\tfrac12\left((m-1)^2+m_H^2\right)\log\Big(1-\frac{1}{r-1}\Big) + (m-1)m_H\log\Big(1+\frac{1}{r-1}\Big)\\
&= -\tfrac12\left((m-1)^2+m_H^2\right)\log(r-2) + (m-1)m_H\log(r)\\
&\quad + \tfrac12(m-1-m_H)^2\log(r-1).\end{aligned}$$

Consequently, assuming $r \geq 3$, (A3) holds, and (A4) says

$$\frac{\mathbb{E}\,Y_n^2}{(\mathbb{E}\,Y_n)^2} \to \frac{(r-1)^{(m-1-m_H)^2/2}\,r^{(m-1)m_H}}{(r-2)^{((m-1)^2+m_H^2)/2}}. \tag{9.33}$$

If (9.33) holds, then Theorem 9.12 yields $Y_n/\mathbb{E}\,Y_n \xrightarrow{d} W$ for a certain random variable W with $\mathbb{E}\,W = 1$. It is easily seen that $\mu_k > 0$ and thus $\delta_k > -1$ for all $k \geq 1$, except when $k = 1$ and all constituents $\mathbb{G}_i$ are either $\mathbb{G}^*(n,1) = \mathbb{G}(n,1)$ or $\mathbb{H}(n)$; in this case loops clearly are impossible. Except in the loopless case we thus have $W > 0$ a.s., and Proposition 9.49 shows that $\mathbb{G}$ and $\mathbb{G}^*(n,r)$ are contiguous; in the loopless case we condition on $X_{1n} = 0$, and obtain as in the proof of Theorem 9.30 that $\mathbb{G}$ and $\mathbb{G}'(n,r)$ are contiguous.

In other words, we have shown that contiguity holds in all cases when (9.33) holds. While it seems quite difficult to prove (9.33) in general by present methods, it has been verified in a number of special cases by arguments similar to the proof of Lemma 9.10 above. (Some of the references below actually treat only the corresponding results for $\mathbb{G}(n,r)$, but then the case $\mathbb{G}^*(n,r)$ can be shown by similar methods.)

We begin with the simplest case.

Theorem 9.34. *If $r \geq 3$, then $\mathbb{G}^*(n,r-1) + \mathbb{G}^*(n,1) \approx \mathbb{G}^*(n,r)$.*

Proof. In this case, the number Y_n of decompositions equals the number of perfect matchings in $\mathbb{G}^*(n,r)$, and the condition (9.33) is

$$\frac{\mathbb{E}\,Y_n^2}{(\mathbb{E}\,Y_n)^2} \to \frac{(r-1)^{1/2}}{(r-2)^{1/2}}. \tag{9.34}$$

This has been proved by Bollobás and McKay (1986), which completes the proof. ∎

Another case is the following theorem, proved by Robalewska (1996).

Theorem 9.35. *If $r \geq 3$, then $\mathbb{G}^*(n,r-2) + \mathbb{G}^*(n,2) \approx \mathbb{G}^*(n,r)$.*

Proof. This is proved as Theorem 9.34; we now have to verify

$$\frac{\mathbb{E}\,Y_n^2}{(\mathbb{E}\,Y_n)^2} \to \frac{(r-1)^{1/2}}{(r-2)^{1/2}}, \tag{9.35}$$

where Y_n is the number of 2-factors of $\mathbb{G}^*(n,r)$, that is, the number of 2-regular submultigraphs. This can be verified by long calculations similar to those in the proof of Lemma 9.10; we omit the details (Robalewska 1996). ∎

As a simple application, we obtain the following theorem which gives partial justification for the intuitive feeling that increasing the degree of a random regular graph adds edges and makes it easier to, for example, find a specified subgraph. (*Cf.* Section 1.3, where other models of random graphs are covered.)

Theorem 9.36. *Let $2 \leq r \leq s$.*

(i) *Any increasing property that holds* a.a.s. *for $\mathbb{G}^*(n,r)$ holds* a.a.s. *for $\mathbb{G}^*(n,s)$ too.*

(ii) *Any increasing property that holds* a.a.s. *for $\mathbb{G}(n,r)$ holds* a.a.s. *for $\mathbb{G}(n,s)$ too.*

Proof. First consider even n only. Any increasing property that holds *a.a.s.* for $\mathbb{G}^*(n,r)$ holds *a.a.s.* for $\mathbb{G}^*(n,r) + \mathbb{G}^*(n,1)$ too; hence it holds *a.a.s.* also for $\mathbb{G}^*(n,r+1)$ by Theorem 9.34. The assertion (i) follows by induction. For odd n, r and s have to be even, and we argue using Theorem 9.35 instead.

The statement (ii) is proved similarly, using $\mathbb{G}(n,r) \oplus \mathbb{G}(n,1) \approx \mathbb{G}(n,r+1)$ and $\mathbb{G}(n,r) \oplus \mathbb{G}(n,2) \approx \mathbb{G}(n,r+2)$, which follow from Theorem 9.34 and Theorem 9.35 by conditioning. ∎

Remark 9.37. Theorem 9.36 is not true for $r=1$. For example, $\mathbb{G}^*(n,1) = \mathbb{G}(n,1)$ contains always a perfect matching (in fact, it *is* one), while $\mathbb{G}^*(n,2)$ or $\mathbb{G}(n,2)$ contains a perfect matching only if it contains no cycles of odd length, and it follows easily from Theorem 9.5 that the probability of this tends to 0.

Remark 9.38. Unlike the simple monotonicity result in Lemma 1.10 for $\mathbb{G}(n,p)$ and $\mathbb{G}(n,M)$, it is, in general, *not* true that if $\mathcal{P}$ is an increasing property and $2 \leq r \leq s$, then $\mathbb{P}(\mathbb{G}^*(n,r)$ satisfies $\mathcal{P}) \leq \mathbb{P}(\mathbb{G}^*(n,s)$ satisfies $\mathcal{P})$. A counter-example is obtained if we take $r=2$, $s=3$, $n=2$ and let $\mathcal{P}$ be the property that the multigraph contains a multiple edge. Then $\mathbb{P}(\mathbb{G}^*(2,2)$ satisfies $\mathcal{P}) = 2/3$ while $\mathbb{P}(\mathbb{G}^*(2,3)$ satisfies $\mathcal{P}) = 2/5$. We do not know whether the corresponding statement for the random graphs $\mathbb{G}(n,r)$ and $\mathbb{G}(n,s)$ also may fail; nor do we know if the statement is always true asymptotically, taking $\limsup_{n\to\infty}$ of both sides.

Remark 9.39. The original proof by Robinson and Wormald (1992) that a Hamilton cycle *a.a.s.* exists in $\mathbb{G}(n,3)$ was by (a version of) the method used in the proof of Theorem 9.20 above. For $\mathbb{G}(n,r)$, $r \geq 4$ (Robinson and Wormald 1994), where the variance calculation as seen above is much harder, they preferred a slightly different method, establishing (implicitly) Theorem 9.34 and Theorem 9.36 for n even by the method above, which uses the easier variance estimate (9.34) for the number of perfect matchings. The case $r \geq 4$ then follows from the case $r = 3$ (with an extra argument when n is odd).

We continue with further instances of contiguity.

Theorem 9.40. $\mathbb{G}^*(n,1) + \mathbb{G}^*(n,1) + \mathbb{G}^*(n,1) \approx \mathbb{G}'(n,3)$.

Proof. This time the condition (9.33) is

$$\frac{\mathbb{E}\, Y_n^2}{(\mathbb{E}\, Y_n)^2} \to 4,$$

where Y_n is the number of partitions of the edge set of $\mathbb{G}^*(n,3)$ into three disjoint perfect matchings. This is verified by Janson (1995b) and Molloy, Robalewska, Robinson and Wormald (1997). ∎

Theorem 9.41. $\mathbb{H}(n) + \mathbb{H}(n) \approx \mathbb{G}'(n,4)$.

Proof. This time (9.33) says

$$\frac{\mathbb{E}\, Y_n^2}{(\mathbb{E}\, Y_n)^2} \to \sqrt{24},$$

where Y_n is the number of partitions of the edge set of $\mathbb{G}^*(n,4)$ into two Hamilton cycles. This is verified by Kim and Wormald (2000+). ∎

The theorems above are so far the only instances where (9.33) has been verified. Nevertheless, by combining them, we obtain contiguity in all remaining cases too. We begin with the case of $\mathbb{G}^*(n,r_1) + \mathbb{G}^*(n,r_2)$.

Lemma 9.42. *If $r_1 \geq 1$, $r_2 \geq 1$ and $r_1 + r_2 \geq 3$, then $\mathbb{G}^*(n,r_1) + \mathbb{G}^*(n,r_2) \approx \mathbb{G}^*(n,r_1+r_2)$.*

Proof. We may assume $r_1 \geq r_2$. The cases $r_2 = 1$ and 2 are Theorems 9.34 and 9.35. For $r_1 \geq r_2 \geq 3$, we use induction on r_2 and assume that the result is true for smaller r_2 (and any r_1).

We observe first that if $G_1 \approx G_1'$ and $G_2 \approx G_2'$, then $G_1 + G_2 \approx G_1' + G_2'$; this is a consequence of Proposition 9.50(ii)(iii). Recall also that contiguity is an equivalence relation.

Hence, if $r_1 \geq r_2 \geq 3$ and at least one of them is odd (which forces n even), using Theorem 9.34 twice and the induction hypothesis,

$$\begin{aligned}\mathbb{G}^*(n,r_1) + \mathbb{G}^*(n,r_2) &\approx \mathbb{G}^*(n,r_1) + \mathbb{G}^*(n,1) + \mathbb{G}^*(n,r_2-1)\\ &\approx \mathbb{G}^*(n,r_1+1) + \mathbb{G}^*(n,r_2-1)\\ &\approx \mathbb{G}^*(n,r_1+r_2),\end{aligned}$$

which verifies the induction step. If both r_1 and r_2 are even (which allows odd n too), we argue similarly using Theorem 9.35 instead. ■

Finally, we prove the full result.

Theorem 9.43. *Let $\mathbb{G}$ be a union as in Theorem 9.32, and assume that $r \geq 3$. Then $\mathbb{G}$ is contiguous to $\mathbb{G}^*(n,r)$ or $\mathbb{G}'(n,r)$ (the latter in the cases in which each $\mathbb{G}_i$ is either $\mathbb{G}^*(n,1)$ or $\mathbb{H}(n)$ and thus loops are impossible).*

Proof. Consider first the case in which all constituents are $\mathbb{G}^*(n,1)$ or $\mathbb{H}(n)$. If there are constituents of both these types we begin by combining one $\mathbb{G}^*(n,1)$ and one $\mathbb{H}(n)$ using Theorem 9.30; if all are $\mathbb{G}^*(n,1)$, and there thus are $r \geq 3$ of them, we combine three of them using Theorem 9.40; if all are $\mathbb{H}(n)$ we combine two of them using Theorem 9.41. In all three subcases this yields a new sum, contiguous to the original one, with one constituent $\mathbb{G}'(n,s)$ ($s = 3$ or 4). Now we observe that by Theorem 9.34 and conditioning on no loops, using Proposition 9.50(i), $\mathbb{G}'(n,s) + \mathbb{G}^*(n,1) \approx \mathbb{G}'(n,s+1)$ when $s \geq 2$; similarly by Theorem 9.29 and conditioning, $\mathbb{G}'(n,s) + \mathbb{H}(n) \approx \mathbb{G}'(n,s+2)$ when $s \geq 2$. Consequently we may absorb all remaining $\mathbb{G}^*(n,1)$ and $\mathbb{H}(n)$ one by one into the $\mathbb{G}'(n,s)$; eventually we reach $\mathbb{G}'(n,r)$.

Secondly, if there is at least one constituent $\mathbb{G}^*(n,r_i)$ with $r_i \geq 2$, we may absorb all other constituents into it one by one using Lemma 9.42 and Theorem 9.29; eventually we reach $\mathbb{G}^*(n,r)$. ■

Corollary 9.44. *Let $\mathbb{G} = \mathbb{G}_1 \oplus \ldots \oplus \mathbb{G}_m$, where $m \geq 1$ and $\mathbb{G}_1, \ldots, \mathbb{G}_m$ are independent random regular graphs such that $\mathbb{G}_i$ is a copy of either $\mathbb{G}(n,r_i)$ (with $r_i \geq 1$) or $\mathbb{H}(n)$, in the latter case we let $r_i = 2$. Let $r = r_1 + \cdots + r_m$; thus $\mathbb{G}$ is an r-regular graph. If $r \geq 3$, then $\mathbb{G} \approx \mathbb{G}(n,r)$.*

Proof. By Theorem 9.43, conditioning on the multigraphs being simple; see Proposition 9.50(i). ■

Remark 9.45. Another model (with a different distribution) of a random 2-regular (multi)graph is obtained by taking a (uniformly distributed) random permutation π of $[n]$ and drawing the n edges $i\pi(i)$. We denote this model by $\mathbb{P}(n)$.

It is well known (and easily shown) that for this model, the cycle counts $Z_k \to \text{Po}(1/k)$ (with independent limits); hence Corollary 9.54 shows that $\mathbb{P}(n)$ is not contiguous to any of the three 2-regular multigraphs considered in Example 9.33.

On the other hand, we conjecture that Theorem 9.43 can be extended to allow summands $\mathbb{P}(n)$ too. Indeed, by an argument similar to the proof of Theorem 9.29, $\mathbb{P}(n) + \mathbb{G}^*(n, r-2) \approx \mathbb{G}^*(n, r)$ and by induction as above, the conjecture holds at least when there is one summand $\mathbb{G}^*(n, r_i)$. To settle the remaining cases, with unions of $\mathbb{H}(n)$ and $\mathbb{P}(n)$ only, it would suffice to show that $\mathbb{H}(n) + \mathbb{P}(n)$ and $\mathbb{P}(n) + \mathbb{P}(n)$ both are contiguous to $\mathbb{G}^*(n, 4)$, but this remains an open problem.

It is not difficult to extend the small cycle count Theorem 9.32 to allow also m_P summands $\mathbb{P}(n)$, with m_H replaced by $m_H - m_P$ in the formula for μ_k. Hence condition (9.33) extends too, with the same modification, but it has not yet been verified for the relevant cases. (The right-hand sides of (9.33) are $\sqrt{3/2}$ and $3^{9/2}2^{-13/2}$, respectively.)

The model $\mathbb{P}(n) + \cdots + \mathbb{P}(n)$ of a random $2d$-regular multigraph has occasionally been used. A proof of the conjecture would imply that results, in the form of *a.a.s.* properties, that can be proved for this model hold for $\mathbb{G}^*(n, 2d)$ and $\mathbb{G}(n, 2d)$ too.

9.6 A BRIEF COURSE IN CONTIGUITY

We end this chapter with a discussion of contiguity in general. The purpose of this purely probabilistic section is to provide an easily accessible reference for some facts used in the preceding section. Readers who are either experts in probability theory or completely uninterested in it may skip this section.

The notion of contiguity was introduced in statistics by Le Cam (1960); see also Le Cam (1969, 1986) and Roussas (1972). It is defined as an equivalence relation among sequences of probability measures on a sequence of space (Ω_n).

Definition. Let $(P_n)_1^\infty$ and $(Q_n)_1^\infty$ be two sequences of probability measures, such that for each n, P_n and Q_n both are defined on the same measurable space $(\Omega_n, \mathcal{F}_n)$. We then say that the sequences are *contiguous* if for every sequence of measurable sets $A_n \subset \Omega_n$,

$$\lim_{n\to\infty} P_n(A_n) = 0 \iff \lim_{n\to\infty} Q_n(A_n) = 0.$$

There are many equivalent definitions of contiguity; some are given below and several others are given in the references given above.

Contiguity is mainly used in statistics, but it seems to be a natural and useful property also in the study of random combinatorial structures. In that case, typically, Ω_n is a (finite) set of some combinatorial objects of size n, $\mathcal{F}_n$ is the σ-field of all subsets of Ω_n, and P_n and Q_n are probability measures corresponding to two different ways of selecting an element of Ω_n "at random". In this connection, we also say that two random objects X_n and Y_n, depending on a parameter n, are contiguous if the corresponding sequences of distributions $(\mathcal{L}(X_n))$ and $(\mathcal{L}(Y_n))$ are contiguous. (This entails that X_n and Y_n take values in a common space Ω_n.)

Several examples of contiguity for random regular graphs are given earlier in this chapter. Another combinatorial example was given by Winkler (1991), who showed that for two-dimensional partial orders, the uniform distribution is contiguous to the distribution of the intersection of two independent random linear orders.

It is important to realize that contiguity is an asymptotic property of two *sequences* of probability measures; if we say that two probability measures (or distributions, or random models) are contiguous, we really mean that there is some parameter n (although perhaps not explicitly mentioned) that tends to infinity. If, as is often useful, we informally regard asymptotic results as statements about a fictitious infinite limiting model, then contiguity can be interpreted as mutual absolute continuity of the two probability measures. (In this context it may be observed that in the special case that Ω_n, $\mathcal{F}_n$, P_n and Q_n do not depend on n, contiguity reduces to mutual absolute continuity.)

Note also that the definition says nothing about the rates of convergence of $P_n(A_n)$ and $Q_n(A_n)$; these may be quite different. It is, nevertheless, possible to restate the definition in terms of estimates, which, however, use unknown functions to relate the rates. (In certain examples, it may, of course, be possible to replace these by explicit functions.) We given two such reformulations, which we think may be useful for a better understanding the contiguity concept.

Proposition 9.46. *The sequences (P_n) and (Q_n) are contiguous if and only if there exist a sequence $\varepsilon_n \to 0$ and a continuous function $\varphi : [0,1] \to [0,1]$ with $\varphi(0) = 0$, such that for every n and $A_n \in \mathcal{F}_n$*

$$P_n(A_n) \le \varepsilon_n + \varphi(Q_n(A_n)) \qquad \textit{and} \qquad Q_n(A_n) \le \varepsilon_n + \varphi(P_n(A_n)).$$

Proof. If (P_n) and (Q_n) are contiguous, then for every $\varepsilon > 0$ there exist $n(\varepsilon)$ and $\delta(\varepsilon)$ such that $n \ge n(\varepsilon)$ and $P_n(A_n) \le \delta(\varepsilon) \Rightarrow Q_n(A_n) \le \varepsilon$, and similarly with P_n and Q_n interchanged. The existence of (ε_n) and φ now follows easily (Exercise!). The converse is obvious. ∎

For the next results, we recall that each Q_n has a Lebesgue decomposition $Q_n = Q_n^a + Q_n^s$ where Q_n^a is absolutely continuous with respect to P_n, while Q_n^s and P_n are mutually singular, and that there exists a function $dQ_n/dP_n \ge 0$ (the Radon–Nikodym derivative) such that

$$Q_n^a(A_n) = \int_{A_n} \frac{dQ_n}{dP_n}\, dP_n, \qquad A_n \in \mathcal{F}_n.$$

(If Ω_n is countable (e.g., finite), the measures are given by probability functions $p_n(w)$ and $q_n(w)$; in this case Q_n^s is the restriction of Q_n to the set $\{w : p_n(w) = 0\}$, while $dQ_n/dP_n = q_n(w)/p_n(w)$ when $p_n(w) \ne 0$ (and arbitrary otherwise).)

Proposition 9.47. *The sequences (P_n) and (Q_n) are contiguous if and only if for every $\varepsilon > 0$ there exist $n(\varepsilon)$ and $K(\varepsilon)$, such that for every $n \geq n(\varepsilon)$ there exists a set $B_n \in \mathcal{F}_n$ with $P_n(B_n^c) \leq \varepsilon$ and $Q_n(B_n^c) \leq \varepsilon$, such that*

$$K(\varepsilon)^{-1} \leq \frac{Q_n(A_n)}{P_n(A_n)} \leq K(\varepsilon) \qquad \text{for all } A_n \subseteq B_n. \tag{9.36}$$

Remark 9.48. Here we may replace (9.36) by the equivalent condition that the restriction of Q_n to B_n is absolutely continuous with respect to P_n, with

$$K(\varepsilon)^{-1} \leq \frac{dQ_n}{dP_n} \leq K(\varepsilon) \qquad P_n\text{-a.s. on } B_n.$$

Proof. Suppose that (P_n) and (Q_n) are contiguous. Fix a large number K, let C_n be a set such that $P_n(C_n) = 0 = Q_n^s(C_n^c)$ (this is possible by the definition of singular measures), and define

$$D_n = \{w \in \Omega_n : dQ_n/dP_n > K\}.$$

Then

$$P_n(C_n \cup D_n) = \int_{D_n} dP_n \leq \frac{1}{K}\int_{D_n} \frac{dQ_n}{dP_n}\, dP_n = \frac{1}{K} Q_n^a(D_n) \leq \frac{1}{K}.$$

Hence Proposition 9.46 yields $Q_n(C_n \cup D_n) \leq \varepsilon_n + \varphi(K^{-1})$. Furthermore, for any $A_n \subseteq (C_n \cup D_n)^c$,

$$Q_n(A_n) = Q_n^a(A_n) = \int_{A_n} \frac{dQ_n}{dP_n}\, dP_n \leq K P_n(A_n).$$

We similarly define C_n' and D_n' with the roles of P_n and Q_n interchanged, and let $B_n = (C_n \cup D_n \cup C_n' \cup D_n')^c$. Then $P_n(B_n^c)$, $Q_n(B_n^c) \leq K^{-1} + \varepsilon_n + \varphi(K^{-1})$, which is less than ε for $n \geq n(\varepsilon)$ if $n(\varepsilon)$ and $K = K(\varepsilon)$ are large enough.

Conversely, if such sets B_n exist, then for large n and every $A_n \in \mathcal{F}_n$,

$$Q_n(A_n) \leq Q_n(A_n \cap B_n) + Q_n(B_n^c) \leq K(\varepsilon) P_n(A_n) + \varepsilon.$$

In particular, if $P_n(A_n) \to 0$, then $\limsup Q_n(A_n) \leq \varepsilon$, for every $\varepsilon > 0$, and thus $Q_n(A_n) \to 0$. By symmetry, we obtain also the converse implication and thus the sequences are contiguous. ∎

For the next result, we consider for simplicity only the case in which dQ_n/dP_n converges in distribution.

Proposition 9.49. *Suppose that $L_n = dQ_n/dP_n$, regarded as a random variable on $(\Omega_n, \mathcal{F}_n, P_n)$, converges in distribution to some random variable L as $n \to \infty$. Then (P_n) and (Q_n) are contiguous if and only if $L > 0$ a.s. and $\mathbb{E}\, L = 1$.*

Proof. Suppose that (P_n) and (Q_n) are contiguous. Let $\varepsilon > 0$, and let $K(\varepsilon)$ and B_n, $n \geq n(\varepsilon)$, be as in Proposition 9.47. In particular, by Remark 9.48, $K(\varepsilon)^{-1} \leq L_n \leq K(\varepsilon)$ on B_n. Thus

$$\begin{aligned}\mathbb{P}(L=0) &\leq \mathbb{P}(L < K(\varepsilon)^{-1}) \leq \limsup_{n\to\infty} P_n(L_n < K(\varepsilon)^{-1}) \\ &\leq \limsup_{n\to\infty} P_n(B_n^c) \leq \varepsilon.\end{aligned}$$

Moreover, since $L_n \wedge K(\varepsilon) \xrightarrow{d} L \wedge K(\varepsilon)$ as $n \to \infty$, dominated convergence yields

$$\begin{aligned}\mathbb{E}\,L \geq \mathbb{E}(L \wedge K(\varepsilon)) &= \lim_{n\to\infty} \int_{\Omega_n} L_n \wedge K(\varepsilon)\, dP_n \\ &\geq \limsup_{n\to\infty} \int_{B_n} L_n\, dP_n = \limsup_{n\to\infty} Q_n(B_n) \geq 1-\varepsilon.\end{aligned}$$

Since ε is arbitrary, and by Fatou's lemma $\mathbb{E}\,L \leq \liminf \mathbb{E}\,L_n = 1$, this shows that $\mathbb{P}(L=0) = 0$ and $\mathbb{E}\,L = 1$.

Conversely, suppose that $\mathbb{P}(L=0) = 0$ and $\mathbb{E}\,L = 1$, and let $\varepsilon > 0$. There exists $\delta > 0$ such that $\mathbb{P}(L \leq \delta) < \varepsilon$; we may furthermore choose δ such that $\delta < 1$ and $\mathbb{P}(L = \delta) = 0$. Then $P_n(L_n \leq \delta) \to \mathbb{P}(L \leq \delta)$, and thus $P_n(L_n \leq \delta) < \varepsilon$ for large n, which also yields $Q_n^a(L_n \leq \delta) = \int_{L_n \leq \delta} L_n\, dP_n < \varepsilon\delta \leq \varepsilon$.

We can also find $K > 1/\varepsilon$ such that $\mathbb{E}(L \wedge K) > 1 - \varepsilon$ and thus

$$\lim_{n\to\infty} \int (L_n \wedge K)\, dP_n = \mathbb{E}(L \wedge K) > 1 - \varepsilon.$$

Hence, for large n, say $n \geq n(\varepsilon)$,

$$\begin{aligned}\varepsilon > 1 - \int_{\Omega_n} (L_n \wedge K)\, dP_n &= Q_n^s(\Omega_n) + Q_n^a(\Omega_n) - \int_{\Omega_n} (L_n \wedge K)\, dP_n \\ &= Q_n^s(\Omega_n) + \int_{\Omega_n} (L_n - L_n \wedge K)\, dP_n \geq Q_n^s(\Omega_n) + \tfrac{1}{2} \int_{L_n > 2K} L_n\, dP_n \\ &= Q_n^s(\Omega_n) + \tfrac{1}{2} Q_n^a(L_n > 2K).\end{aligned}$$

Let N_n be a set with $P_n(N_n) = Q_n^s(N_n^c) = 0$, and define $B_n = \{w \in N_n^c : \delta \leq L_n \leq 2K\}$. Then, for $n \geq n(\varepsilon)$,

$$P_n(B_n^c) = P_n(L_n < \delta) + P_n(L_n > 2K) < \varepsilon + \int L_n\, dP_n/2K < 2\varepsilon$$

and

$$Q_n(B_n^c) = Q_n^s(\Omega_n) + Q_n^a(L_n < \delta) + Q_n^a(L_n > 2K) < 3\varepsilon.$$

Hence Proposition 9.47 implies that (P_n) and (Q_n) are contiguous. ∎

Contiguity is preserved by some natural operations, as the next proposition shows. In (ii) and (iii) we suppose that $(\Omega_n', \mathcal{F}_n')$ is another sequence of measure spaces.

Proposition 9.50. *Suppose that (P_n) and (Q_n) are contiguous.*

(i) *If (A_n) is any sequence of events such that* $\liminf P_n(A_n) > 0$, *then the conditioned measures $P_n(\cdot \mid A_n)$ and $Q_n(\cdot \mid A_n)$ are contiguous.*

(ii) *If $f_n : \Omega_n \to \Omega_n'$ are measurable functions, then the induced measures $P_n \circ f_n^{-1}$ and $Q_n \circ f_n^{-1}$ on Ω_n' are contiguous.*

(iii) *If P_n' and Q_n' are contiguous probability measures on Ω_n', then the product measures $P_n \times P_n'$ and $Q_n \times Q_n'$ on $\Omega_n \times \Omega_n'$ are contiguous.*

Proof. (i) and (ii) follow easily from the definition (note that in (i), also $\liminf Q_n(A_n) > 0$ by the contiguity); for (iii) it is perhaps simplest to use Proposition 9.47 with Remark 9.48. We omit the details (Exercise!). ∎

Contiguity of two sequences (P_n) and (Q_n) has several useful consequences for limit theorems. The definition says that any property which holds under P_n with probability tending to 1 as $n \to \infty$ (i.e., asymptotically almost surely) also holds under Q_n with probability tending to 1 (and conversely). An immediate consequence is that if X_n are any random variables such that X_n under P_n converges in probability to some constant c, then X_n converges in probability to the same constant c also under Q_n.

For convergence in distribution we need some further information. In the next two results we suppose that S and S' are two complete separable metric spaces, for example $\mathbb{R}$, $\mathbb{R}^d$ or the space $\mathbb{R}^\infty$ of infinite sequences. (See Billingsley (1968) for details on convergence in metric spaces.) We use $\mathcal{L}(X \mid P)$ to denote the distribution of X under P.

In the first result we assume joint convergence under P_n of the variables X_n and the Radon–Nikodym derivatives dQ_n/dP_n.

Proposition 9.51. *Suppose that (P_n) and (Q_n) are contiguous, and let $L_n = dQ_n/dP_n$. Suppose further that X_n are random variables defined on Ω_n with values in S, such that $\mathcal{L}((X_n, L_n) \mid P_n) \to \mathcal{L}(X, L)$ for some random variables X and L (with values in S and $\mathbb{R}$, respectively), defined on a probability space $(\Omega, \mathcal{F}, P)$. Then $\mathcal{L}(X_n \mid Q_n) \to \mathcal{L}(X \mid Q)$, where $dQ = L\,dP$ (i.e., $Q(A) = \int_A L\,dP$, $A \in \mathcal{F}$).*

Proof. Note that $\int L\,dP = 1$ by Proposition 9.49, so Q is a probability measure. Let f be a bounded continuous real-valued function on S. Then $\mathcal{L}(f(X_n)L_n \mid P_n) \to \mathcal{L}(f(X)L)$. If $f \geq 0$, then Fatou's lemma yields

$$\liminf_{n\to\infty} \int f(X_n)L_n\,dP_n \geq \int f(X)L\,dP. \tag{9.37}$$

Taking first $f = 1$, we see that

$$\liminf_{n\to\infty} Q_n^a = \liminf_{n\to\infty} \int L_n\,dP_n \geq \int L\,dP = 1, \tag{9.38}$$

and thus $\int L_n\,dP_n \to \int L\,dP$. Hence it follows, by adding a constant to f, that (9.37) actually holds for every bounded continuous f. But then we may also substitute $-f$ in (9.37), and obtain

$$\int f(X_n)\,dQ_n^a = \int f(X_n)L_n\,dP_n \to \int f(X)L\,dP = \int f(X)\,dQ. \tag{9.39}$$

Moreover, by (9.38), $Q_n^s(\Omega_n) = 1 - Q_n^a(\Omega_n) \to 0$, and we obtain

$$\int f(X_n)\,dQ_n \to \int f(X)\,dQ$$

for every such f, which gives the result. ∎

The next result gives a necessary condition for contiguity.

Proposition 9.52. *Suppose that (P_n) and (Q_n) are contiguous, and that X_n are random variables defined on Ω_n with values in S. Suppose further that X_n converges in distribution under both P_n and Q_n:*

$$\mathcal{L}(X_n \mid P_n) \to P_X \qquad \text{and} \qquad \mathcal{L}(X_n \mid Q_n) \to Q_X$$

for two probability measures P_X and Q_X on S. Then P_X and Q_X are mutually absolutely continuous.

Proof. Let $L_n = dQ_n/dP_n$. Since $\mathbb{E}\,L_n \leq 1$, the sequence (L_n) is tight and we may, by restricting attention to a subsequence, assume that $\mathcal{L}((X_n, L_n) \mid P_n)$ converges. Proposition 9.51 now implies that Q_X is absolutely continuous with respect to P_X. The converse holds by symmetry. ∎

We combine Proposition 9.52 with the following standard result. (In fact, the given conditions are equivalent to mutual absolute continuity.)

Proposition 9.53. *Suppose that the measure $\prod_i \mathrm{Po}(\lambda_i)$ and $\prod_i \mathrm{Po}(\lambda_i')$ on $\mathbb{Z}^\infty$ are mutually absolutely continuous. Then*

$$\lambda_i = 0 \iff \lambda_i' = 0,$$

and

$$\sum_i \frac{(\lambda_i - \lambda_i')^2}{\lambda_i + \lambda_i'} < \infty.$$

Proof. The first conclusion is trivial. For the second, we may use the Hellinger integral, defined for any two probability measures μ and μ' on the same space by $I(\mu, \mu') = \int \left(\frac{d\mu}{d\nu}\frac{d\mu'}{d\nu}\right)^{1/2} d\nu$, where ν is any σ-finite measure with $\mu, \mu' \ll \nu$.

It is easily seen that μ and μ' are mutually singular if and only if $I(\mu, \mu') = 0$, and that in our case

$$I\left(\prod_i \mathrm{Po}(\lambda_i), \prod_i \mathrm{Po}(\lambda_i')\right) = \prod_i I(\mathrm{Po}(\lambda_i), \mathrm{Po}(\lambda_i')) = \prod_i e^{-(\sqrt{\lambda_i}-\sqrt{\lambda_i'})^2/2}.$$

Hence the measures are mutually singular unless

$$\infty > \sum(\sqrt{\lambda_i} - \sqrt{\lambda_i'})^2 = \sum(\lambda_i - \lambda_i')^2/(\sqrt{\lambda_i} + \sqrt{\lambda_i'})^2$$
$$\geq \tfrac{1}{2}\sum(\lambda_i - \lambda_i')^2/(\lambda_i + \lambda_i'). \qquad \blacksquare$$

Corollary 9.54. *Suppose that (P_n) and (Q_n) are contiguous, and that X_{kn}, $k \geq 1$, are integer-valued random variables defined on Ω_n. Suppose further that X_{kn} is asymptotically Poisson distributed under both P_n and Q_n:*

$$\mathcal{L}(X_{kn} \mid P_n) \to \mathrm{Po}(\lambda_k) \qquad \textit{and} \qquad \mathcal{L}(X_{kn} \mid Q_n) \to \mathrm{Po}(\lambda_k')$$

for some λ_k and λ_k', for each k, as $n \to \infty$, and that these hold jointly for all k with independent limits, for both P_n and Q_n. Then

$$\sum_i \frac{(\lambda_i - \lambda_i')^2}{\lambda_i + \lambda_i'} < \infty. \qquad \blacksquare$$

10

Zero-One Laws

10.1 PRELIMINARIES

Most of the problems we have studied so far followed a similar scheme: given a graph property $\mathcal{A}$ estimate the asymptotic probability that $\mathcal{A}$ holds for a random graph $\mathbb{G}(n,p)$, where the edge probability $p = p(n)$ varies. In this chapter we consider questions of a different type: we fix a function $p = p(n)$ and study the asymptotic behavior of the probability that $\mathbb{G}(n,p)$ has $\mathcal{A}$, where $\mathcal{A}$ ranges over a prescribed class of graph properties $\mathfrak{A}$. Thus, in the following section we identify functions $p = p(n)$ for which the *zero-one law* holds, that is, the probability that $\mathbb{G}(n,p)$ has $\mathcal{A}$ tends to either zero or one as $n \to \infty$ for every property $\mathcal{A}$ from $\mathfrak{A}$. If $p = p(n)$ is close to the threshold function of some property from $\mathfrak{A}$, then the behavior of $\mathbb{G}(n,p)$ becomes more complicated. In particular, another phenomenon may happen: for each $\mathcal{A} \in \mathfrak{A}$ the probability that $\mathbb{G}(n,p)$ has $\mathcal{A}$ converges (but, maybe, to a non-trivial limit). Several instances of $p(n)$ satisfying such a *weak zero-one law* are given in Section 10.3. Then we introduce sum schemes of models – a fairly general approach which can be used to show zero-one laws for random discrete structures. Finally, we ask when the behavior of the probability that $\mathbb{G}(n,p)$ has some property can be effectively determined. In particular, if for a given function $p = p(n)$, there is a procedure which separates all properties in $\mathfrak{A}$ which hold *a.a.s.* from those which *a.a.s.* do not hold for $\mathbb{G}(n,p)$.

In order to start our considerations we need to decide first what class of graph properties $\mathfrak{A}$ we are going to study. Indeed, it would be naive to hope that for some non-trivial $p = p(n)$ the probability that $\mathbb{G}(n,p)$ has $\mathcal{A}$ converges for every graph property $\mathcal{A}$; the probability that $\mathbb{G}(n,p)$ has an even number of vertices oscillates between 0 and 1 as $n \to \infty$, no matter what function $p = p(n)$ we choose. Hence, we have to find a class of properties $\mathfrak{A}$ which is wide enough to include many interesting graph properties yet does not contain "odd" properties such as "the number of vertices of a graph is even".

A natural way of describing different families of properties comes from mathematical logic: we specify a language L, treat all graphs as models of L and consider the class of all properties which can be written as sentences of L. Therefore, in this chapter we employ notions and facts from mathematical logic which have not appeared in this book so far. In the remaining part of this section, we list basic terminology and results we will need for our considerations.

We will deal only with *first-order languages*, whose vocabulary, besides Boolean symbols $\neg$, $\wedge$, $\vee$, quantifiers $\forall$, $\exists$, and brackets, consists of a countable number of logical variables denoted by $x, y, z, \ldots$, possibly with indices, the binary predicate of equality "$=$", and a finite number of additional predicates. Two such languages will be of special interest to us: the first-order language of graphs $L_\sim$, which contains only one additional predicate "$\sim$", and the first-order language of linearly ordered graphs $L_\sim^{\mathrm{ord}}$, which, besides "$\sim$", contains also a binary predicate "$\leq$". The expression "$x \sim y$" is interpreted as "the vertices x and y are adjacent", and "$\leq$" should be understood as the standard ordering relation in the set of natural numbers. Thus, in this part of the book, graphs defined on subsets of natural numbers are viewed as models of the languages $L_\sim$ and $L_\sim^{\mathrm{ord}}$. Consequently, we will be mainly concerned with graph properties which can be expressed by sentences of $L_\sim$ or $L_\sim^{\mathrm{ord}}$. For example, the property that a graph contains a cycle of length 1729, and the property that any two vertices of a graph are connected by path of length 123 can be expressed by sentences in both $L_\sim$ and $L_\sim^{\mathrm{ord}}$. The property that a graph contains a "monotone" path of length two,

$$\exists_x \exists_y \exists_z (x \leq y \wedge y \leq z \wedge x \sim y \wedge y \sim z)\,,$$

can be expressed in $L_\sim^{\mathrm{ord}}$ but not in $L_\sim$. Note that the symbol "$\in$" is not a part of the vocabulary described above and consequently all logical variables of first-order languages must be interpreted as elements of a model, in our case the vertices of a graph. Hence, there is no obvious way to write in $L_\sim$ (or $L_\sim^{\mathrm{ord}}$) that a graph is connected; as a matter of fact we will soon see that this property cannot be stated in $L_\sim$. Note also that neither $L_\sim$ nor $L_\sim^{\mathrm{ord}}$ contains logical constants, so we cannot use the labels of vertices, and no sentence of $L_\sim$ (or $L_\sim^{\mathrm{ord}}$) can express the property which, for a graph G with vertex set $[n]$, means: "vertices $\lfloor n/2 \rfloor$ and $\lfloor n/2 \rfloor + 1$ are adjacent". In $L_\sim^{\mathrm{ord}}$ we can implicitly define vertices 1 and n as those without predecessors and successors, respectively, so it is possible to say that, for example, vertex 7 is adjacent to vertex $n-5$; in $L_\sim$ we cannot do even that.

Finally, the *quantifier depth* $\mathrm{qd}(\psi)$ of a formula ψ tells us, roughly speaking, how many times quantifiers are nested inside ψ. More formally, the quantifier depth can be defined recursively with respect to the structure of a formula by setting:

- $\mathrm{qd}(\psi) = 0$ for a formula ψ without quantifiers;
- $\mathrm{qd}(\psi) = \mathrm{qd}(\neg\psi)$;
- $\mathrm{qd}(\psi) = \max_i\{\mathrm{qd}(\psi_i)\}$ if either $\psi = \bigvee_i \psi_i$ or $\psi = \bigwedge_i \psi_i$;
- $\mathrm{qd}(\exists_x \psi(x)) = \mathrm{qd}(\psi(x)) + 1$ and $\mathrm{qd}(\forall_x \psi(x)) = \mathrm{qd}(\psi(x)) + 1$.

Thus, for example, the property that a graph contains a cycle of length six but none of length four can be in an obvious way written as a sentence of $L_\sim$ of quantifier depth six. However, it is not hard to see that the same property can be expressed by the following slightly more complicated sentence of $L_\sim$ of quantifier depth four:

$$\begin{aligned}
&\forall_{x_1}\forall_{x_2}\forall_{x_3}\forall_{x_4}\big(\neg(\neg x_1 = x_3 \wedge \neg x_2 = x_4 \wedge x_1 \sim x_2 \wedge x_2 \sim x_3 \\
&\qquad\qquad\qquad\qquad \wedge x_3 \sim x_4 \wedge x_1 \sim x_4)\big) \\
&\wedge \Big(\exists_{x_1}\exists_{x_2}\exists_{x_3}\big(\neg x_1 = x_2 \wedge \neg x_1 = x_3 \wedge \neg x_2 = x_3 \qquad (10.1)\\
&\qquad \wedge \exists_{x_4}(\neg x_1 = x_4 \wedge \neg x_1 \sim x_4 \wedge x_2 \sim x_4 \wedge x_4 \sim x_3) \\
&\qquad \wedge \exists_{x_4}(\neg x_2 = x_4 \wedge x_1 \sim x_4 \wedge \neg x_2 \sim x_4 \wedge x_3 \sim x_4) \\
&\qquad \wedge \exists_{x_4}(\neg x_3 = x_4 \wedge x_1 \sim x_4 \wedge x_2 \sim x_4 \wedge \neg x_3 \sim x_4)\big)\Big) .
\end{aligned}$$

If a model M satisfies the property described by a sentence ψ, we sometimes say that M is a model for ψ, and write $M \models \psi$. For a model M of L, we denote by $\mathrm{Th}_k^L(M)$, or simply by $\mathrm{Th}_k(M)$, the set of all sentences of L of quantifier depth at most k which hold for M, that is,

$$\mathrm{Th}_k(M) = \mathrm{Th}_k^L(M) = \{\psi \in L : M \models \psi \text{ and } \mathrm{qd}(\psi) \le k\} .$$

Let us also mention that for any first-order language L and every k there are only finite number of non-equivalent properties expressed by sentences from $\mathrm{Th}_k^L(M)$, that is, if for each such property we choose one sentence which describes it, then the set $\mathrm{Th}_k^L(M)$ becomes finite.

10.2 EHRENFEUCHT GAMES AND ZERO-ONE LAWS

In the study of combinatorial properties of discrete structures viewed as models of some languages it is sometimes convenient to use elements of the theory of games. In many cases such an approach provides combinatorial insight into otherwise purely logical considerations and often simplifies the statement of the problem. For analysis of first-order theories the Ehrenfeucht game, introduced by Ehrenfeucht (1961), has proved to be particularly useful.

Ehrenfeucht games

Let M' and M'' denote two models of a first-order language L and let k be a natural number. The *k-round Ehrenfeucht game*, denoted by $\mathrm{Ehr}_k(M', M'')$, is a perfect information game, played on models M' and M'' by two players. In each round the first player chooses one of the models and an element in it. The second player must answer by picking an element in the other model. This procedure is repeated k times until, at the end of the game, two sequences $(x'_1, \ldots, x'_k)$ and $(x''_1, \ldots, x''_k)$ of elements of M' and M'', respectively, are obtained. Let us emphasize that in each step of the game the first player is free to change the model from which he picks an element; if in the first step he decides to choose x'_1 from M', in the second round he might either pick $x''_2 \in M''$, or move again at M'. Thus, some of the vertices $x'_1, \ldots, x'_k$ might have been selected by the first player and the others by the second one. The second player wins $\mathrm{Ehr}_k(M', M'')$ if the substructures spanned by the sequences $(x'_1, \ldots, x'_k)$ and $(x''_1, \ldots, x''_k)$ are isomorphic as models of L; otherwise the game goes to the first player. The isomorphism should map x'_i into x''_i, thus, in the case of the language $L_\sim$ the second player should assure that for all $1 \le i, j \le k$ we have $x'_i = x'_j$ if and only if $x''_i = x''_j$, and $x'_i \sim x'_j$ if and only if $x''_i \sim x''_j$, whereas in the case of $L^{\mathrm{ord}}_\sim$ we additionally require that $x'_i \le x'_j$ if and only if $x''_i \le x''_j$.

Example 10.1. Let ψ be the property that a graph has diameter at most two, that is,

$$\forall_x \forall_y \exists_z (x = y \vee x \sim y \vee (x \sim z \wedge z \sim y)),$$

and suppose that we have two graphs G' and G'' such that ψ holds for G' but not for G''. It is not hard to see that in such a case the first player has a sure win in the game $\mathrm{Ehr}_3(G', G'')$. In the first two moves he should pick two vertices x''_1 and x''_2 which are *not* connected by a path of length at most two in G''. The second player must answer by selecting non-adjacent vertices x'_1 and x'_2 in G'. Then, in his last move, the first player should choose as x'_3 a common neighbor of vertices x'_1 and x'_2 in G', so the second player cannot duplicate his move in G'' because of the choice of x''_1 and x''_2.

Example 10.2. Let G' and G'' be two graphs such that in G' the shortest cycle has length six whereas G'' contains no cycles of length smaller than seven. Then, although the first player cannot build in G'' a cycle of length six in four moves, nonetheless he can win $\mathrm{Ehr}_4(G', G'')$. Indeed, in the first three moves he should choose vertices x'_1, x'_2, x'_3 which belong to a cycle of length six in G' and are such that the distance between any two of them is two. In response to that the second player must pick three non-adjacent vertices x''_1, x''_2, x''_3 in G'' but, since G'' contains no short cycles, either all three of them are adjacent to one vertex, or the distance between at least one pair of vertices, say x''_2 and x''_3, must be larger than two. In either case the first player wins if he chooses as x'_4 the common neighbor of x'_2 and x'_3.

One might suspect that in Example 10.1 the strategy of the first player, who in the first two rounds picks elements from G'' and in the third one switches to G', has something to do with the fact that in the sentence ψ two general quantifiers are followed by the existential one. Furthermore, the reader patient enough to study the formula (10.1) can easily notice that the strategy of the first player in Example 10.2 in a way "reflects" the structure of the sentence (10.1) of depth four which holds for G' but not for G''. More generally, it is not hard to see that if the first player can win $\text{Ehr}_k(M', M'')$ then, knowing his winning strategy, one can write down a sentence ψ of quantifier depth at most k which holds for M' but not for M''. On the other hand, if there is a sentence of quantifier depth at most k which holds for only one of M' and M'', the first player can use it to create his winning strategy for $\text{Ehr}_k(M', M'')$. This observation, stated below as Lemma 10.3, can be used to replace an argument from mathematical logic by purely combinatorial considerations.

Lemma 10.3. *The second player has a winning strategy in the game* $\text{Ehr}_k(M', M'')$ *if and only if* $\text{Th}_k(M') = \text{Th}_k(M'')$. ∎

One can use the above fact to prove that certain properties cannot be expressed in the first-order language of graphs. We give just one such example; the reader can easily show using a similar argument that the properties that a graph contains a cycle, or a perfect matching, or that it is bipartite, or, finally, that it contains an even number of vertices, cannot be expressed in the first-order language of graph theory (Exercise!).

Corollary 10.4. *The property that a graph is connected cannot be written as a sentence from* $L_\sim$.

Proof. Suppose that there is a sentence ψ in $L_\sim$ of quantifier depth k, which says that a graph is connected. Furthermore, let G' be a cycle of length 3^{k+1} and let G'' consist of two copies of G'. Then there exists a simple winning strategy for the second player in $\text{Ehr}_k(G', G'')$: if in the i-th move the first player picks a vertex which is within distance 3^{k-i} from some vertex chosen previously, the second player should select a "corresponding" vertex in the second model, otherwise he may answer with any vertex which lies far enough from vertices already chosen. The reader is encouraged to verify that this strategy really works, i.e., that, basically, if the distances between x_1' and x_2', and x_1'' and x_2'' are both larger than 3^{k-1}, then the first player cannot effectively use the fact that the distances might be different to win the game $\text{Ehr}_k(G', G'')$. Thus, $\text{Th}_k(G') = \text{Th}_k(G'')$, contradicting the fact that $\psi \in \text{Th}_k(G')$ while $\psi \notin \text{Th}_k(G'')$. ∎

Zero-one laws for very dense and very sparse random graphs

Throughout the chapter we will repeatedly use Lemma 10.3 to establish the zero-one law in $\mathbb{G}(n, p)$ for some particular function $p = p(n)$. Since for this type of problem it does not matter whether we study properties of a random graph or properties of its complement, we may and will always assume that $p = p(n) \leq 1/2$.

We start with a classical result on dense random graphs by Glebski, Kogan, Liagonkii and Talanov (1969), independently proved also by Fagin (1976).

Theorem 10.5. *If* $1/2 \geq p = p(n) = n^{-o(1)}$ *then for every sentence* ψ *from* $L_\sim$ *the probability that* $\mathbb{G}(n,p)$ *satisfies* ψ *tends either to* 0 *or to* 1 *as* $n \to \infty$.

Proof. For a natural number $k \geq 1$ let φ_k be the property that for every pair of disjoint subsets of vertices S_1 and S_2, such that $|S_1| + |S_2| \leq k-1$, there exists a vertex which is adjacent to all vertices from S_1 and no vertices from S_2. (Note, by the way, that since both S_1 and S_2 have prescribed size, φ_k is a first-order sentence of quantifier depth at most k.) Then, for every $k \geq 1$, we have

$$\mathbb{P}(\mathbb{G}(n,p) \models \neg\varphi_k) \leq \sum_{\ell_1=0}^{k-1} \sum_{\ell_2=0}^{k-1-\ell_1} n^{\ell_1+\ell_2}(1-p^{\ell_1}(1-p)^{\ell_2})^{n-k}$$
$$\leq k^2 n^{k-1} \exp(k - np^{k-1}(1-p)^{k-1}) \leq k^2 e^k n^k \exp(-n^{1-o(1)}) = o(1),$$

and so *a.a.s.* $\mathbb{G}(n,p)$ satisfies φ_k. On the other hand, for any two graphs G' and G'' for which φ_k holds, the second player can win $\mathrm{Ehr}_k(G', G'')$ by playing an obvious strategy: the fact that both φ_k holds for both G' and G'' guarantees that in each of the first k steps he will be able to imitate the moves of his opponent. Thus, *a.a.s.* $\mathrm{Th}_k(G(n,p)) = \mathrm{Th}_k(G_k)$, where G_k is any graph for which φ_k holds (the existence of G_k follows from the probabilistic part of our proof). Consequently, for any sentence ψ from $L_\sim$ of quantifier depth k we have either $\psi \in \mathrm{Th}_k(G_k)$, and then

$$\lim_{n\to\infty} \mathbb{P}(\mathbb{G}(n,p) \models \psi) = 1,$$

or $\psi \notin \mathrm{Th}_k(G_k)$, which implies that

$$\lim_{n\to\infty} \mathbb{P}(\mathbb{G}(n,p) \models \psi) = 0. \qquad \blacksquare$$

It is worthwhile to note that the above argument proves that for every sentence ψ of $L_\sim$ of quantifier depth at most k either ψ or $\neg\psi$ is a consequence of the axiom set which consists of one sentence φ_k, in other words, we showed that $\{\varphi_k\}$ is a complete system of axioms for the theory which consists of the sentences of depth at most k from $L_\sim$. Needless to say, this can be verified directly without invoking Ehrenfeucht games. As a matter of fact, from the remarks which led us to Lemma 10.3, it should be clear that virtually every proof which employs Ehrenfeucht games can be transformed into a purely logical argument, and it is only a matter of personal preference whether one decides to use the logical or the combinatorial approach.

Now let us consider the behavior of a very sparse random graph $\mathbb{G}(n,p)$, that is, the case $p = p(n) \le n^{-1+o(1)}$. In order to do that we first identify some functions for which the zero-one law certainly does *not* hold. According to Theorem 3.19, if

$$p_\ell = cn^{-1-1/\ell}$$

for some natural number $\ell \ge 1$ and a constant $c > 0$, then the probability of the first-order property saying that $\mathbb{G}(n,p_\ell)$ contains a tree of size $\ell+1$ converges to a limit which is neither zero nor one. Furthermore, from the same result we infer that if

$$np_0 = c'$$

for some constant $c' > 0$, then the probability that $\mathbb{G}(n,p_0)$ contains a triangle also tends to a non-trivial limit. Finally, let r, s be integers such that $s \ge r-1 \ge 0$ and

$$rnp_{r,s} = \log n + s\log\log n + c'',$$

where this time c'' is any constant, not necessarily positive, and let $\mathcal{A}_{r,s}$ denote the first-order property that a graph contains a path $v_1\cdots v_r$ such that the total number of edges incident to the vertices $v_1,\ldots,v_r$ is s. Elementary application of the method of moments (Exercise!) shows that the probability that $\mathbb{G}(n,p_{r,s})$ has $\mathcal{A}_{r,s}$ tends to a non-trivial limit, which rules out $p_{r,s}$ as a possible candidate for a function for which the zero-one law holds. We will show that, as long as $p \le n^{-1+o(1)}$, the functions p_ℓ, p_0 and $p_{r,s}$ are, in a way, the only ones which do not obey the zero-one law (see Shelah and Spencer (1988) and Łuczak and Spencer (1991), where Theorem 10.6 is proved using a somewhat different method).

Theorem 10.6. *Let $p = p(n)$ be a function for which one of the following conditions holds:*

(i) $n^{-1-1/\ell} \ll p \ll n^{-1-1/(\ell+1)}$ *for some natural number* $\ell \ge 1$,

(ii) $p \ll n^{-1}$ *but* $p = n^{-1-o(1)}$,

(iii) $p \gg n^{-1}$ *but* $p = n^{-1+o(1)}$ *and for every given pair of integers r and s such that $s \ge r-1 \ge 0$ we have either $rnp - \log n - s\log\log n \to -\infty$, or $rnp - \log n - s\log\log n \to \infty$.*

Then, for every sentence ψ from $L_\sim$ the probability that $\mathbb{G}(n,p)$ satisfies ψ tends either to 0 or to 1 as $n \to \infty$.

Proof. Let ψ be a sentence from $L_\sim$ of quantifier depth k. Consider first the simplest case $n^{-1-1/\ell} \ll p \ll n^{-1-1/(\ell+1)}$. In this early period of the evolution, *a.a.s.* $\mathbb{G}(n,p)$ consists of isolated trees of size at most $\ell+1$ and every tree of size at most $\ell+1$ appears as a component of $\mathbb{G}(n,p)$ at least k times (see Theorems 3.4 and 3.19). Clearly, this property uniquely determines all properties of a graph described by the sentences of length at most k, since in the Ehrenfeucht game on two such graphs the second player may always choose a vertex which lies in a tree isomorphic to one picked by the first player.

The case when $p \ll n^{-1}$ but $p = n^{-1-o(1)}$ is only slightly more difficult. Set

$$\Xi_k = \{\mathrm{Th}_k(T) : T \text{ is a finite tree}\}$$

and let

$$\mathcal{T}_k = \{T_1, T_2, \ldots, T_{|\Xi_k|}\}$$

be a collection of trees such that for every finite tree T there is a tree $T' \in \mathcal{T}_k$ for which $\mathrm{Th}_k(T) = \mathrm{Th}_k(T')$. Furthermore, let G_k be a graph which consists of k copies of each tree from $\mathcal{T}_k$. If $p = p(n) = n^{-1-o(1)}$ and $np \to 0$ then *a.a.s.* $G' = \mathbb{G}(n,p)$ is a forest which contains at least k copies of each given tree as components (Exercise!), in particular, it contains at least k copies of each element from $\mathcal{T}_k$. It is easy to see that this property ensures that the second player can win the game $\mathrm{Ehr}_k(G_k, G')$. Indeed, if the first player chooses a vertex in some component T of G', the second one will pick a component T' of G_k such that $\mathrm{Th}_k(T') = \mathrm{Th}_k(T)$, and select a vertex in T' which guarantees him a win in $\mathrm{Ehr}_k(T, T')$. Needless to say, if at some stage of the game the first player decides to choose again a vertex from either T or T' the second player will play according to his winning strategy for $\mathrm{Ehr}_k(T, T')$.

The last part of the proof is slightly more complicated. Roughly speaking, now the second player will use the fact that for $p = n^{-1+o(1)}$ the typical random graph $\mathbb{G}(n,p)$ "locally" resembles a tree, in which every vertex has large degree. The only problem he might face is the presence of sparsely distributed short cycles and some number of vertices of small degree.

In order to deal with short cycles, let us recall that the distance between two subsets of vertices V' and V'' of a graph is measured by the number of edges in the shortest path joining a vertex from V' to a vertex from V''. Note that Theorem 3.4 implies that if $p = n^{-1+o(1)}$ then *a.a.s.* any two cycles of $\mathbb{G}(n,p)$ shorter than 3^{k+1} lie at a distance larger than 3^{k+2}. Elementary first moment calculations show also that if $p \gg n^{-1}$ then *a.a.s.* every vertex of $\mathbb{G}(n,p)$ which lies within distance 3^{k+2} from such a short cycle must have degree at least 3^{k+1} (Exercise!). Furthermore, *a.a.s.* $\mathbb{G}(n,p)$ contains at least k cycles of each length ℓ, where $3 \le \ell \le 3^{k+1}$, provided $pn \to \infty$ (Exercise!). Thus, although $\mathbb{G}(n,p)$ contains a lot of short cycles, they lie far apart from each other and from other "critical" structures and can be dealt with without much problem (see the proof of Corollary 10.4).

Unfortunately, unlike short cycles, vertices of small degree can appear in clusters, so we need to classify them according to their neighborhood. We say that two vertices of a graph are of the same 0-*type* if either they have the same degree smaller than k, or both of them have degree at least k. Clearly, this is an equivalence relation which partitions the vertices, according to their degrees, into $k+1$ classes. Furthermore, for $i = 1, \ldots, k$, we say that vertices w and w' of the graph have the same i-*type* if for each $(i-1)$-type the number of neighbors of w and w' which belong to that type are either the same, or both at least k. Note that for every given i there are only finitely many i-types. Moreover, to determine the i-type of a vertex v one needs only to check if v is the root of some finite subgraph F of $\mathbb{G}(n,p)$ of depth $i+1$ such that the size of F is bounded by a function of i and k and its vertices have certain prescribed degrees in $\mathbb{G}(n,p)$. It turns out that, whenever a function $p = p(n)$ fulfills the assumptions of Theorem 10.6 for any given natural numbers i and k, the class of all i-types, where $1 \le i \le 3^{k+1}$, can be split into two subclasses of "likely" and "unlikely" i-types, where the division depends on the behavior of p with respect to $p_{r,s}$. It can be shown that *a.a.s.* $\mathbb{G}(n,p)$ contains no vertices of unlikely types; on the other hand *a.a.s.* $\mathbb{G}(n,p)$ contains at least k vertices of each likely type, which can be chosen in such a way that they lie at distance at least 3^{k+1} from each other. A rigorous verification of this fact relies on the computation of the expectation and variance of the number of vertices of a given i-type. Thus, if the expected number of vertices of a given i-type tends to 0, then *a.a.s.* such vertices do not appear in $\mathbb{G}(n,p)$. On the other hand, if the expected number of vertices of an i-type tends to infinity, then one can use Chebyshev's inequality (1.2) to verify that, under the assumption of the theorem, $\mathbb{G}(n,p)$ *a.a.s.* contains many of them. Since this is a standard, but quite tedious argument, we neither elaborate on it, nor dare to recommend it to the reader as an exercise.

Now let us consider the k-round Ehrenfeucht game on two graphs such that both of them

- have many sparsely distributed short cycles,
- contain no vertices of unlikely i-types, for $i = 0, 1, \ldots, 3^{k+1}$,
- contain a rich sparsely distributed system of representatives of likely i-types, for $i = 0, 1, \ldots, 3^{k+1}$,

where each of the above items should be understood in the context of the probabilistic results on properties of $\mathbb{G}(n,p)$ we mentioned. In such a game the second player has the following winning strategy. If in the i-th round of the game the first player chooses a vertex, the second one checks whether it lies within distance 3^{k-i} from either a cycle shorter than 3^{k+1}, or some vertex chosen previously: if this is the case, then he answers with an "analogous" vertex in the second graph (in the case of a cycle this means any vertex which lies at the same distance from the corresponding cycle in the other graph); otherwise he chooses any vertex of the same 3^{k-i+1}-type which lies at distance at least 3^{k-i+1} from all vertices selected so far. ∎

Sophisticated winning strategies: The closure

Let us assume now that $p = p(n) = n^{-\alpha+o(1)}$ for some constant $\alpha \in (0,1)$. It turns out that in this case the behavior of $\mathbb{G}(n,p)$ depends strongly on whether α is rational or irrational. The analysis of the case in which α is rational is somewhat involved: we postpone it until the next section and concentrate on the result of Shelah and Spencer (1988), stating that the zero-one law holds for every irrational $\alpha \in (0,1)$.

Theorem 10.7. *Let $p = p(n) = n^{-\alpha+o(1)}$, where $\alpha \in (0,1)$ is irrational. Then, for every sentence ψ from $L_{\sim}$ the probability that $\mathbb{G}(n,p)$ satisfies ψ tends either to 0 or to 1 as $n \to \infty$.*

As in the case of Theorems 10.5 and 10.6, we prove Theorem 10.7 by showing that a certain Ehrenfeucht game can be won by the second player; however now his strategy must be much more sophisticated. Indeed, for $p(n) = n^{-o(1)}$ the random graph $\mathbb{G}(n,p)$ is *a.a.s.* so dense that it contains every possible configuration which can emerge during the Ehrenfeucht game. Thus, the second player can pick in each move any vertex which is appropriate at this stage of the game and not worry about his opponent's further moves. Now this is no longer the case, if $p(n) = n^{-\alpha+o(1)}$ and $\alpha \in (0,1)$, then one can easily find a graph H such that *a.a.s.* $\mathbb{G}(n,p)$ contains a copy of H, but not more than $o_p(n)$ of them. Hence, if the first player chooses in his first move a vertex in a copy of such an H, the second player must choose a vertex which also belongs to a copy of H, and at each stage of the game he must constantly analyze all possible strategies of his adversary. Furthermore, unlike in the game played on sparse graphs containing only a few "critical" structures lying far away from each other, here the diameter of the graph is bounded from above by $\lceil 1/(1-\alpha) \rceil$ (see Bollobás (1985, Chapter X)). Thus, there are no "isolated" structures and if the Ehrenfeucht game lasts long enough the first player can connect by a path any two vertices of the graph.

In order to get some feeling for what the strategy of the second player should be, let us pick, say $\alpha = \sqrt{3}/2 = 0.86602\ldots$, and consider the Ehrenfeucht game of length k played on two graphs G', G'' such that if a sentence ψ of quantifier depth at most k holds *a.a.s.* in $\mathbb{G}(n,p)$ with $p = p(n) = n^{-\alpha}$, then it holds also for both G' and G''.

First set $k = 3$. Then $\mathbb{G}(n,p)$ contains $\Theta_p(n^{3-3\alpha})$ triangles (see Remark 3.7) and, since $0 < 3 - 3\alpha < 1$, we may assume that both G' and G'' contain some triangles as well as vertices which do not belong to them. Thus, before his first move the second player should check at least whether the vertex selected by the first player lies on a triangle, and pick his vertex accordingly.

Now suppose that $k = 4$. Then the first move of the second player should depend not only on whether the vertex v picked by the first player belongs to a triangle: he certainly must consider also larger structures containing v, such as cycles of size four, five, six (see Example 10.2) and, as one can easily check, even seven. Furthermore, he must examine whether v is adjacent to some triangle and, if this is the case, choose a vertex which also has this property since otherwise the first player can win easily.

What should be the first move of the second player when k becomes even larger, say $k = 15$? Certainly, he must, in particular, take into account all aspects of the choice of the first player we mentioned for $k = 4$, since otherwise he will lose after the first four moves. But should he care whether the vertex v chosen by the first player lies within distance, say, ten from some triangle? It does not seem to be absolutely necessary: *a.a.s.* the diameter $\mathbb{G}(n,p)$ is eight and so is the diameter of G' and G'', thus the second player might hope that he can always join the first vertex to a triangle when he notices that the first player is trying to do so. Hence, he must concentrate on the fact that the vertex v selected by the first player either belongs to a "rare" structure (like, in the case of $\alpha = \sqrt{3}/2$, a short cycle), or at least lies close to it. More specifically, he should compute the "closure" of v, which consists of such structures, and choose in the second graph a vertex with the same closure as v. Now he might try to play according to the following strategy: if the first player chooses a vertex in the closure of the vertices chosen so far, then the second one picks a corresponding vertex in the closure of the set of vertices chosen in the other graph; if the first player decides to move outside the closure, then, hopefully, the second player will be able to find an "equivalent" vertex outside the closure in the second graph.

Hence, the closure of a vertex or a set of vertices, should consist, roughly speaking, of all "rare" structures containing this vertex or set and, in the random graph, "rare" typically means "denser than average". Note, however, that our sketchy analysis of the case $\alpha = \sqrt{3}/2$, $k = 4$, has suggested that a triangle which lies at distance one from a vertex should belong to its closure, while, on the other hand, if a vertex v belongs to a triangle it is clearly not necessary to include all vertices adjacent to it in the closure of v. Thus, to find the closure of a vertex v we should not only examine all copies of a graph H in $\mathbb{G}(n,p)$ which contain v – we must also take into account how v is placed in them.

Thus, instead of graphs, we will consider *rooted graphs* (R,H), which consist of a graph H and a subset of its vertices $R \subseteq V(H)$ (see Section 3.4; although now, for technical reasons, we allow edges between vertices of R, they are irrelevant for our argument). The elements of R are *roots*, the vertices from $V(H) \setminus R$ are called *extension vertices*, and we say that an edge of H is *proper* if it does not join two roots. The number of extension vertices for a rooted graph (R,H) is denoted by $v(R,H)$, whereas $e(R,H)$ stands for the number of proper edges. We say that (R,H) is *sparse* if $v(R,H) > \alpha e(R,H)$, and *dense* otherwise, where, throughout the rest of this section, α denotes a fixed irrational number from the interval $(0,1)$. Thus, (R,H) is dense if either $R = V(H)$, or $v(R,H) < \alpha e(R,H)$. We say that (R,H) is *rigid* if (S,H) is dense for every S with $R \subseteq S \subseteq V(H)$, and *safe* if (R,H') is sparse for every subgraph H' of H with $R \subset V(H') \subseteq V(H)$. Note that if (R,H) is rigid, then any rooted graph (R',H') such that $R \subseteq R'$, $V(H') = R' \cup V(H)$, and $H \subseteq H'$, is rigid as well.

An extensive list of elementary properties of safe and rigid rooted graphs can be found in Shelah and Spencer (1988); here we will mention only three of them. Throughout this section $H \cup H'$ $[H \cap H']$ denotes the graph with vertex set $V(H) \cup V(H')$ $[V(H) \cap V(H')]$ and edge set $E(H) \cup E(H')$ $[E(H) \cap E(H')]$.

Fact 10.8. *If both (R, H) and $(V(H), H')$ are rigid, then (R, H') is rigid.*

Proof. Let $R \subseteq S \subseteq V(H')$. Then, since both (R, H) and $(V(H), H')$ are rigid, we get

$$\begin{aligned} v(S, H') &= v(S \cap V(H), H) + v(S \cup V(H), H') \\ &\leq \alpha e(S \cap V(H), H) + \alpha e(S \cup V(H), H') \leq \alpha e(S, H'). \end{aligned}$$ ■

Fact 10.9. *If both (R, H) and (R, H') are rigid, then $(R, H' \cup H)$ is rigid.*

Proof. Note that $(V(H), H \cup H') = (V(H) \cup R, V(H) \cup H')$ is rigid and apply Fact 10.8. ■

Fact 10.10. *If (R, H) is not safe, then there exists a subgraph H' of H such that $R \subset V(H')$ and the rooted graph (R, H') is rigid.*

Proof. Let H' be a minimal subgraph of H such that $R \subset V(H')$ and the rooted graph (R, H') is dense. Furthermore, let S'' be any set such that $R \subset S'' \subset V(H')$ and let H'' denote the subgraph spanned by S'' in H'. Because of the choice of H', the rooted graph (R, H'') is sparse, and so

$$v(S'', H') = v(R, H') - v(R, H'') < \alpha e(R, H') - \alpha e(R, H'') = \alpha e(S'', H').$$

Thus, (R, H') is rigid. ■

Let (R, H) be a given rooted graph and R' be a subset of vertices of a graph G, such that $|R| = |R'|$. A pair (R', H') is an *(R, H)-extension of R in G* if H' is a subgraph of G such that $R' \subseteq V(H') \subseteq V(G)$ and there exists a bijection $f : V(H) \to V(H')$ which maps R into R' and transforms all proper edges of (R, H) into edges of H', that is, if $\{v', v''\} \in E(H)$ and either $v', v'' \in V(H) \setminus R$, or $v' \in R$, $v'' \in V(H) \setminus R$, then also $\{f(v'), f(v'')\} \in E(H')$. Note that we do not care about edges joining two roots, nor do we prohibit additional edges in H'.

Finally, we introduce the crucial notion of t-closure. Let $t \geq 0$, G be a graph and $W \subseteq V(G)$. The *t-closure of W* is the minimal set $\bar{W} \supseteq W$ such that for every rigid rooted graph (R, H) with $v(R, H) \leq t$, each (R, H)-extension of a subset R' of $\bar{W}$ is contained in $\bar{W}$. The t-closure of W we denote by $\mathrm{cl}_t(W)$, or, if $W = \{w_1, \ldots, w_r\}$, we put simply $\mathrm{cl}_t(w_1, \ldots, w_r)$. It is easy to see that $\mathrm{cl}_t(W)$ is uniquely defined, and that for all W and t there exists a sequence $W = W_0 \subset W_1 \subset \cdots \subset W_\ell = \mathrm{cl}_t(W)$ such that for $i = 1, \ldots, \ell$, the set W_i is obtained by appending to W_{i-1} the vertices of some rigid (R_i, H_i)-extension of a subset of W_{i-1}, where $v(R_i, H_i) \leq t$. Thus, due to Facts 10.8 and 10.9, $(W, \mathrm{cl}_t(W))$ is a rigid extension of W.

Our next result states that in a random graph the t-closure of a small set is unlikely to be very large.

Lemma 10.11. *Let $p = p(n) = n^{-\alpha+o(1)}$, where $\alpha \in (0,1)$ is an irrational number. Then for all natural numbers r and t, there exists $M = M(\alpha, r, t)$ such that* a.a.s. *for every choice of vertices $v_1, \ldots, v_r$ of $\mathbb{G}(n,p)$ we have*

$$|\operatorname{cl}_t(v_1, \ldots, v_r)| \leq M.$$

Proof. As we have already observed, for every set W and a natural number t there exists a sequence $W = W_0 \subset W_1 \subset \cdots \subset W_\ell = \operatorname{cl}_t(W)$ such that $W_i = W_{i-1} \cup V(H_i')$, for some rigid (R_i, H_i)-extension (R_i', H_i'), with $v(R_i, H_i) \leq t$ and $R_i' \subseteq W_{i-1}$. Furthermore, it is easy to see that we may assume that all rooted graphs (R_i, H_i) are *minimal rigid graphs*, that is, for each rigid graph $(\hat{R}_i, \hat{H}_i)$ such that $\hat{R}_i \subseteq R_i$, $\emptyset \neq V(\hat{H}_i) \setminus \hat{R}_i \subseteq V(H_i) \setminus R_i$ and $\hat{H}_i \subseteq H_i$ we have $\hat{R}_i = R_i$ and $\hat{H}_i = H_i$. Note also that for each minimal rigid rooted graph (R_i, H_i) with $v(R_i, H_i) \leq t$ we must have $|R_i| \leq t/\alpha + t$. Indeed, otherwise either there exists a vertex in R_i with no neighbors in $V(H_i) \setminus R_i$, or some vertex $v \in V(H_i) \setminus R_i$ has more than $1 + 1/\alpha$ neighbors in R_i, contradicting the minimality of (R_i, H_i). Thus, since there are only finitely many non-isomorphic rooted graphs (R, H) with $|R| \leq t/\alpha + t$ and $v(R, H) \leq t$, we have

$$\begin{aligned} \varepsilon_t = \min\{\alpha e(R,H) - v(R,H) : (R,H) \text{ is dense} \\ \text{and } 0 < v(R,H) \leq t,\ |R| \leq t/\alpha + t\} > 0. \end{aligned}$$

We will show that the assertion holds with $M = r + t\lceil r/\varepsilon_t \rceil$.

The idea of the proof is rather simple. Suppose that $W = \{v_1, \ldots, v_r\}$ is a subset of vertices of a graph G such that $|\operatorname{cl}_t(W)| \geq M$. Let $W = W_0 \subset W_1 \subset \cdots \subset W_m$ be a sequence of subsets of G such that for $i = 1, 2, \ldots, m$ the set W_i is the result of an (R_i, H_i)-expansion of W_{i-1}, where (R_i, H_i) is a minimal rigid rooted graph with $v(R_i, H_i) \leq t$, and finally $|W_m| > M$ while for $i = 0, \ldots, m-1$ we have $|W_i| \leq M$. We will show that the subgraph induced by W_m contains at least $|W_m|/\alpha$ edges, whereas, due to Theorem 3.4, *a.a.s.* $\mathbb{G}(n,p)$ contains no subgraphs on at most $M + t$ vertices with density larger than $1/\alpha$.

Thus, let G_i denote the subgraph of G spanned by W_i, $i = 0, 1, \ldots, m$. Since G_i is the $(W_{i-1} \cup R_i, W_{i-1} \cup H_i)$-extension of W_{i-1}, where the rooted graph $(W_{i-1} \cup R_i, W_{i-1} \cup H_i)$ is rigid, and furthermore, (R_i, H_i) is a minimal rigid rooted graph, we have

$$|W_i| - |W_{i-1}| = v(W_{i-1} \cup R_i, W_{i-1} \cup H_i) \leq \alpha e(W_{i-1} \cup R_i, W_{i-1} \cup H_i) + \varepsilon_t .$$

Then, for $i = 1, 2, \ldots, m$,

$$e(G_i) - e(G_{i-1}) \geq \frac{|W_i| - |W_{i-1}|}{\alpha} + \frac{\varepsilon_t}{\alpha} .$$

Consequently, since $|W_i| - |W_{i-1}| \leq t$ for $i = 1, 2, \ldots, m$, and $|W_m| > M = r + t\lceil r/\varepsilon_t \rceil$, we have $m > \lceil r/\varepsilon_t \rceil$, and so

$$\frac{e(G_m)}{v(G_m)} > \frac{1}{|W_m|}\left(\frac{|W_m| - r}{\alpha} + \left\lceil \frac{r}{\varepsilon_t} \right\rceil \frac{\varepsilon_t}{\alpha}\right) \geq \frac{1}{\alpha} .$$ ∎

Now we state two probabilistic lemmas on $\mathbb{G}(n,p)$ which will be crucial for constructing the winning strategy for the second player. The first one, Lemma 10.12, guarantees that in the first move the second player can choose a vertex whose closure is identical with the closure of the vertex selected by the first player. Lemma 10.13 assures the existence of a suitable move for the second player when the first one selects a vertex outside the closure.

Lemma 10.12. *Let $p = p(n) = n^{-\alpha+o(1)}$, where $\alpha \in (0,1)$ is an irrational number, let $(\{w\}, H)$ be a rooted rigid graph such that $m(H) < 1/\alpha$, and let $t \geq v(\{w\}, H)$. Then* a.a.s. *$\mathbb{G}(n,p)$ contains a vertex v such that $\mathrm{cl}_t(v)$ is isomorphic to $(\{w\}, H)$, where the isomorphism maps v into w.* ∎

Lemma 10.13. *Let $t \geq 1$, $p = p(n) = n^{-\alpha+o(1)}$, where $\alpha \in (0,1)$ is an irrational number and let (R, H) be a rooted safe graph with $|R| = r$. Then* a.a.s. *for every choice of vertices $v_1, \ldots, v_r$ of $\mathbb{G}(n,p)$ there exists an (R,H)-extension $(\{v_1, \ldots, v_r\}, H')$ such that $\mathrm{cl}_t(V(H')) = \mathrm{cl}_t(v_1, \ldots, v_r) \cup V(H')$.* ∎

Unfortunately, the proofs of the above results are somewhat technical, so we just briefly comment on them. The existence of the required extensions in both lemmas is not very hard to show: in the case of Lemma 10.12 it is an immediate consequence of Theorem 3.4, for Lemma 10.13 it follows from Theorem 3.27. However, one needs also to prove that at least one such extension contains no vertices which can be used as the roots of rigid rooted graphs to extend the closure of the set we are dealing with. This can be done using a similar technique as employed in the proof of Lemma 10.11, but since the argument is long and not very instructive, we decided to omit it here.

Proof of Theorem 10.7. Let k be a fixed natural number. Define a sequence $t_1, \ldots, t_k$ setting $t_k = 0$ and $t_i = M(\alpha, k, t_{i+1})$ for $i = 1, 2, \ldots, k-1$, where $M(\alpha, r, t)$ is a function which fulfills the assertion of Lemma 10.11. Let G', G'' be two graphs such that for all $r \leq k$, $t \leq t_1$, and every graph H with at most $M(\alpha, k, t_1)$ vertices, all graph properties which, according to Theorem 3.4, Lemmas 10.11, 10.12 and 10.13, *a.a.s.* hold for $\mathbb{G}(n,p)$, hold also for both G' and G''. We will show that the second player can win the Ehrenfeucht game $\mathrm{Ehr}_k(G', G'')$, which, due to Lemma 10.3, will imply the theorem.

A winning strategy for the second player for $\mathrm{Ehr}_k(G', G'')$ has already been anticipated in our remarks following the statement of Theorem 10.7. We will show that, for $i = 1, \ldots, k$, in the i-th step of the game the second player can make his move in such a way that the subgraph G'_i induced in G' by $\mathrm{cl}_{t_i}(x'_1, \ldots, x'_i)$, and G''_i spanned in G'' by $\mathrm{cl}_{t_i}(x''_1, \ldots, x''_i)$, are isomorphic, where the isomorphism maps x'_j into x''_j for $j = 1, \ldots, i$. (Here and below $x'_j \in V(G')$ and $x''_j \in V(G'')$ stand for the vertices chosen in the j-th step of the game $\mathrm{Ehr}_k(G', G'')$.)

Since neither G' nor G'' contains subgraphs H with $|V(H)| \leq M(\alpha, k, t_1)$ and $m(H) \geq 1/\alpha$ (Theorem 3.4), the existence of a vertex with appropriate t_1-closure which can be chosen in the first move by the second player is assured by Lemma 10.12. Thus, let us assume that for some i, $i = 1, \ldots, k-1$, the subgraphs G'_i and G''_i are isomorphic. Furthermore, without loss of generality we may assume that the first player decides to pick a vertex x'_{i+1} from G'. Our aim is to show that there exists a vertex x''_{i+1} in G'' such that the subgraph G'_{i+1} spanned in G' by $\mathrm{cl}_{t_{i+1}}(x'_1, \ldots, x'_{i+1})$ is isomorphic to the subgraph G''_{i+1} induced by $\mathrm{cl}_{t_{i+1}}(x''_1, \ldots, x''_{i+1})$ in G''. We consider the following two cases.

Case 1: $x'_{i+1} \in \mathrm{cl}_{t_i}(x'_1, \ldots, x'_i)$.
Then, using the isomorphism of G'_i and G''_i, the second player should choose in G''_i the vertex x''_{i+1} corresponding to x'_{i+1}. Since

$$|V(G'_{i+1})| = |\,\mathrm{cl}_{t_{i+1}}(x'_1, \ldots, x'_{i+1})| \le M(\alpha, k, t_{i+1}) = t_i,$$

the definition of t_i-closure and Fact 10.8 imply that

$$V(G'_{i+1}) = \mathrm{cl}_{t_{i+1}}(x'_1, \ldots, x'_{i+1}) \subseteq \mathrm{cl}_{t_i}(x'_1, \ldots, x'_i) = V(G'_i).$$

Similarly, we have $V(G''_{i+1}) \subseteq V(G''_i)$, and so G'_{i+1} and G''_{i+1} are isomorphic as subgraphs of G'_i and G''_i respectively.

Case 2: $x'_{i+1} \notin \mathrm{cl}_{t_i}(x'_1, \ldots, x'_i)$.
Let H'_i denote the maximal subgraph such that

$$x'_1, \ldots, x'_i \in V(H'_i) \subseteq \mathrm{cl}_{t_{i+1}}(x'_1, \ldots, x'_{i+1}) = V(G'_{i+1})$$

and the rooted graph $(\{x'_1, \ldots, x'_i\}, H'_i)$ is rigid. Note that, due to Fact 10.9, H'_i is uniquely defined and contains $\mathrm{cl}_{t_{i+1}}(x'_1, \ldots, x'_i)$. Because H'_i is chosen to be maximal, from Fact 10.10 it follows that the rooted graph $(V(H'_i), G'_{i+1})$ is safe. Moreover, since

$$|V(H'_i)| \le |\,\mathrm{cl}_{t_{i+1}}(x'_1, \ldots, x'_{i+1})| \le M(\alpha, k, t_{i+1}) = t_i,$$

we have

$$V(H'_i) \subseteq \mathrm{cl}_{t_i}(x'_1, \ldots, x'_i) = V(G'_i).$$

and consequently $x'_{i+1} \notin V(H'_i)$.

Now let H''_i be a subgraph of G''_i isomorphic to $H'_i \subseteq G'_i$. Then the second player should use Lemma 10.13 to find a $(V(H'_i), G'_{i+1})$-extension $(V(H''_i), G''_{i+1})$ of H''_i in G'' which does not enlarge the t_{i+1}-closure of $V(H''_i)$. Finally, as x''_{i+1}, he should pick the vertex of $(V(H''_i), G''_{i+1})$ which corresponds to x'_{i+1} in G'_{i+1}. ■

10.3 FILLING GAPS

In this section we complete the "first-order picture" of $\mathbb{G}(n,p)$. Thus, we look at the behavior of $\mathbb{G}(n,p)$ for probability functions $p = p(n)$ which are "close" to the known thresholds, such as $p = n^{-\alpha}$, where either $\alpha = 1 + 1/\ell$ for some natural number ℓ, or α is a positive rational number smaller than one. Finally, we discuss briefly what happens when we consider properties of $\mathbb{G}(n,p)$ which can be expressed in $L^{\mathrm{ord}}_{\sim}$, a much stronger language than $L_{\sim}$ which has been studied so far.

Weak zero-one laws

We start with the case $p \leq n^{-1+o(1)}$ when, as we have already learned from the previous section, the behavior of $\mathbb{G}(n,p)$ is rather easy to study. Theorem 10.6, which specifies the threshold functions for this range of $p(n)$, is nicely supplemented by the following result, which says that at these thresholds the probabilities of sentences from $L_\sim$ converge, that is, the *weak zero-one law* holds. (For a more thorough analysis of the behavior of $\mathbb{G}(n,p)$ at this stage of the evolution see Lynch (1990) and Spencer and Thoma (1999).)

Theorem 10.14. *Let $p = p(n)$ be such that one of the following conditions holds:*

(i) $n^{1+1/\ell}p \to c$, *for some natural number $\ell \geq 1$ and a positive constant c,*

(ii) $np \to c'$, *for some positive constant c',*

(iii) $rnp - \log n - s \log\log n \to c''$ *for some, not necessarily positive, constant c'' and integers r, s such that $0 \leq r - 1 \leq s$.*

Then, for every sentence ψ from $L_\sim$, the probability that $\mathbb{G}(n,p)$ satisfies ψ converges as $n \to \infty$.

Proof. The proof closely follows the way which led us to Theorem 10.6. Let $n^{1+1/\ell}p \to c > 0$ and ψ be a sentence from $L_\sim$ of quantifier depth k. Then, a.a.s. $\mathbb{G}(n,p)$ is a forest, which contains no trees of size larger than $\ell+1$, and at least k copies of each tree of size at most ℓ. Thus, our main concern is trees of size $\ell+1$ which appear in $\mathbb{G}(n,p)$ with probabilities which are bounded away from both 0 and 1 (see Theorem 3.19). Let $T_1, \dots, T_m$ be the family of all pairwise non-isomorphic trees with $\ell+1$ vertices and $r_1, \dots, r_m$ be natural numbers such that $0 \leq r_i \leq k$ for $1 \leq i \leq m$. We say that a graph G has property $\varphi_{k,\ell}(r_1, \dots, r_m)$ if the following hold:

(a) G is a forest;

(b) for any tree T on at most ℓ vertices G contains at least k components isomorphic to T;

(c) G contains no components on $\ell + 2$ vertices;

(d) for all i, $1 \leq i \leq m$, the number of copies of T_i in G is r_i if $0 \leq r_i \leq k-1$, and at least k if $r_i = k$.

A simple calculation of the moments of an appropriate m-dimensional random variable shows that for every sequence $r_1, \dots, r_m$ the limit

$$\lim_{n\to\infty} \mathbb{P}(\mathbb{G}(n,p) \models \varphi_{k,\ell}(r_1, \dots, r_m)) = \lambda(r_1, \dots, r_m)$$

exists (see Remark 3.20); as a matter of fact, $\lambda(r_1, \dots, r_m)$ is a sum of certain probabilities of a multidimensional Poisson distribution. Now let

$G(r_1, \ldots, r_m)$ be a graph for which $\varphi_{k,\ell}(r_1, \ldots, r_m)$ holds. One can see immediately that, for any other graph G which fulfills $\varphi_{k,\ell}(r_1, \ldots, r_m)$, the second player has a winning strategy in the game $\mathrm{Ehr}_k(G, G(r_1, \ldots, r_m))$. Thus,

$$\lim_{n\to\infty} \mathbb{P}(\mathbb{G}(n,p) \models \psi) = \sum_{\substack{(r_1,\ldots,r_m)\\ G(r_1,\ldots,r_m)\models\psi}} \lambda(r_1, \ldots, r_m)$$

and (i) follows.

The proofs of (ii) and (iii) are similar, although slightly longer and more involved, so we omit them here. ■

Zero-one laws and recursive functions

In order to complete the picture of the first-order properties of $\mathbb{G}(n,p)$, it remains to study the behavior of $\mathbb{G}(n,p)$ for $p = p(n) = n^{-\alpha+o(1)}$, where $\alpha \in (0,1)$ is a rational number. We remark that for each rational $\alpha \in (0,1]$ there exists a strictly balanced graph H_α such that $d(H_\alpha) = \alpha$ (Ruciński and Vince 1986). Thus, for $pn^\alpha \to c > 0$, the probability that $\mathbb{G}(n,p)$ contains H_α tends to a constant $a = a(c)$ which is neither 0 nor 1 (see Theorem 3.19) and the zero-one law does not hold. If $p = p(n) = n^{-\alpha+o(1)}$ the behavior of $\mathbb{G}(n,p)$ is much more involved and thus, to simplify its description and avoid many technical details, we will consider only the special case $\alpha = 1/7$.

As observed by Łuczak and Spencer (1991), the argument which led us to Theorem 10.7 can be modified to show that the zero-one law holds when p is slightly larger than $n^{-1/7}$.

Theorem 10.15. *Let $p = p(n)$ be such that*

$$p^7 n \geq 7 \log n + \omega(n) \log\log n\ ,$$

where $\omega = \omega(n) \to \infty$ but $p = p(n) = n^{-1/7+o(1)}$. Then, for every sentence ψ from $L_\sim$ the probability that $\mathbb{G}(n,p)$ satisfies ψ tends either to 0 or to 1 as $n \to \infty$. ■

Remark 10.16. The above result is, in a way, best possible. Indeed, for every natural number ℓ and $np^7 = 7 \log n + \ell \log\log n$, the probability of the property that $\mathbb{G}(n,p)$ contains seven vertices $v_1, \ldots, v_7$ which have precisely ℓ common neighbors tends to a constant λ, where $0 < \lambda < 1$, as $n \to \infty$ (Exercise!). Thus, the function $\omega(n)$ in Theorem 10.15 cannot be replaced by any constant.

What happens when $p = n^{-1/7+o(1)}$ but $np^7 = o(\log n)$? The bold reader who has managed to read the book until this point would probably be willing to bet that now, as in Theorem 10.6, we specify a spectrum of the threshold functions for different properties, and in between them the zero-one law must hold. This is indeed what an experienced graph theorist should expect.

However, as shown by Shelah and Spencer (1988), the truth is much more surprising.

Theorem 10.17. *There exists a sentence ψ from $L_\sim$ such that for every $p = p(n)$ satisfying*

$$n^{-1/\log\log\log\log\log n} \le np^7 \le \frac{\log n}{\log\log\log\log\log n} \tag{10.2}$$

we have

$$\liminf_{n\to\infty} \mathbb{P}(\mathbb{G}(n,p) \models \psi) = 0$$

and

$$\limsup_{n\to\infty} \mathbb{P}(\mathbb{G}(n,p) \models \psi) = 1 .$$

Before we say a few words about the proof of Theorem 10.17 we comment briefly on the content of this very counterintuitive result. First of all, note that the theorem states that the probability of a first-order property can oscillate between 0 and 1 even for such odd functions as $p' = n^{-1/7}\log^{-\pi} n$ or $p'' = n^{-1/7-1/\sqrt{\log\log\log n}}$ which certainly do not resemble any threshold function we have seen so far. Furthermore, there exists one sentence ψ which behaves eccentrically for *all* such functions at the same time. Finally, the inequality (10.2) looks very "asymmetric". Indeed, we already know from Theorem 10.15 that the zero-one law holds if np^7 is substantially larger than $\log n$, hence the upper bound for np^7 cannot be replaced by a much better one (although it is by no means best possible – see Theorem 10.19 and Remark 10.20 below); on the other hand the lower bound for np^7 given in Theorem 10.17 tends to 0 much faster than any power of the logarithm.

Can we significantly extend the range of $p = p(n)$ for which Theorem 10.17 holds? More precisely, can we replace in (10.2) the iterated logarithm by any function which tends to infinity? The following theorem by Łuczak and Spencer (1991) answers this question in the negative.

Theorem 10.18. *There exists a function $\omega = \omega(n)$ which tends to infinity as $n \to \infty$ such that for $p = p(n) = n^{-1/7-1/\omega}$ and every sentence ψ from $L_\sim$ the probability that $\mathbb{G}(n,p)$ satisfies ψ converges either to 0, or to 1.*

Proof. Let a_i be a sequence of irrational numbers which monotonically tends to infinity as $i \to \infty$, say, $a_i = i\sqrt{2}$, and let $A_0 = \{a_1, a_2, \dots\}$ and $p_{a_i} = n^{-1/7-1/a_i}$. Furthermore, let $\{\psi_1, \psi_2, \dots\}$ be the set of all sentences from $L_\sim$. In order to find $\omega(n)$ we first define a decreasing sequence of infinite sets $A_0 \supseteq A_1 \supseteq A_2 \supseteq \dots$ and a sequence of sentences $\hat{\psi}_i$ from $L_\sim$, where for every $i \ge 1$ either $\hat{\psi}_i = \psi_i$ or $\hat{\psi}_i = \neg\psi_i$. Thus, let us suppose that for some $i \ge 0$ the set A_i has already been defined. Then, by Theorem 10.7, we can partition A_i into two classes, according to whether for $a \in A_i$ the formula ψ_{i+1} holds *a.a.s.* for $\mathbb{G}(n, p_a)$, or *a.a.s.* $\mathbb{G}(n, p_a)$ satisfies $\neg\psi_{i+1}$. Let A_{i+1} denote the set of the above partition which is infinite (if both are infinite we can pick either of them) and let $\hat{\psi}_{i+1}$ be the sentence from $\{\psi_{i+1}, \neg\psi_{i+1}\}$ which holds *a.a.s.* for $\mathbb{G}(n, p_a)$, for all $a \in A_{i+1}$. Moreover, choose $\alpha_i \in A_i$ such that $\alpha_1 < \alpha_2 < \dots$ and select n_i in such a way that $1 \leq n_1 \leq n_2 \leq \dots$ and for every $j \leq i$ and $n \geq n_i$

$$\mathbb{P}(\mathbb{G}(n, p_{\alpha_i}) \models \hat{\psi}_j) \geq 1 - 1/i \,.$$

Now, it is enough to set $\omega(n) = \alpha_i$ for each $n_i \leq n < n_{i+1}$, $i \geq 1$. ■

It turns out, however, that each function $\omega(n)$ which fulfills the assumptions of the above theorem tends to zero unimaginably slowly. Before we make this statement precise, let us recall that a recursive function $f : \mathbb{N} \to \mathbb{N}$ is a function such that there exists a procedure which for every n computes the value of $f(n)$ in a finite time using a deterministic Turing machine. (The reader who is not familiar with this notion is advised either to look at one of the equivalent definitions of a recursive function in any book on the theory of computation, or accept Church's thesis and view a recursive function just as a function which can be effectively computed.) Somewhat surprisingly at first sight, recursive functions play an important role in the analysis of the behavior of $\mathbb{G}(n, p)$.

Theorem 10.19. *For every recursive function $\omega = \omega(n)$ which tends to infinity as $n \to \infty$, and every $p = p(n)$, satisfying*

$$n^{-1/\omega} \leq np^7 \leq \log n/\omega \,, \tag{10.3}$$

there exists a sentence ψ from $L_\sim$ such that

$$\liminf_{n\to\infty} \mathbb{P}(\mathbb{G}(n, p) \models \psi) = 0$$

and

$$\limsup_{n\to\infty} \mathbb{P}(\mathbb{G}(n, p) \models \psi) = 1 \,.$$

Remark 10.20. The fact that Theorem 10.18 remains true if we replace $n^{-1/\log\log\log\log\log n}$ by $n^{-1/\omega}$, where the recursive function ω tends to infinity as $n \to \infty$, was observed by Łuczak and Spencer (1991) (note, however, that now ψ does depend on ω). This lower bound for np^7 is best possible in the sense that, as Theorem 10.18 shows, we cannot omit the assumption that ω is recursive. The fact that $\log n/ \log\log\log\log\log n$ in (10.2) can be replaced by $\log n/\omega$, where ω is a recursive function which tends to infinity, follows from the original argument of Shelah and Spencer (1988). Let us also mention that at this moment it is not known whether the assumption that ω is recursive is really necessary. It is even possible that the assertion of Theorem 10.19 holds as long as $np^7 \leq (7 - \varepsilon) \log n$ for some $\varepsilon > 0$ (see Theorem 10.15).

Although we will not give here complete proofs of Theorems 10.17 and 10.19 (which are long and technical), we sketch the main idea of the argument so that the somewhat unexpected appearance of recursive functions will become clear. Let us suppose that $p = n^{-1/7-1/\omega}$, where $\omega = \omega(n)$ tends to infinity slowly, say, slower than $\log\log\log n$. For each subset $\{v_1, \ldots, v_7\}$ of vertices of $\mathbb{G}(n,p)$ let $N(v_1, \ldots, v_7)$ denote the set of common neighbors of $v_1, \ldots, v_7$. It turns out (Exercise!) that *a.a.s.* the maximum of $|N(v_1, \ldots, v_7)|$ over all choices of $v_1, \ldots, v_7$ differs from ω by no more than 4. Furthermore, *a.a.s.* for any two sets $\{v_1, \ldots, v_7\}$ and $\{v'_1, \ldots, v'_7\}$ and every symmetric binary relation R on $N(v_1, \ldots, v_7) \cup N(v'_1, \ldots, v'_7)$ with fewer than 2ω elements, there exist elements x_1, x_2, x_3, x_4 and $y_1, y_2, \ldots, y_{20}$ such that zRz' if and only if the set $N(z, z', x_1, x_2, x_3, x_4, y_i)$ is non-empty for some y_i, where $1 \le i \le 20$. Since, clearly, $|N(v_1, \ldots, v_7)| \ge |N(v'_1, \ldots, v'_7)|$ if and only if there exists a "matching" binary relation between $N(v_1, \ldots, v_7) \setminus N(v'_1, \ldots, v'_7)$ and $N(v'_1, \ldots, v'_7) \setminus N(v_1, \ldots, v_7)$ which saturates each element from $N(v'_1, \ldots, v'_7) \setminus N(v_1, \ldots, v_7)$, it is possible to write a formula in $L_\sim$ which would mean "$|N(v_1, \ldots, v_7)| \ge |N(v'_1, \ldots, v'_7)|$". Consequently, it is possible to "identify" in $L_\sim$ one of the subsets $\{v_1, \ldots, v_7\}$ which maximizes $|N(v_1, \ldots, v_7)|$, in other words, it is possible to define a predicate P in $L_\sim$ such that $P(x)$ holds if and only if x belongs to some $A = N(v_1, \ldots, v_7)$, where for each choice of $v'_1, \ldots, v'_7$ we have $|N(v_1, \ldots, v_7)| \ge |N(v'_1, \ldots, v'_7)|$. In a similar way one can show that *a.a.s.* for each symmetric ternary relation T on A with at most 9ω elements there are vertices x_1, x_2, x_3, and $y_1, y_2, \ldots, y_{90}$ such that $T(z, z', z'')$ holds if and only if $N(z, z', z'', x_1, x_2, x_3, y_i) \ne \emptyset$ for some i, $1 \le i \le 90$. Thus, each such relation T could be, in principle, expressed in $L_\sim$. One can use a few additional technical tricks to show that there exists a subset $S \subseteq A$ in $\mathbb{G}(n,p)$ which can be defined using $L_\sim$, such that $|S| \ge |A|^{1/3}/2 \ge \omega^{1/4}$ and *each* ternary relation on A can be "encoded" in $L_\sim$. Thus, we have arrived at the following result (in the statement we include all functions $p = p(n)$ for which (10.3) holds, although when np^7 does not tend to 0 fast enough the encoding procedure is more involved). Here, for simplicity, we use $\underline{v} = (v_1, \ldots, v_m)$ to denote sequences of vertices of length m, where m does not depend on n.

Lemma 10.21. *There exist formulas $P^*(x, \underline{v})$ and $T^*(x, y, z, \underline{v})$ in $L_\sim$ such that if $\omega = \omega(n) \le \log n$ is any function which tends to infinity as $n \to \infty$ and*

$$n^{-1/\omega} \le np^7 \le \log n/\omega\,,$$

then a.a.s. *for every choice of $\underline{v} = \underline{v}^S$ the set $S(\underline{v}) = \{w : P^*(w, \underline{v})\}$ of vertices of $\mathbb{G}(n,p)$ has the following two properties:*

(i) *either $S(\underline{v}) = \emptyset$, or $\omega^{1/8} \le |S(\underline{v})| < n$,*

(ii) *for every ternary relation T defined on $S(\underline{v})$ there exists $\underline{v}^T$ such that if $x, y, z \in S(\underline{v})$, then $T(x, y, z)$ holds if and only if $T^*(x, y, z, \underline{v}^T)$ is true for $\mathbb{G}(n,p)$.*

Furthermore, a.a.s. *there exists $\underline{v}$ such that $S(\underline{v}) \ne \emptyset$, that is, $|S(\underline{v})| \ge \omega^{1/8}$.* ■

Before we sketch the proofs of Theorems 10.17 and 10.19, we state one more fact about recursive functions.

Lemma 10.22. *Let $g : \mathbb{N} \to \mathbb{N}$ be any function such that for some non-decreasing recursive function $f : \mathbb{N} \to \mathbb{N}$ which tends to infinity as $n \to \infty$ we have $f(n) \le g(n) < n$ for $n \ge 2$. Then there exists a recursive function $h : \mathbb{N} \to \mathbb{N}$ such that $h(g(\mathbb{N}))$ contains all natural numbers.*

Proof. For any non-decreasing unbounded recursive function $f : \mathbb{N} \to \mathbb{N}$ such that $f(n) < n$ for $n \ge 2$, define $f^* : \mathbb{N} \to \mathbb{N}$, setting

$$f^*(n) = \min_k \{\underbrace{f(f(\ldots f}_{k}(n)\ldots)) = 1\} .$$

Obviously, f^* is recursive and non-decreasing. Moreover it tends to infinity as $n \to \infty$ but much slower than f. We leave to the reader the elementary calculations which show that the assertion holds for $h = f^*$ (Exercise!). ■

Sketch of the proofs of Theorems 10.17 and 10.19. Let us suppose that the condition (10.3) holds for $p = p(n)$. Furthermore, to simplify the argument slightly, let us assume that the recursive function $\omega(n)$ which appears in the assumptions of Theorem 10.19 is non-decreasing, and set $f(n) = [\omega(n)]^{1/8}$. Then, by Lemma 10.21, *a.a.s.* $\mathbb{G}(n,p)$ contains a set S, such that $f(n) \le |S| < n$, which can be encoded by $P^*(\cdot, \underline{v}^S)$ for some $\underline{v}^S$. Furthermore, any ternary relation T on S can be encoded in $L_\sim$ by the predicate $T^*(\cdot,\cdot,\cdot,\underline{v}^T)$ for some sequence $\underline{v}^T$. Thus, for instance, there exists $\underline{v}^{\le}$ and w such that for all x, y, z, for which $P^*(x,\underline{v}^S)$, $P^*(y,\underline{v}^S)$, $P^*(z,\underline{v}^S)$, we have:

(i) if $T^*(x,y,w,\underline{v}^{\le})$ and $T^*(y,z,w,\underline{v}^{\le})$, then $T^*(x,z,w,\underline{v}^{\le})$;

(ii) if $T^*(x,y,w,\underline{v}^{\le})$ and $T^*(y,x,w,\underline{v}^{\le})$, then $x = y$;

(iii) either $T^*(x,y,w,\underline{v}^{\le})$, or $T^*(y,x,w,\underline{v}^{\le})$.

Hence, $\underline{v}^{\le}$ and w define on S a linear order. Now, we can use Lemma 10.21 to find an element $\underline{v}^+$ such that for x, y, z, for which $P^*(x,\underline{v}^S)$, $P^*(y,\underline{v}^S)$, $P^*(z,\underline{v}^S)$, the relation $T^*(x,y,z,\underline{v}^+)$ holds if and only if $\hat{x} + \hat{y} = \hat{z}$, where $\hat{x}$, $\hat{y}$, $\hat{z}$ denote the positions of elements x, y and z, respectively, in the order determined by $\underline{v}^{\le}$ and w. Indeed, it is enough to require that the relation $T^*(x,y,z,\underline{v}^+)$ has all the properties which hold for the sum of two natural numbers; for instance, if $T^*(x,y,z,\underline{v}^+)$, then $T^*(x,F(y),F(z),\underline{v}^+)$, where $F(y)$ and $F(z)$ stand for the elements which in the linear order determined by $\underline{v}^{\le}$ and w succeed y and z, respectively. Clearly, in a similar manner we can encode in such a linearly ordered set S *any* recursive function. In particular, if h is a function whose existence is assured by Lemma 10.22, there exists $\underline{v}^h$ such that for x, y, z with $P^*(x,\underline{v}^S)$, $P^*(y,\underline{v}^S)$, $P^*(z,\underline{v}^S)$ we have $T^*(x,y,z,\underline{v}^h)$ if and only if $y = h(x)$. Now let s be a maximal element from S, that is,

$P^*(s, \underline{v}^S)$ and for each x such that $P^*(x, \underline{v}^S)$ and $T^*(x, s, w, \underline{v}^{\leq})$ we have $x = s$. Furthermore, let ψ be the sentence

$$\exists_{\underline{v}^S} \exists_{\underline{v}^{\leq}} \exists_w \exists_{\underline{v}^+} \exists_{\underline{v}^h} \exists_s \exists_x \exists_y \; \alpha(\underline{v}^S, \underline{v}^{\leq}, w, \underline{v}^+, \underline{v}^h, s) \\ \wedge P^*(x, \underline{v}^S) \wedge P^*(y, \underline{v}^S) \wedge T^*(s, x, x, \underline{v}^h) \wedge T^*(y, y, x, \underline{v}^+) \,,$$

where α is a long and complicated formula which contains the definition of S, the axioms of linear ordering, the axioms of addition, the definitions of h and s, and so on. The above ψ belongs to $L_{\sim}$ and expresses the property "$h(|S|)$ is even". Hence, by Lemmas 10.21 and 10.22, as $n \to \infty$ the probability that ψ holds for $\mathbb{G}(n, p)$ oscillates between 0 and 1. ∎

The idea of encoding the initial segments of the natural numbers by substructures of a random structure has been widely used in the literature to show that the zero-one law does not occur. Typically, in such cases some sort of "recursive bound behavior", similar to that described by Theorems 10.18 and 10.19, can be observed. A good example of such a phenomenon is the following theorem of Dolan and Lynch (1993), who ingeniously used the strength of the language $L_{\sim}^{\mathrm{ord}}$ to encode the initial segments of the natural numbers even in random graphs with very few edges.

Theorem 10.23.

(i) *If $p = p(n)$ is such that $n^2 p \to c$ as $n \to \infty$ for some constant c, then for every ψ from $L_{\sim}^{\mathrm{ord}}$ the probability that $\mathbb{G}(n, p)$ satisfies ψ converges as $n \to \infty$.*

(ii) *There exists a function $p = p(n)$ such that $n^2 p \to \infty$ and for every ψ from $L_{\sim}^{\mathrm{ord}}$ the probability that $\mathbb{G}(n, p)$ satisfies ψ converges.*

(iii) *If $p = p(n) \leq 1/2$ but $n^2 p \geq f(n)$ for some unbounded recursive function $f : \mathbb{N} \to \mathbb{N}$, then there exists a sentence ψ of $L_{\sim}^{\mathrm{ord}}$ such that*

$$\liminf_{n \to \infty} \mathbb{P}(\mathbb{G}(n, p) \models \psi) = 0$$

and

$$\limsup_{n \to \infty} \mathbb{P}(\mathbb{G}(n, p) \models \psi) = 1 \,. \quad ∎$$

10.4 SUMS OF MODELS

In this part of the chapter we will study the behavior of random structures using yet another approach: a sum of models. Although it does not contribute much to the picture of $\mathbb{G}(n, p)$, which is now nearly complete, it is a simple yet very convenient tool which can be used to obtain different types of zero-one laws for various random structures. Because of the character of the book, we present here only the simplest version of a more sophisticated notion (see Łuczak and Shelah (1995), where more general sum schemes were used to study the behavior of the random graph $\overline{\mathbb{G}}(n, \bar{p})$ defined below).

Sum schemes

Throughout this section L will denote a first-order language with predicates $P_1, \ldots, P_t$, all of them binary, and $\mathfrak{M}$ will stand for a family of models of L with binary relations $R_1, \ldots, R_t$. Suppose that we are given a t-tuple of pairs of zeros and ones $\underline{\tau} = ((\tau_1^-, \tau_1^+), \ldots, (\tau_t^-, \tau_t^+))$. A binary operation $\oplus : \mathfrak{M} \times \mathfrak{M} \to \mathfrak{M}$ is a *sum scheme for* $\mathfrak{M}$ *with signature* $\underline{\tau}$ if for all $M_1, M_2 \in \mathfrak{M}$ the set of elements of $M = M_1 \oplus M_2$ is a disjoint sum of the set U_1 of elements of M_1 and the set U_2 of elements of M_2, and for every $i = 1, 2, \ldots, t$ the following hold:

(i) if $x, y \in U_r$, $r = 1, 2$, then xR_iy in M if and only if xR_iy in M_r;

(ii) if $x \in U_1$ and $y \in U_2$, then xR_iy when $\tau_i^+ = 1$, and $\neg xR_iy$ whenever $\tau_i^+ = 0$. Similarly, for such x, y, we have yR_ix for $\tau_i^- = 1$, and $\neg yR_ix$ otherwise.

Note that the operation $\oplus$ is associative but typically it is not commutative (unless $\tau_i^+ = \tau_i^-$ for all $i = 1, \ldots, t$).

Example 10.24. Let G_1 and G_2 be graphs with vertex sets $[n_1]$ and $[n_2]$, respectively. Define $G = G_1 \dotplus G_2$ as a graph with vertex set $[n_1 + n_2]$, such that $\{x, y\}$ is an edge of G if and only if either

- $1 \leq x, y \leq n_1$ and $\{x, y\}$ is an edge of G_1,

or

- $n_1 + 1 \leq x, y \leq n_1 + n_2$ and $\{x - n_1, y - n_1\}$ is an edge of G_2.

Thus, roughly speaking, to obtain G one should put G_2 behind G_1 preserving the order of vertices in each of the graphs. Clearly, the operation $\dotplus$ is a sum scheme for graphs viewed as the models of either $L_\sim$ or $L_\sim^{\mathrm{ord}}$, with signature $((0,0))$ for $L_\sim$ and $((0,0),(0,1))$ for $L_\sim^{\mathrm{ord}}$.

The relevance of sum schemes for the study of first-order properties is based on the following simple observation.

Lemma 10.25. *If* $\oplus : \mathfrak{M} \times \mathfrak{M} \to \mathfrak{M}$ *is a sum scheme for* $\mathfrak{M}$, *then for all* $M_1, M_2 \in \mathfrak{M}$ *the set* $\mathrm{Th}_k(M_1 \oplus M_2)$ *can be computed from* $\mathrm{Th}_k(M_1)$ *and* $\mathrm{Th}_k(M_2)$, *i.e.,* $\mathrm{Th}_k(M_1) = \mathrm{Th}_k(M_1')$ *and* $\mathrm{Th}_k(M_2) = \mathrm{Th}_k(M_2')$ *implies* $\mathrm{Th}_k(M_1 \oplus M_2) = \mathrm{Th}_k(M_1' \oplus M_2')$.

Proof. Suppose that $\mathrm{Th}_k(M_1) = \mathrm{Th}_k(M_1')$ and $\mathrm{Th}_k(M_2) = \mathrm{Th}_k(M_2')$. Thus, according to Lemma 10.3, the second player has winning strategies in both $\mathrm{Ehr}_k(M_1, M_1')$ and $\mathrm{Ehr}_k(M_2, M_2')$. But then, he can win also $\mathrm{Ehr}_k(M_1 \oplus M_2, M_1' \oplus M_2')$ using his winning strategy for $\mathrm{Ehr}_k(M_1, M_1')$ whenever the first player selects a vertex from either M_1 or M_1', and playing according to his winning strategy for $\mathrm{Ehr}_k(M_2, M_2')$ whenever the first player picks a vertex from M_2 or M_2'. Thus, the assertion follows immediately from Lemma 10.3. ■

Let k be a natural number and $\oplus : \mathfrak{M} \times \mathfrak{M} \to \mathfrak{M}$ be a sum scheme for a family of models $\mathfrak{M}$. A model $M \in \mathfrak{M}$ is $(\mathfrak{M}, \oplus, k)$-*persistent*, or simply *persistent*, if for every $M' \in \mathfrak{M}$ we have $\mathrm{Th}_k(M \oplus M' \oplus M) = \mathrm{Th}_k(M)$. We will show that for every sum scheme such a persistent model can be found.

Theorem 10.26. *Every family of models $\mathfrak{M}$ with a sum scheme $\oplus : \mathfrak{M}\times\mathfrak{M} \to \mathfrak{M}$ contains at least one $(\mathfrak{M}, \oplus, k)$-persistent model for every natural number k.*

The proof of Theorem 10.26 will rely on the following property of the operation $\oplus$.

Lemma 10.27. *For every k there exists $m = m(k)$ such that if $\oplus : \mathfrak{M}\times\mathfrak{M} \to \mathfrak{M}$ is a sum scheme for $\mathfrak{M}$, then for every model $M \in \mathfrak{M}$ we have*

$$\mathrm{Th}_k(\underbrace{M \oplus \cdots \oplus M}_{m \text{ times}}) = \mathrm{Th}_k(\underbrace{M \oplus \cdots \oplus M}_{m+1 \text{ times}}).$$

Proof. We will show that the assertion holds if we put, rather crudely, $m(k) = 3^{k+1}$. Thus, let $G' = \bigoplus_{i=1}^{m} M_i'$ and $G'' = \bigoplus_{i=1}^{m+1} M_i''$, where $M_1', \ldots, M_m'$ and $M_1'', \ldots, M_{m+1}''$ are identical copies of M. We will describe a winning strategy for the second player in the game $\mathrm{Ehr}_k(G', G'')$. Let us suppose that in the first i steps of the game the players choose vertices from copies $M_{r_1}', \ldots, M_{r_i}'$ and $M_{s_1}'', \ldots, M_{s_i}''$, respectively. Then if the first player in the $(i+1)$-th step picks a vertex v from $M_{r_{i+1}}'$ [or $M_{s_{i+1}}''$] the second player should select the corresponding vertex v from $M_{s_{i+1}}''$ [$M_{r_{i+1}}'$] chosen in such a way that

(a) $s_{i+1} = r_{i+1}$ if $r_{i+1} \le 3^{k-i}$ [or $s_{i+1} \le 3^{k-i}$];

(b) $m + 1 - s_{i+1} = m - r_{i+1}$ if $m - r_{i+1} \le 3^{k-i}$ [or $m + 1 - s_{i+1} \le 3^{k-i}$];

(c) $s_{i+1} - s_{i'} = r_{i+1} - r_{i'}$ if for some $1 \le i' \le i$ we have $|r_{i+1} - r_{i'}| \le 3^{k-i}$ [or $|s_{i+1} - s_{i'}| \le 3^{k-i}$];

(d) $s_{i+1} \le s_{i'}$ if and only if $r_{i+1} \le r_{i'}$, for all $i' = 1, 2, \ldots, i$.

As in the proof of Corollary 10.4, one can check (Exercise!) that such a strategy is consistent, that is, if the second player follows it from the beginning of the game he can always find a suitable element s_{i+1} [or r_{i+1}] in the $(i+1)$-th move of the game. Thus, at the end of the game, for all $1 \le i, j \le k$ we have $r_i \le r_j$ if and only if $s_i \le s_j$, and the second player wins. ■

We will need also the following elementary result on finite semigroups.

Lemma 10.28. *Let S be a finite semigroup for which there exists a natural number m such that $s^m = s^{m+1}$ for every $s \in S$. Then S contains an element w such that $wuw = w$ for every $u \in S$.*

Proof. For $x, y \in S$ let us write $x \rightsquigarrow y$ if for some $s \in S$ we have $y = xsx$. Set $A_z = \{y \in S : z \rightsquigarrow y\}$ and let A be a minimal subset in the family $\{A_z : z \in S\}$. Since the relation "$\rightsquigarrow$" is transitive, for every $x \in A$, we have $x \rightsquigarrow y$ if and only if $y \in A$, and so for every $x \in A$ we have $A = A_x$. Thus, to verify the assertion, it is enough to show that A consists of only one element.

Since we may assume that $m \geq 3$, for every $x \in A$ we have $xx^{m-2}x = x^m = x^{m+1} \in A$. Observe also that the choice of A ensures that for each $x \in A$ the function $f_x : A \to A : y \mapsto x^m y x^m$ is a bijection, that is, for every $x \in A$ we have

$$A = x^m A x^m = \{x^m y x^m : y \in A\}. \tag{10.4}$$

Indeed, clearly $x^m A x^m \subseteq A$, and for every $z \in A$, there exists $s \in S$ such that

$$z = x^m s x^m = x^{m+1} s x^{m+1} = x^m (xsx) x^m = x^m y x^m,$$

where $y = xsx \in A$.

Now let $x, y \in A$. Then, $y = xsx$ for some $s \in S$, and so $xy = x(xs)x \in A$. Furthermore,

$$x^m y x^m = x^{m+1} y x^m = x^m (xy) x^m$$

and, because of (10.4), $y = xy$. Similarly,

$$y^m x y^m = y^m (xy) y^m$$

and so $y = xy = x$. Hence, A consists of only one element w and the assertion follows. ∎

Proof of Theorem 10.26. Let us notice first that, due to Lemma 10.25, for a given k, the sum scheme $\oplus$ induces an operation $\underline{\oplus}$ acting on the set $\Xi_k(\mathfrak{M}) = \{\mathrm{Th}_k(M) : M \in \mathfrak{M}\}$. Since the operation $\oplus$ is associative, so is $\underline{\oplus}$, and consequently $(\Xi_k(\mathfrak{M}), \underline{\oplus})$ is a finite semigroup. Due to Lemma 10.27, this semigroup fulfills the assumption of Lemma 10.28 and thus contains an element $w = \mathrm{Th}_k(M)$ such that for every $M' \in \mathfrak{M}$

$$\mathrm{Th}_k(M) = \mathrm{Th}_k(M) \underline{\oplus} \mathrm{Th}_k(M') \underline{\oplus} \mathrm{Th}_k(M) = \mathrm{Th}_k(M \oplus M' \oplus M). \quad ∎$$

Remark 10.29. For a language L in which all predicates are symmetric, as in the case of the first-order language of graphs $L_\sim$, Theorem 10.26 can be proved in a much simpler way. Indeed, one can easily see that then Lemma 10.27 holds with $m(k) = k$. Furthermore, the semigroup $(\Xi_k(\mathfrak{M}), \underline{\oplus})$ is commutative, and so as a persistent element one can choose

$$\Big(\prod_{x \in \Xi_k(\mathfrak{M})} x\Big)^m = \Big(\prod_{x \in \Xi_k(\mathfrak{M})} x\Big)^k.$$

The random graph $\overline{\mathbb{G}}(n, \overline{p})$

In the previous sections of this chapter we have characterized quite precisely for which $p = p(n)$ in $\mathbb{G}(n, p)$ the zero-one law holds; now we use sum schemes to study the behavior of a slightly different type of random graph, denoted by $\overline{\mathbb{G}}(n, \overline{p})$. Let $\overline{p}$ be a sequence of probabilities $\{p_i\}_{i=1}^{\infty}$. Then, $\overline{\mathbb{G}}(n, \overline{p})$ is a graph on the vertex set $[n]$ obtained by joining each pair of vertices $1 \leq k < \ell \leq n$ with probability $p_{\ell-k}$, independently for each such pair. Thus, in particular, $\mathbb{G}(n, p)$ is a special case of $\overline{\mathbb{G}}(n, \overline{p})$ when all terms of the sequence $\overline{p}$ are equal to p. On the other hand, we will consider asymptotic properties of $\overline{\mathbb{G}}(n, \overline{p})$ when n tends to infinity but the sequence $\overline{p}$ does *not* depend on n, thus it is not possible to deduce our previous results about $\mathbb{G}(n, p)$ from those given below on $\overline{\mathbb{G}}(n, \overline{p})$, unless p does not depend on n.

In order to simplify slightly our considerations we will study the behavior of $\overline{\mathbb{G}}(n, \overline{p})$ only for sequences $\overline{p} = \{p_i\}_{i=1}^{\infty}$ for which $0 < p_i \leq 1/2$. The assumption that $\overline{p} > 0$ helps to avoid some pathological cases when, for example, $\overline{p}$ is selected in such a way that no triangles are allowed in $\overline{\mathbb{G}}(n, \overline{p})$; let us mention, however, that Theorems 10.31, 10.33 and 10.34 below remain true also for sequences which contain zero terms (Łuczak and Shelah 1995). The upper bound for p_i takes care of another troublesome situation when $\overline{p}$ contains some ones or terms which quickly approach one as $i \to \infty$. It is easy to see that, under the above assumptions, *a.a.s.* the random graph $\overline{\mathbb{G}}(n, \overline{p})$ contains many copies of any given subgraph H. Following Łuczak and Shelah (1995) we will show that if the terms of $\overline{p}$ tend to 0 fast enough, then a stronger result holds: *a.a.s.* $\overline{\mathbb{G}}(n, \overline{p})$ contains a copy of H of a very special type.

Thus, let H and G be graphs with vertex sets $[m]$ and $[n]$, respectively. We say that a subgraph H' of G spanned by vertices $i + 1, \ldots, i + m$ is an *exact copy* of H if for all $1 \leq k < \ell \leq m$ the pair $\{k, \ell\}$ is an edge of H if and only if $\{i + k, i + \ell\}$ is an edge of G, and no vertex from H' is adjacent to vertices outside H'. (Thus, H' is an exact copy if it is an order-preserving induced copy of H on consecutive vertices which, moreover, is a sum of components of G.) Furthermore, such an exact copy of H' is *separating* if no vertex $i' \leq i$ of G is adjacent to a vertex $i'' \geq i + m + 1$, in other words, one can write the graph G as a sum $G' \dotplus H' \dotplus G''$, or, maybe, $G' \dotplus H'$ or $H' \dotplus G''$ (for the definition of $G_1 \dotplus G_2$ see Example 10.24).

Lemma 10.30. *Let $\overline{p} = \{p_i\}_{i=1}^{\infty}$ be such that $0 < p_i \leq 1/2$ for every $i \geq 1$, and let H be a graph with vertex set $[m]$.*

(i) *If $\sum_{i=1}^{n} p_i = o(\log n)$, then* a.a.s. *$\overline{\mathbb{G}}(n, \overline{p})$ contains an exact copy of H.*

(ii) *If the series $\sum_{i=1}^{\infty} ip_i$ converges, then* a.a.s. *$\overline{\mathbb{G}}(n, \overline{p})$ contains an exact separating copy of H which contains no vertices with labels larger than $m\lceil n^{0.9} \rceil$.*

Proof. Let us consider a family $\mathcal{A}$ of disjoint sets $A_\ell = \{(\ell - 1)m + 1, \ldots, \ell m\}$, where $\ell = 1, \ldots, \lceil n^{0.9} \rceil$. The probability that a given set A_ℓ from $\mathcal{A}$ spans an exact copy of H is given by $p(A_\ell) = p(H)p'(A_\ell)$, where the first factor,

$$p(H) = \prod_{e=\{i,j\}\in E(H)} p(|i-j|) \prod_{e'=\{i',j'\}\notin E(H)} (1 - p(|i'-j'|)) > 0\,,$$

remains the same for all ℓ, while the second one,

$$p'(A_\ell) = \prod_{r=(\ell-1)m+1}^{\ell m} \prod_{s=1}^{(\ell-1)m} (1 - p_{r-s}) \prod_{t=\ell m+1}^{n} (1 - p_{t-s})\,,$$

may depend on ℓ. However, from our assumptions it follows that

$$p'(A_\ell) \geq \left(\prod_{i=1}^{n}(1-p_i)\right)^{2m} \geq \exp\left(-2m\sum_{i=1}^{n} p_i\right) \geq n^{-0.05},$$

so there exists a subfamily $\mathcal{A}'$ of $\mathcal{A}$ with, say, $\lceil n^{0.7}\rceil$ elements, such that for every $A \in \mathcal{A}'$ the probability $p'(A)$ is roughly the same, that is, for some function $f(n)$, where $n^{-0.05} \leq f(n) \leq 1$, and every $A \in \mathcal{A}'$ we have

$$f(n)(1 - o(n^{-0.1})) \leq p'(A) \leq f(n)\,.$$

Let X denote the number of those sets from $\mathcal{A}'$ which span exact copies of H. Then, the expectation of X is given by

$$\mathbb{E}\,X = \sum_{A\in\mathcal{A}'} p(A) = (1 + o(n^{-0.1}))p(H)f(n)n^{0.7} \geq n^{0.6}\,,$$

whereas, for the second factorial moment of X, we get

$$\mathbb{E}[X(X-1)] = \sum_{\substack{A,B\in\mathcal{A}'\\ A\neq B}} p(A)p(B)\Big/\prod_{i\in A}\prod_{j\in B}(1 - p_{|i-j|})\,.$$

In order to estimate $\mathbb{E}[X(X-1)]$, note that among the first n terms of every sequence $\overline{p}$ which fulfills the assumptions of Lemma 10.30 one can find at most $n^{0.25}$ terms larger than $n^{-0.2}$. Thus, for all except at most $2mn^{0.25}|\mathcal{A}'| \leq n$ pairs $A, B \in \mathcal{A}'$ we have

$$\prod_{i\in A}\prod_{j\in B}(1 - p_{|i-j|}) \geq (1 - n^{-0.2})^m \geq 1 - n^{-0.1}\,.$$

Furthermore, for all $A, B \in \mathcal{A}'$

$$\prod_{i\in A}\prod_{j\in B}(1 - p_{|i-j|}) \geq \prod_{i=1}^{n}(1 - p_i)^{2m} \geq \exp\left(-4m\sum_{i=1}^{n} p_i\right) \geq n^{-0.05}\,.$$

Thus,

$$\mathbb{E}[X(X-1)] \leq n^{1.4}f^2(n)(1+O(n^{-0.1}))+nf^2(n)n^{0.05} = (\mathbb{E}\,X)^2(1+O(n^{-0.1}))\,,$$

and so for the variance of X we get

$$\operatorname{Var} X = \mathbb{E}[X(X-1)] + \mathbb{E}\, X - (\mathbb{E}\, X)^2 = o((\mathbb{E}\, X)^2)\,.$$

Consequently, *a.a.s.* $X > \mathbb{E}\, X/2 > 0$, due to Chebyshev's inequality.

The second part of the lemma can be shown in a similar way, but now the probability that $A_\ell \in \mathcal{A}$ is an exact separating copy of H is the product of three factors $p(H)$, $p'(A_\ell)$ and $p''(A_\ell)$, where

$$p''(A_\ell) = \prod_{i=1}^{(\ell-1)m} \prod_{j=\ell m+1}^{n} (1 - p_{j-i}) \geq \prod_{i=1}^{n} (1 - p_i)^i \geq \exp\Bigl(-2\sum_{i=1}^{\infty} i p_i\Bigr)$$

is bounded away from zero whenever the series $\sum_{i=1}^{\infty} i p_i$ converges. As we have done before, one must choose a family $\mathcal{A}''$ of sets for which the probability $p(H)p'(A)p''(A)$ is roughly the same for all $A \in \mathcal{A}''$ and use Chebyshev's inequality to show that *a.a.s.* at least one set from $\mathcal{A}''$ spans an exact separating copy of H. Since the proof follows very closely the previous argument, we omit it here (for details see Łuczak and Shelah (1995)). ∎

Zero-one laws for $\overline{\mathbb{G}}(n, \overline{p})$

One can easily deduce from Theorem 10.26 and Lemma 10.30 that the zero-one law holds for $\mathbb{G}(n, \overline{p})$ whenever the sequence $\overline{p}$ tends to 0 quickly enough.

Theorem 10.31. *Let a sequence $\overline{p} = \{p_i\}_{i=1}^{\infty}$ be such that $0 < p_i \leq 1/2$ for $i \geq 1$ and*

$$\sum_{i=1}^{n} p_i = o(\log n)\,.$$

Then, for every sentence ψ from $L_\sim$ the probability that ψ holds for $\overline{\mathbb{G}}(n, \overline{p})$ tends either to 0 or to 1 as $n \to \infty$.

Proof. Let $\overline{p}$ fulfill the assumptions of the theorem and ψ be a sentence of quantifier depth k from $L_\sim$. From Theorem 10.26 we know that there exists a graph H_k such that for every graph G we have $\operatorname{Th}_k(H_k \dotplus G \dotplus H_k) = \operatorname{Th}_k(H_k)$. But Lemma 10.30 implies that *a.a.s.* $\overline{\mathbb{G}}(n, \overline{p})$ contains an exact copy of $H_k \dotplus H_k$ and thus *a.a.s.* $\operatorname{Th}_k(\overline{\mathbb{G}}(n, \overline{p})) = \operatorname{Th}_k(H_k)$, that is, if H_k satisfies ψ, then

$$\lim_{n\to\infty} \mathbb{P}(\overline{\mathbb{G}}(n, \overline{p}) \models \psi) = 1$$

otherwise

$$\lim_{n\to\infty} \mathbb{P}(\overline{\mathbb{G}}(n, \overline{p}) \models \psi) = 0\,.$$ ∎

Remark 10.32. As was observed by Łuczak and Shelah (1995), the above result is best possible in the following sense. Let $\varepsilon > 0$ and ψ' be a sentence from $L_\sim$ which says that a graph contains no components with $\lceil 10/\varepsilon \rceil$ vertices. Then, there exists a sequence $\bar{p}' = \{p'_i\}_{i=1}^{\infty}$ such that $0 < p'_i \leq 1/2$ and for every $n \geq 2$ we have $\sum_{i=1}^{n} p'_i \leq \varepsilon \log n$, but $\liminf_{n\to\infty} \mathbb{P}(\overline{\mathbb{G}}(n,\bar{p}') \models \psi') = 0$ while $\limsup_{n\to\infty} \mathbb{P}(\overline{\mathbb{G}}(n,\bar{p}') \models \psi') = 1$.

What happens when, instead of $L_\sim$, we consider the stronger language $L_\sim^{\text{ord}}$? Observe that if ψ is the first-order statement which says that the vertex without predecessors (i.e., 1) is adjacent to its immediate successor (i.e., 2), then for every $n \geq 2$

$$0 < \mathbb{P}(\overline{\mathbb{G}}(n,\bar{p}) \models \psi) = p_1 \leq 1/2 ,$$

in other words, for $L_\sim^{\text{ord}}$ the zero-one law does not hold. However, if the terms of $\bar{p}$ tend to 0 fast enough, then for every sentence ψ from $L_\sim^{\text{ord}}$ the probability that $\overline{\mathbb{G}}(n,\bar{p})$ satisfies ψ converges.

Theorem 10.33. *Let $\bar{p} = \{p_i\}_{i=1}^{\infty}$ denote a sequence such that $0 < p_i \leq 1/2$ for every $i \geq 1$ and the series $\sum_{i=1}^{\infty} ip_i$ converges. Then, for every sentence ψ from $L_\sim^{\text{ord}}$, the probability that ψ holds for $\overline{\mathbb{G}}(n,\bar{p})$ converges as $n \to \infty$.*

Proof. As usual, let ψ denote a sentence of $L_\sim^{\text{ord}}$ of quantifier depth k and let $\bar{p}$ fulfill the assumptions of the theorem. We will show that $\{\mathbb{P}(\overline{\mathbb{G}}(n,\bar{p}) \models \psi)\}_{n=1}^{\infty}$ is a Cauchy sequence, and so it must converge.

Let $\{\overline{\mathbb{G}}'(n,\bar{p})\}_{n=1}^{\infty}$ be a sequence of random graphs constructed in the following way. Take $\overline{\mathbb{G}}'(1,\bar{p})$ to be a graph with a single vertex 1. For $n \geq 1$, define $\overline{\mathbb{G}}'(n+1,\bar{p})$ in the following way:

(i) if $1 \leq i < j \leq \lceil n/2 \rceil$, then the pair $\{i,j\}$ is an edge of $\overline{\mathbb{G}}'(n+1,\bar{p})$ if and only if $\{i,j\}$ is an edge of $\overline{\mathbb{G}}'(n,\bar{p})$;

(ii) if $\lceil n/2 \rceil + 1 \leq i < j \leq n+1$, then the pair $\{i,j\}$ is an edge of $\overline{\mathbb{G}}'(n+1,\bar{p})$ if and only if $\{i-1, j-1\}$ is an edge of $\overline{\mathbb{G}}'(n,\bar{p})$;

(iii) if $1 \leq i \leq \lceil n/2 \rceil < j \leq n+1$ then $\{i,j\}$ is an edge of $\overline{\mathbb{G}}'(n+1,\bar{p})$ with probability p_{j-i}, independently for each such pair.

Clearly, the n-th element of the Markov chain $\{\overline{\mathbb{G}}'(n,\bar{p})\}_{n=1}^{\infty}$ can be identified with the graph $\overline{\mathbb{G}}(n,\bar{p})$, and so we may assume that $\overline{\mathbb{G}}(n,\bar{p}) = \overline{\mathbb{G}}'(n,\bar{p})$. Let us also remark that, of course, a much easier way of obtaining $\overline{\mathbb{G}}(n+1,\bar{p})$ from $\overline{\mathbb{G}}(n,\bar{p})$ is to add to $\overline{\mathbb{G}}(n,\bar{p})$ a new vertex $n+1$ and join it to the vertices of $\overline{\mathbb{G}}(n,\bar{p})$ with appropriate probabilities. Nonetheless, our method of inserting a new vertex in the middle of $\overline{\mathbb{G}}(n,\bar{p})$ has one crucial advantage: the subgraphs induced in $\overline{\mathbb{G}}(n,\bar{p})$ and $\overline{\mathbb{G}}(n+1,\bar{p})$ by the vertices with small labels are the same and, furthermore, the subgraphs spanned by the vertices with large labels in both graphs are identical as well.

Let H_k denote a persistent graph with, say, m vertices, whose existence is assured by Theorem 10.26; thus, for every graph G we have $\text{Th}_k(H_k \dotplus G \dotplus H_k) = \text{Th}_k(H_k)$. From Lemma 10.30 we know that the probability that a separating exact copy of H_k appears somewhere at the beginning of $\overline{\mathbb{G}}(n,\overline{p})$ tends to 1 as $n \to \infty$ and, similarly, by symmetry, the probability that one can find a separating exact copy of H_k somewhere near the end of $\overline{\mathbb{G}}(n,\overline{p})$ tends to 1 as well. More precisely, with probability larger than $1-\varepsilon'(n)$, for some function $\varepsilon'(n)$ such that $\lim_{n\to\infty}\varepsilon'(n)=0$, $\overline{\mathbb{G}}(n,\overline{p})$ can be represented as $G_1 \dotplus H_k' \dotplus G_2 \dotplus H_k'' \dotplus G_3$, where H_k' and H_k'' are exact separating copies of H_k, and H_k' contains no vertex with label larger than $m\lceil n^{0.9}\rceil$, while no vertex with label smaller than $n+1-m\lceil n^{0.9}\rceil$ belongs to H_k''. Now, for $n'>n$, let us consider the random graph $\overline{\mathbb{G}}(n',\overline{p})$ viewed as a stage of the Markov chain described above, that is, we assume that $\overline{\mathbb{G}}(n',\overline{p})$ was obtained from $\overline{\mathbb{G}}(n,\overline{p})$ in a sequence of "middle vertex insertions". Clearly, if in this process we have inserted no edge $\{i,j\}$ such that $j-i \geq n/2-m\lceil n^{0.9}\rceil$ and either $i \leq m\lceil n^{0.9}\rceil$ or $j \geq n+1-m\lceil n^{0.9}\rceil$, then $\overline{\mathbb{G}}(n,\overline{p})$ can be represented as $G_1 \dotplus H_k' \dotplus G_2' \dotplus H_k'' \dotplus G_3$ where G_1 and H_k' are the same as in the decomposition of $\mathbb{G}(n,\overline{p})$, and H_k'' and G_3 are shifted to the right by $n'-n$. Furthermore, the probability that $\overline{\mathbb{G}}(n',\overline{p})$ contains edges $\{i,j\}$ which can corrupt this decomposition is bounded from above by

$$2m\lceil n^{0.9}\rceil \sum_{i=\lceil n/2\rceil - m\lceil n^{0.9}\rceil}^{n'} p_i \leq \sum_{i=\lfloor n/3\rfloor}^{\infty} ip_i = \varepsilon''(n),$$

where $\varepsilon''(n)$ tends to 0 as $n \to \infty$.

Hence, using the fact that H_k is persistent, with probability at least $1-\varepsilon'(n)-\varepsilon''(n)$, we have

$$\begin{aligned}\mathrm{Th}_k(\overline{\mathbb{G}}(n',\overline{p})) &= \mathrm{Th}_k(G_1 \dotplus H_k' \dotplus G_2' \dotplus H_k'' \dotplus G_3) = \mathrm{Th}_k(G_1 \dotplus H_k \dotplus G_3) \\ &= \mathrm{Th}_k(G_1 \dotplus H_k' \dotplus G_2 \dotplus H_k'' \dotplus G_3) = \mathrm{Th}_k(\overline{\mathbb{G}}(n,p)).\end{aligned}$$

In particular, for $n'>n$,

$$|\,\mathbb{P}(\overline{\mathbb{G}}(n',\overline{p}) \models \psi) - \mathbb{P}(\overline{\mathbb{G}}(n,\overline{p}) \models \psi)| \leq \varepsilon'(n)+\varepsilon''(n),$$

where $\varepsilon'(n)+\varepsilon''(n) \to 0$ as $n\to\infty$. Thus, the sequence $\{\mathbb{P}(\overline{\mathbb{G}}(n,\overline{p}) \models \psi)\}_{n=1}^{\infty}$ is Cauchy and so it must converge. ■

Let us mention one more result on the first-order properties of $\overline{\mathbb{G}}(n,\overline{p})$. Łuczak and Shelah (1995) showed that for every $\varepsilon>0$ and every sequence $\overline{p}=\{p_i\}_{i=1}^{\infty}$ there is a sentence ψ from $L_{\sim}^{\mathrm{ord}}$, and a sequence $\overline{p}'=\{p_i'\}_{i=1}^{\infty}$ obtained from $\overline{p}$ by adding to it some number of zero terms, such that

$$\limsup_{n\to\infty} \mathbb{P}(\overline{\mathbb{G}}(n,\overline{p}') \models \psi) - \liminf_{n\to\infty} \mathbb{P}(\overline{\mathbb{G}}(n,\overline{p}') \models \psi) > 1-\varepsilon\,.$$

However, for their construction one cannot put $\varepsilon=0$ unless the series $\sum_i p_i$ diverges. It is somewhat surprising that, in fact, if $\sum_i p_i < \infty$, then for no sentence ψ of $L_{\sim}^{\mathrm{ord}}$ the sequence $\{\mathbb{P}(\overline{\mathbb{G}}(n,\overline{p}) \models \psi\}_{n=1}^{\infty}$ has both 0 and 1 as concentration points; that is, the following *defect zero-one law* holds.

Theorem 10.34. *Let $\overline{p} = \{p_i\}_{i=1}^{\infty}$ be a sequence such that $0 < p_i \leq 1/2$ for every $i \geq 1$ and the series $\sum_{i=1}^{\infty} p_i$ converges. Then, for every sentence ψ from $L_{\sim}^{\mathrm{ord}}$,*

$$\limsup_{n\to\infty} \mathbb{P}(\overline{\mathbb{G}}(n,\overline{p}) \models \psi) - \liminf_{n\to\infty} \mathbb{P}(\overline{\mathbb{G}}(n,\overline{p}) \models \psi) < 1 .$$

Proof. Let ψ be a sentence of quantifier depth k and H_k be a persistent graph on m vertices whose existence is implied by Theorem 10.26. The probability that $\overline{\mathbb{G}}(n,\overline{p})$ can be represented as $H_k' \dotplus G \dotplus H_k''$, where H_k' and H_k'' are exact separating copies of H_k, is, for $n \geq 3m$, given by

$$\begin{aligned}\left(\mathbb{P}(\overline{\mathbb{G}}(m,\overline{p}) = H_k)\right)^2 &\prod_{i=1}^{n-1} (1-p_i)^{\min\{2m,2i,n-i\}} \\ &\geq \left[\mathbb{P}(\overline{\mathbb{G}}(m,\overline{p}) = H_k)\left(\prod_{i=1}^{\infty}(1-p_i)\right)^m\right]^2 \\ &\geq \left[\mathbb{P}(\overline{\mathbb{G}}(m,\overline{p}) = H_k)\exp\Big(-2m\sum_{i=1}^{\infty}p_i\Big)\right]^2 ,\end{aligned}$$

and so it is larger than a suitably chosen constant $\varepsilon > 0$ which depends on both ψ and $\overline{p}$ but not on n. Hence, because of the persistence of H_k, if ψ holds for H_k, then

$$\liminf_{n\to\infty} \mathbb{P}(\overline{\mathbb{G}}(n,\overline{p}) \models \psi) > \varepsilon,$$

and if H_k is not a model for ψ, then

$$\limsup_{n\to\infty} \mathbb{P}(\overline{\mathbb{G}}(n,\overline{p}) \models \psi) < 1 - \varepsilon .$$ ■

Finally, we remark that there are other types of zero-one laws we did not consider in this chapter. For instance, one can ask when the difference $\mathbb{P}(\overline{\mathbb{G}}(n+1,\overline{p}) \models \psi) - \mathbb{P}(\overline{\mathbb{G}}(n,\overline{p}) \models \psi)$ tends to 0 as $n \to \infty$ for every sentence ψ of $L_{\sim}$ or $L_{\sim}^{\mathrm{ord}}$ (Shelah 1996), or study the convergence of $\{\mathbb{P}(\overline{\mathbb{G}}(n,\overline{p}) \models \psi)\}_{n=1}^{\infty}$, when n is chosen from some dense subset of the natural numbers (Lynch 1993). We should also mention that we have emphasized methods of combinatorial and probabilistic flavor which are not, by any means, the only ways to study zero-one laws for combinatorial structures (see, e.g., the survey of Compton (1988)).

10.5 SEPARABILITY AND THE SPEED OF CONVERGENCE

If we can show that a random structure satisfies the zero-one law another problem immediately emerges: is it possible to determine *effectively* which properties hold for the random structure *a.a.s.* and which *a.a.s.* do not? As a matter of fact, such a separability question can be asked even when the zero-one law does not hold. Thus, we would like to find an algorithm which, given a function $p = p(n)$ and a sentence ψ from L, puts ψ into one of two disjoint classes: one which contains all sentences of L which hold for $\mathbb{G}(n,p)$ *a.a.s.*, and the other, which contains all sentences which *a.a.s.* do not hold for $\mathbb{G}(n,p)$. The sentences that are not of the above two types may be placed in either class, provided only that the decision will be reached after a finite number of steps – the algorithm is not allowed to run forever.

Does there exist a separating procedure for $\mathbb{G}(n,p)$ and the language $L_\sim$, in the case in which $p = p(n)$ is one of the functions listed in Theorems 10.5, 10.6, 10.7 or 10.15, where we know that the zero-one law holds? At first sight the positive answer seems to follow directly from the method we used in the proofs. Indeed, in each case we have employed a similar scheme. Given a sentence ψ of quantifier depth k we first determined a finite family of sentences $\Phi_k = \{\varphi_i\}_{i \in I}$ such that each sentence from Φ_k holds *a.a.s.* for $\mathbb{G}(n,p)$. Then we showed that for any two graphs G' and G'' such that both of them possess all properties from Φ_k we had $\mathrm{Th}_k(G') = \mathrm{Th}_k(G'')$, that is, we verified, using Lemma 10.3, that Φ_k was a complete system of axioms for the theory containing all sentences of L of quantifier depth at most k. Now let G' be a graph such that G' satisfies φ_i for every $\varphi_i \in \Phi_k$; clearly G' can be found by examining one by one all graphs on n vertices, as n ranges from one to infinity. Then, it is enough to check whether ψ holds for G': if the answer is positive, then *a.a.s.* ψ holds for $\mathbb{G}(n,p)$, otherwise *a.a.s.* $\mathbb{G}(n,p)$ is a model for $\neg\psi$.

Before we state the above result in a rigorous form, we need to make a comment of a somewhat technical nature. In order to decide on the limit probability that $\mathbb{G}(n,p)$ is a model for ψ, the algorithm must be given both ψ and the *function* $p = p(n)$ in a form such that one can study the behavior of $p(n)$ for large values of n. However, instead of struggling with the problem of how to present a function $p : \mathbb{N} \to \mathbb{R}$ in a finite and accessible way, we will cheat a little bit and ***assume*** that we know the asymptotic behavior of $p = p(n)$. Thus, for instance, in the case of a function $p = p(n)$ for which the assumptions of Theorem 10.6(i) hold, we assume that there is a way to compute ℓ such that $n^{-1-1/\ell} \ll p \ll n^{-1-1/(\ell+1)}$, and use the value of ℓ to construct the appropriate complete axiom system $\Phi_k^{(\ell)}$ in order to check whether *a.a.s.* ψ is satisfied in $\mathbb{G}(n,p)$. This idea works well for functions $p = p(n)$ which fulfill the assumptions of one of Theorems 10.5, 10.6, or 10.15. When $p(n) = n^{-\alpha+o(1)}$ for an irrational $\alpha \in (0,1)$ we still need to find a compact description of α. However, a quick inspection of the proof of Theorem 10.7 reveals that we can study the behavior of $\mathbb{G}(n,p)$ for $p = p(n) = n^{-\alpha+o(1)}$, where $\alpha \in (0,1)$ is irrational, provided we can estimate expressions like $v - \alpha e$, for all natural numbers v and e. This, in turn, can be accomplished if we are given a subprocedure Oracle_α which, for any rational r, can tell whether $\alpha < r$. Thus we arrive at the following result.

Theorem 10.35. *Let $p = p(n)$ be a function for which we can decide that it fulfills the assumptions of either Theorem 10.5 or Theorem 10.15, or one of the assumptions of Theorem 10.6. Then there exists an algorithm which, given a sentence ψ from $L_\sim$, computes* $\lim_{n\to\infty} \mathbb{P}(\mathbb{G}(n,p) \models \psi)$.

Furthermore, for every irrational $\alpha \in (0,1)$ and $p = p(n) = n^{-\alpha+o(1)}$ there exists an algorithm using Oracle_α *as a subprocedure which, given a sentence ψ from $L_\sim$, computes* $\lim_{n\to\infty} \mathbb{P}(\mathbb{G}(n,p) \models \psi)$. ■

The above result might suggest that finding a separating procedure should be possible at least when we can prove that the zero-one law holds. We will soon show that it is not the case. Our argument will be based on the following consequence of Trahtenbrot's theorem (Trahtenbrot 1950).

Theorem 10.36. *There is no procedure which can decide, for every sentence ψ from $L_\sim$, whether there exists a finite graph G such that G satisfies ψ.* ■

Theorem 10.36 implies, in particular, that the minimal size of the model in which a sentence holds can grow very fast with the depth of the sentence.

Corollary 10.37. *There exists a sequence $\{\psi_k\}_{k=1}^{\infty}$ of sentences from $L_\sim$ such that for every $k = 1, 2, \ldots$, the sentence ψ_k has quantifier depth k and ψ_k holds for some finite graph G_k, but the function*

$$g(k) = \min\{v(G_k) : G_k \models \psi_k\}$$

is not bounded from above by any recursive function of k.

Proof. Suppose that the assertion does not hold and for every sentence ψ from $L_\sim$ of quantifier depth k, either ψ holds for no finite graph, or it is valid for some finite graph of size bounded from above by a recursive (i.e., computable) function $g(k)$. Consider a procedure which computes $g(k)$, constructs all graphs on the set $\{1, 2, \ldots, \ell\}$ for all $1 \le \ell \le k$, and checks one by one whether ψ holds for one of them. Such an algorithm would verify whether ψ holds for a finite graph, contradicting Theorem 10.36. ■

Let us look again at the random graph $\overline{\mathbb{G}}(n, \overline{p})$ we studied in the previous section. In Theorem 10.31 we proved that if $\sum_{i=1}^{n} p_i = o(\log n)$, then for every sentence ψ from $L_\sim$ the zero-one law holds. We will show that, somewhat surprisingly, no procedure can decide whether the limit probability of a sentence is zero or one. Moreover, the speed of convergence for some sentences from $L_\sim$ can be very, very slow.

Theorem 10.38. *Let $\overline{p} = \{p_i\}_{i=1}^{\infty}$ be such that $0 < p_i \le 1/2$ and $\sum_{i=1}^{n} p_i = o(\log n)$. Then there is no procedure which for every sentence ψ from $L_\sim$ can decide whether* a.a.s. *ψ holds for $\overline{\mathbb{G}}(n, \overline{p})$.*

Furthermore, there exists a sequence $\{\psi_k\}_{k=1}^{\infty}$ of sentences from $L_\sim$ such that, for every $k = 1, 2, \ldots$, the sentence ψ_k has quantifier depth k and

$$\lim_{n\to\infty} \mathbb{P}(\overline{\mathbb{G}}(n, \overline{p}) \models \psi_k) = 1 ,$$

but the function

$$f(k) = \min\{n : \mathbb{P}(\overline{\mathbb{G}}(n,\overline{p}) \models \psi_k) > 0\}$$

is not bounded from above by any recursive function of k.

Proof. Let $\overline{p}$ be a sequence of probabilities which fulfills the assumptions of the theorem and let ψ be any sentence of $L_\sim$. Denote by $\hat{\psi}$ the sentence saying that ψ holds in the neighborhood of some vertex of a graph, that is, in order to construct $\hat{\psi}$ from ψ one needs to replace each occurrence of $y \sim z$ by $(x \sim y \wedge x \sim z \wedge y \sim z)$, each $\neg(y \sim z)$ by $(x \sim y \wedge x \sim z \wedge \neg(y \sim z))$, and add at the beginning of the formula the existential quantifier $\exists_x$. Then, obviously, $\hat{\psi}$ is a sentence from $L_\sim$ of quantifier depth at most one larger than ψ, and ψ has a finite model if and only if there is a finite model for $\hat{\psi}$. Furthermore, a graph is a model for $\hat{\psi}$ if and only if one of its components satisfies $\hat{\psi}$.

Clearly, if $\hat{\psi}$ holds *a.a.s.* for $\overline{\mathbb{G}}(n,\overline{p})$, then it has a finite model. Note however, that Lemma 10.30 implies that the converse is also true: if $\hat{\psi}$ holds for some graph H, then *a.a.s.* $\overline{\mathbb{G}}(n,\overline{p})$ contains this graph as a component, and thus *a.a.s.* $\overline{\mathbb{G}}(n,\overline{p})$ is a model for $\hat{\psi}$. Hence, if we could effectively compute the limit probability for every sentence from $L_\sim$, we could also decide for every ψ from $L_\sim$ whether it has a finite model, which would contradict Theorem 10.36.

In order to show the second part of the proof, it is enough to consider sentences $\hat{\psi}_k$ corresponding to sentences ψ_k from $L_\sim$ chosen as in Corollary 10.37. ■

Finally, let us return to $\mathbb{G}(n,p)$ and consider the case in which $p = p(n) = n^{-1/7+o(1)}$, where np^7 is of an order smaller than $\log n$, so Theorem 10.15 does not apply and, unless np^7 tends to 0 very fast, the zero-one law does not hold (Theorem 10.19). Then, as observed by Dolan (1992), one can use the idea employed in the proof of Theorem 10.19 to show that for such a $p = p(n)$ no separating procedure can be found.

Theorem 10.39. *Let* $p = p(n) = n^{-1/7+o(1)}$ *be such that* $np^7 = o(\log n)$. *Then there exists no procedure which can separate those sentences of* $L_\sim$ *which* a.a.s. *hold in* $\mathbb{G}(n,p)$ *from those whose probability tends to* 0 *as* $n \to \infty$.

Furthermore, there exists a sequence $\{\psi_k\}_{k=1}^{\infty}$ *of sentences from* $L_\sim$ *such that, for every* $k = 1, 2, \ldots$, *the sentence* ψ_k *has quantifier depth* k *and*

$$\lim_{n\to\infty} \mathbb{P}(\mathbb{G}(n,p) \models \psi_k) = 1 ,$$

but the function

$$f(k) = \min\{n : \mathbb{P}(\mathbb{G}(n,p) \models \psi_k) > 0\}$$

is not bounded from above by any recursive function of k.

Proof. Let ψ be any sentence from $L_\sim$ and let $\hat{\psi}$ denote the sentence from $L_\sim$ stating that there exists an element x such that ψ holds in the set of all elements y for which $x \sim y$. Then, as in the proof of Theorem 10.38, we infer that ψ holds for some finite model if and only if for each m large enough there exists a model for $\hat{\psi}$ of m elements.

From Lemma 10.21 we know that one can find formulas P^* and T^* from $L_\sim$ such that for some function $\omega = \omega(n)$, where $\omega(n) \to \infty$ as $n \to \infty$, the following holds: *a.a.s.* $\mathbb{G}(n,p)$ contains a set of vertices S, $|S| \geq \omega$, which is defined by P^* and some sequence $\underline{v}^S$, and such that every ternary relation on S can be encoded by T^*. Now let $\hat{\psi}^*$ be the sentence obtained from $\hat{\psi}$ in the following way: replace each occurrence of "$x \sim y$" by "$T^*(x,y,z,\underline{v}^T) \wedge T^*(y,x,z,\underline{v}^T)$", for each logical variable x_i used in $\hat{\psi}$, add to it "$\wedge P^*(x_i,\underline{v}^S)$", and put at the beginning the existential quantifiers $\exists_{\underline{v}^S}\exists_{\underline{v}^T}\exists_z$. Then, clearly, $\hat{\psi}^*$ is a sentence from $L_\sim$. Moreover, $\hat{\psi}^*$ states that for some subset S of vertices of $\mathbb{G}(n,p)$, such that $|S| \geq \omega$, there exists a symmetric binary relation "$\sim'$" on S such that S with "$\sim'$" is a finite model for $\hat{\psi}$. (Note that "$\sim'$" is *not* the adjacency relation of $\mathbb{G}(n,p)$.) Thus, if *a.a.s.* $\mathbb{G}(n,p)$ satisfies $\hat{\psi}^*$, then $\hat{\psi}$ and thus ψ have finite models. On the other hand, if ψ has a finite model, then for every set S large enough there exists a symmetric binary relation "$\sim'$" such that S with "$\sim'$" is a model for $\hat{\psi}$, and so $\hat{\psi}^*$ *a.a.s.* holds for $\mathbb{G}(n,p)$. Consequently, an algorithm separating properties which *a.a.s.* hold in $\mathbb{G}(n,p)$ from those which *a.a.s.* do not hold, could be used to decide if ψ has a finite model, contradicting Trahtenbrot's theorem.

The second part of the statement follows from Corollary 10.37 in a similar manner. ■

References

Page numbers in italics show where the works are cited in this book.

D. ACHLIOPTAS & E. FRIEDGUT (1999), A sharp threshold for k-colorability. *Random Structures Algorithms* **14**, 63–70. (*23, 196*)

D. ACHLIOPTAS & M. MOLLOY (1997), The analysis of a list-coloring algorithm on a random graph. In *38th Annual Symposium on Foundations of Computer Science (FOCS '97)*, 204–212. (*107, 196*)

D. ACHLIOPTAS & M. MOLLOY (1999), Almost all graphs with $2.522n$ edges are not 3-colorable. *Electronic J. Comb.* **6**, R29. (*196*)

R. AHLSWEDE & D.E. DAYKIN (1978), An inequality for the weights of two families of sets, their unions and intersections. *Z. Wahrscheinl. Geb.* **43**, 183–185. (*30*)

D. ALDOUS (1990), A random tree model associated with random graphs. *Random Structures Algorithms* **1**, 383–402. (*5*)

D. ALDOUS (1997), Brownian excursions, critical random graphs and the multiplicative coalescent. *Ann. Probab.* **25**, 812–854. (*127*)

D.J. ALDOUS & J. PITMAN (1994), Brownian bridge asymptotics for random mappings. *Random Structures Algorithms* **5**, 487–512. (*3*)

N. ALON & Z. FÜREDI (1992), Spanning subgraphs of random graphs. *Graphs and Combinatorics (Research problem)* **8**, 91–94. (*94, 96*)

N. ALON & M. KRIVELEVICH (1997), The concentration of the chromatic number of random graphs. *Combinatorica* **17**, 303–313. (*189*)

N. ALON, M. KRIVELEVICH & B. SUDAKOV (1998), Finding a large hidden clique in a random graph. *Random Structures Algorithms* **13**, 457–466. (*187*)

N. ALON & J. SPENCER (1992), *The Probabilistic Method*. John Wiley & Sons, New York. (*12, 27, 29, 38, 49, 50*)

N. ALON & R. YUSTER (1993), Threshold functions for H-factors. *Combin. Probab. Comput.* **2**, 137–144. (*91, 92*)

P. ANDERSSON (1998), On the asymptotic distribution of subgraphs counts in a random tournament. *Random Structures Algorithms* **13**, 249–260. (*3, 176*)

K.B. ATHREYA & P.E. NEY (1972), *Branching processes*. Die Grundlehren der mathematischen Wissenschaften, Band 196, Springer, Berlin. (*107*)

K. AZUMA (1967), Weighted sums of certain dependent variables. *Tôhoku Math. J.* **3**, 357–367. (*37*)

L. BABAI, M. SIMONOVITS & J. SPENCER (1990), Extremal subgraphs of random graphs. *J. Graph Theory* **14**, 599–622. (*231*)

A.D. BARBOUR (1982), Poisson convergence and random graphs. *Math. Proc. Camb. Phil. Soc.* **92**, 349–359. (*152, 157, 162*)

A.D. BARBOUR, L. HOLST & S. JANSON (1992), *Poisson Approximation*. Oxford University Press, Oxford. (*153, 154, 156*)

A. D. BARBOUR, S. JANSON, M. KAROŃSKI & A. RUCIŃSKI (1990), Small cliques in random graphs. *Random Structures Algorithms* **1**, 403–434. (*176, 245*)

A.D. BARBOUR, M. KAROŃSKI & A. RUCIŃSKI (1989), A central limit theorem for decomposable random variables with applications to random graphs. *J. Combin. Theory Ser. B* **47**, 125–145. (*157, 158, 161, 162, 172*)

E.A. BENDER & E.R. CANFIELD (1978), The asymptotic number of labeled graphs with given degree sequences. *J. Combin. Theory Ser. A* **24**, 296–307. (*234, 235*)

E.A. BENDER, E.R. CANFIELD & B.D. MCKAY (1990), The asymptotic number of labeled connected graphs with a given number of vertices and edges. *Random Structures Algorithms* **1**, 127–169. (*121*)

G. BENNETT (1962), Probability inequalities for the sum of independent random variables. *J. Amer. Statist. Assoc.* **57**, 33–45. (*29*)

C. BERGE (1973), *Graphs and Hypergraphs*. North Holland, Amsterdam. (*76*)

S. BERNSTEIN (1924), On a modification of Chebyshev's inequality and of the error formula of Laplace. *Ann. Sci. Inst. Sav. Ukraine, Sect. Math.* **1**, 38–49 (Russian). (*25*)

P. BILLINGSLEY (1968), *Convergence of Probability Measures*. John Wiley & Sons, New York. (*8, 9, 68, 169, 268*)

B. BOLLOBÁS (1978), *Extremal Graph Theory*. Academic Press, London. (*48, 94, 205, 230*)

B. BOLLOBÁS (1979), *Graph Theory*. Springer, New York. (*13, 14*)

B. BOLLOBÁS (1980), A probabilistic proof of an asymptotic formula for the number of labelled regular graphs. *Europ. J. Combinatorics* **1**, 311–316. (*235, 236*)

B. BOLLOBÁS (1981a), The diameter of random graphs. *Trans. Amer. Math. Soc.* **267**, 41–52. (*105*)

B. BOLLOBÁS (1981b), Random Graphs. In *Combinatorics*, Proceedings, Swansea 1981, London Math. Soc. Lecture Note Ser. 52, Cambridge Univ. Press, Cambridge, 80–102. (*53, 56, 58, 66, 241, 244*)

B. BOLLOBÁS (1983), Almost all regular graphs are Hamiltonian. *European J. Combin.* **4**, 97–106. (*241*)

B. BOLLOBÁS (1984a), The evolution of random graphs. *Trans. Amer. Math. Soc.* **286**, 257–274. (*103, 112, 119, 120*)

B. BOLLOBÁS (1984b), The evolution of sparse graphs. In *Graph Theory and Combinatorics*, Proceedings, Cambridge Combinatorial Conf. in honour of Paul Erdős, ed. B. Bollobás, Academic Press, 35–57. (*105, 106*)

B. BOLLOBÁS (1985), *Random Graphs*. Academic Press, London. (*4, 14, 17, 71, 105, 119, 180, 235, 236, 244, 280*)

B. BOLLOBÁS (1986), *Combinatorics*. Cambridge Univ. Press, Cambridge. (*30*)

B. BOLLOBÁS (1988a), The chromatic number of random graphs. *Combinatorica* **8**, 49–56. (*37, 39, 190*)

B. BOLLOBÁS (1988b), Martingales, isoperimetric inequalities and random graphs. In *Combinatorics*, Proceedings, Eger 1987, eds. A. Hajnal, L. Lovász & V.T. Sós, Colloq. Math. Soc. János Bolyai 52, North-Holland, Amsterdam, 113–139. (*39, 134*)

B. BOLLOBÁS (1998), *Modern Graph Theory*. Springer, New York. (*60, 184, 209, 212, 214*)

B. BOLLOBÁS & P. ERDŐS (1976), Cliques in random graphs. *Math. Proc. Camb. Phil. Soc.* **80**, 419–427. (*180*)

B. BOLLOBÁS, T.I. FENNER & A.M. FRIEZE (1990), Hamilton cycles in random graphs of minimal degree at least k. In *A tribute to Paul Erdős*, eds. A. Baker, B. Bollobás & A. Hajnal, Cambridge Univ. Press, Cambridge, 59–95. (*105*)

B. BOLLOBÁS & A.M. FRIEZE (1985), On matchings and Hamiltonian cycles in random graphs. In *Random Graphs '83*, Proceedings, Poznań, 1983, eds. M. Karoński & A. Ruciński, North-Holland, Amsterdam–New York, 23–46. (*89, 105*)

B. BOLLOBÁS & B.D. MCKAY (1986), The number of matchings in random regular graphs and bipartite graphs. *J. Combin. Theory Ser. B* **41**, 80–91. (*260*)

B. BOLLOBÁS & A. THOMASON (1985), Random graphs of small order. In *Random Graphs '83*, Proceedings, Poznań, 1983, eds. M. Karoński & A. Ruciński, North-Holland, Amsterdam-New York, 47–97. (*4, 81, 84, 105, 187*)

B. BOLLOBÁS & A. THOMASON (1987), Threshold functions. *Combinatorica* **7**, 35–38. (*18*)

B. BOLLOBÁS & A. THOMASON (1995), Generalized chromatic numbers of random graphs. *Random Structures Algorithms* **6**, 353–356. (*196*)

B. BOLLOBÁS & A. THOMASON (1997), Hereditary and monotone properties of graphs. In *The Mathematics of Paul Erdős, II*, eds. R.L. Graham & J. Nešetřil, Algorithms and Combinatorics, 14, Springer, Berlin, 70–78. (*196*)

B. BOLLOBÁS & J.C. WIERMAN (1989), Subgraph counts and containment probabilities of balanced and unbalanced subgraphs in a large random graph. In *Graph Theory and Its Applications: East and West*, Proceedings, Jinan, 1986, Ann. New York Acad. Sci. 576, New York Acad. Sci., New York, 63–70. (*68*)

R. Boppana & J. Spencer (1989), A useful elementary correlation inequality. *J. Combin. Theory Ser. A* **50**, 305–307. (*33*)

S. Boucheron, G. Lugosi & P. Massart (2000+), A sharp concentration inequality with applications. *Random Structures Algorithms*, to appear. (*43*)

J. Bourgain (1999), On sharp thresholds of monotone properties. Appendix to Friedgut (1999), *J. Amer. Math. Soc.* **12**, 1046–1053. (*23*)

J. Bourgain & G. Kalai (1997), Influences of variables and threshold intervals under group symmetries. *Geom. Funct. Anal.* **7**, 438–461. (*22, 23*)

Yu.D. Burtin (1973), Asymptotic estimates of the diameter and independence and domination numbers of a random graph. *Dokl. Akad. Nauk SSSR* **209**, 765–768 (Russian); English transl. *Soviet Math. Dokl.* **14**, 497–501. (*105*)

L.H.Y. Chen (1975), Poisson approximation for dependent trials. *Ann. Probab.* **3**, 534–545. (*152, 153*)

H. Chernoff (1952), A measure of asymptotic efficiency for tests of a hypothesis based on the sum of observations. *Ann. Math. Statistics* **23**, 493–507. (*26*)

K.L. Chung (1974), *A Course in Probability Theory*. 2nd ed., Academic Press, New York. (*140, 247*)

V. Chvátal (1991), Almost all graphs with $1.44n$ edges are 3-colorable. *Random Structures Algorithms* **2**, 11–18. (*196*)

K.J. Compton (1988), $0-1$ Laws in Logic and Combinatorics. In *Algorithms and Order*, ed. I. Rival, NATO ASI series, Kluwer Academic Publishers, Dordrecht, 353–383. (*301*)

C. Cooper, A.M. Frieze & M.J. Molloy (1994), Hamilton cycles in random regular digraphs. *Combin. Probab. Comput.* **3**, 39–50. (*233*)

C. Cooper, A.M. Frieze, M. Molloy & B. Reed (1996), Perfect matchings in random r-regular, s-uniform hypergraphs. *Combin. Probab. Comput.* **5**, 1–14. (*241*)

P. DE JONG (1996), A central limit theorem with applications to random hypergraphs. *Random Structures Algorithms* **8**, 105–120. (*176*)

A. DEMBO (1997), Information inequalities and concentration of measure. *Ann. Probab.* **25**, 927–939. (*40*)

R. DIESTEL (1996), *Graphentheorie* (German), Springer, New York; English transl. *Graph Theory* (1997), Springer, New York. (*60, 66, 184, 212, 214*)

P. DOLAN (1992), Undecidable statements and random graphs. *Ann. Math. Artificial Intelligence* **6**, 17–25. (*304*)

P. DOLAN & J.F. LYNCH (1993), The logic of ordered random structures. *Random Structures Algorithms* **4**, 429–446. (*292*)

A. EHRENFEUCHT (1961), An application of games to the completeness problem for formalized theories. *Fund. Math.* **49**, 129–141. (*273*)

P. ERDŐS (1947), Some remarks on the theory of graphs. *Bull. Amer. Math. Soc.* **53**, 292–294. (*1, 2, 3*)

P. ERDŐS & L. LOVÁSZ (1975), Problems and results on 3-chromatic hypergraphs and some related questions. In *Infinite and Finite Sets, Vol. II*, eds. A. Hajnal, R. Rado & V.T. Sós, Colloq. Math. Soc. J. Bolyai 10, North-Holland, Amsterdam, 609–627. (*12*)

P. ERDŐS & A. RÉNYI (1959), On random graphs I. *Publ. Math. Debrecen* **6**, 290-297. (*4, 81, 105*)

P. ERDŐS & A. RÉNYI (1960), On the evolution of random graphs. *Publ. Math. Inst. Hungar. Acad. Sci.* **5**, 17–61. (*53, 66, 80, 103, 104, 111, 128, 156, 157, 162*)

P. ERDŐS & A. RÉNYI (1961), On the strength of connectedness of a random graph. *Acta Math. Acad. Sci. Hungar.* **12**, 261–267. (*81, 105*)

P. ERDŐS & A. RÉNYI (1964), On random matrices. *Publ. Math. Inst. Hung. Acad. Sci.* **8**, 455-461. (*81, 83*)

P. ERDŐS & A. RÉNYI (1966), On the existence of a factor of degree one of a connected random graph. *Acta Math. Acad. Sci. Hungar.* **17**, 359–368. (*81, 85, 89, 105*)

P. ERDŐS & A. RÉNYI (1968), On random matrices II. *Studia Sci. Math. Hungar.* **3**, 459-464. (*81*)

P. ERDŐS & M. SIMONOVITS (1966), A limit theorem in graph theory. *Studia Sci. Math. Hungar.* **1**, 51-57. (*60, 205*)

P. ERDŐS & A.H. STONE (1946), On the structure of linear graphs. *Bull. Amer. Math. Soc.* **52**, 1087-1091. (*60, 205*)

P. ERDŐS, S. SUEN & P. WINKLER (1995), On the size of a maximal random graph. *Random Structures Algorithms* **6**, 309–318. (*5*)

P. ERDŐS & P. TETALI (1990), Representations of integers as the sum of k terms. *Random Structures Algorithms* **1**, 245–261. (*50*)

R. FAGIN (1976), Probabilities in finite models. *J. Symbolic Logic* **41**, 50–58. (*276*)

T.I. FENNER & A.M. FRIEZE (1984), Hamiltonian cycles in random regular graphs. *J. Combin. Theory Ser. B* **37**, 103–112. (*241*)

P. FLAJOLET, D.E. KNUTH & B. PITTEL (1989), The first cycles in an evolving graph. *Discrete Math.* **75**, 167–215. (*135*)

C.M. FORTUIN, P.W. KASTELEYN & J. GINIBRE (1971), Correlation inequalities on some partially ordered sets. *Comm. Math. Phys.* **22**, 89–103. (*30*)

P. FRANKL & V. RÖDL (1986), Large triangle-free subgraphs in graphs without K_4. *Graphs and Combinatorics* **2**, 135–144. (*208, 209*)

E. FRIEDGUT (1999), Sharp thresholds of graph properties, and the k-sat problem. *J. Amer. Math. Soc.* **12**, 1017–1054. (*22, 23*)

E. FRIEDGUT & G. KALAI (1996), Every monotone graph property has a sharp threshold. *Proc. Amer. Math. Soc.* **124**, 2993–3002. (*22, 23*)

E. FRIEDGUT & M. KRIVELEVICH (2000), Sharp thresholds for Ramsey properties of random graphs. *Random Structures Algorithms*, to appear. (*23, 198, 204*)

A.M. FRIEZE (1988), Finding Hamilton cycles in sparse random graphs. *J. Combin. Theory Ser. B* **44**, 230–250. (*241*)

A.M. FRIEZE (1990), On the independence number of random graphs. *Discrete Math.* **81**, 171–175. (*183*)

A.M. FRIEZE & S. JANSON (1995), Perfect matchings in random s-regular hypergraphs. *Random Structures Algorithms* **7**, 41–57. (*100, 101*)

A.M. FRIEZE, M.R. JERRUM, M. MOLLOY, R. ROBINSON & N.C. WORMALD (1996), Generating and counting Hamilton cycles in random regular graphs. *J. Algorithms* **21**, 176–198. (*240, 252, 256*)

A.M. FRIEZE & T. ŁUCZAK (1992), On the independence and chromatic numbers of random regular graphs. *J. Combin. Theory Ser. B* **54**, 123–132. (*233*)

A.M. FRIEZE & C. MCDIARMID (1997), Algorithmic theory of random graphs. *Random Structures Algorithms* **10**, 5–42. (*187*)

Z. FÜREDI (1994), Random Ramsey graphs for the four-cycle. *Discrete Math.* **126**, 407–410. (*62, 232*)

H. GARMO (1999), The asymptotic distribution of long cycles in random regular graphs. *Random Structures Algorithms* **15**, 43–92. (*256*)

Y.V. GLEBSKII, D.I. KOGAN, M.I. LIAGONKII & V.A. TALANOV, (1969), Range and degree of realizability of formulas in the restricted predicate calculua. *Kibernetica* **5**, 17–27 (Russian); English transl. *Cybernetics* **5**, 142–154. (*276*)

A.W. GOODMAN (1959), On sets of acquaintances and strangers at any party. *Amer. Math. Mon.* **66**, 778–783. (*209*)

R.L. GRAHAM, D.E. KNUTH & O. PATASHNIK (1989), *Concrete Mathematics*. Addison-Wesley, Reading, Mass. (*144*)

R.L. GRAHAM, V. RÖDL & A. RUCIŃSKI (1996), On Schur properties of random subsets of integers. *J. Number Theory* **61**, 388–408. (*208*)

R.L. GRAHAM, B. ROTHSCHILD & J. SPENCER (1992), *Ramsey Theory*. 2nd ed., Wiley, New York. (*217*)

G. GRIMMETT (1992a), *Percolation*. Springer, New York. (*3*)

G. GRIMMETT (1992b), Weak convergence using higher-order cumulants. *J. Theoret. Probab.* **5**, 767–773. (*151*)

G. GRIMMETT & C.J.H. MCDIARMID (1975), On colouring random graphs. *Math. Proc. Camb. Phil. Soc.* **77**, 313–324. (*186*)

G. GRIMMETT & D.R. STIRZAKER (1992), *Probability and Random Processes: Problems and Solutions*. Oxford University Press, Oxford. (*30*)

J. GRUSZKA, T. ŁUCZAK & A. RUCIŃSKI (1996), On the evolution of a random tournament. *Discrete Math.* **148**, 311–316. (*3*)

A. GUT (1995), *An intermediate course in probability*. Springer, New York. (*8, 9*)

E. GYŐRI, B. ROTHSCHILD & A. RUCIŃSKI (1985), Every graph is contained in a sparsest possible balanced graph. *Math. Proc. Camb. Phil. Soc.* **98**, 397–401. (*58*)

A. HAJNAL & E. SZEMERÉDI (1970), Proof of a conjecture of Erdős. In *Combinatorial Theory and its Applications, Vol. II*, eds. P. Erdős, A. Rényi & V.T. Sós, Colloq. Math. Soc. János Bolyai 4, North-Holland, Amsterdam, 601–623. (*48, 94*)

P.E. HAXELL, Y. KOHAYAKAWA & T. ŁUCZAK (1995), Turán's extremal problem in random graphs: forbidding even cycles. *J. Combin. Theory Ser. B* **64**, 273–287. (*62, 208, 232*)

P.E. HAXELL, Y. KOHAYAKAWA & T. ŁUCZAK (1996), Turán's extremal problem in random graphs: forbidding odd cycles. *Combinatorica* **16**, 133–163. (*208*)

W. HOEFFDING (1963), Probability inequalities for sums of bounded random variables. *J. Amer. Statist. Assoc.* **58**, 13–30. (*29, 37*)

S. JANSON (1987), Poisson convergence and Poisson processes with applications to random graphs. *Stochastic Process. Appl.* **26**, 1–30. (*68, 134*)

S. JANSON (1988), Normal convergence by higher semi-invariants with applications to sums of dependent random variables and random graphs. *Ann. Probab.* **16**, 305–312. (*151*)

S. JANSON (1990a), A functional limit theorem for random graphs with applications to subgraph count statistics. *Random Structures Algorithms* **1**, 15–37. (*166, 172*)

S. JANSON (1990b), Poisson approximation for large deviations. *Random Structures Algorithms* **1**, 221–230. (*31, 49*)

S. JANSON (1993), Multicyclic components in a random graph process. *Random Structures Algorithms* **4**, 71–84. (*117*)

S. JANSON (1994a), *Orthogonal Decompositions and Functional Limit Theorems for Random Graph Statistics*. Memoirs Amer. Math. Soc. **534**, Amer. Math. Soc., Providence, R.I. (*166, 173, 175, 245*)

S. JANSON (1994b), The numbers of spanning trees, Hamilton cycles and perfect matchings in a random graph. *Combin. Probab. Comput.* **3**, 97–126. (*176, 245*)

S. JANSON (1995a), A graph Fourier transform and proportional graphs. *Random Structures Algorithms* **6**, 341–351. (*172, 175*)

S. JANSON (1995b), Random regular graphs: asymptotic distributions and contiguity. *Combin. Probab. Comput.* **4**, 369–405. (*233, 235, 242, 262*)

S. JANSON (1997), *Gaussian Hilbert Spaces*. Cambridge Univ. Press, Cambridge. (*167, 169*)

S. JANSON (1998), New versions of Suen's correlation inequality. *Random Structures Algorithms* **13**, 467–483. (*34, 35*)

S. JANSON, D.E. KNUTH, T. ŁUCZAK & B. PITTEL (1993), The birth of the giant component. *Random Structures Algorithms* **3**, 233–358. (*122, 128, 136, 137, 177*)

S. JANSON & J. KRATOCHVÍL (1991), Proportional graphs. *Random Structures Algorithms* **2**, 209–224. (*172*)

S. JANSON, T. ŁUCZAK & A. RUCIŃSKI (1990), An exponential bound for the probability of nonexistence of a specified subgraph in a random graph. In *Random Graphs '87*, Proceedings, Poznań, 1987, eds. M. Karoński, J. Jaworski & A. Ruciński, John Wiley & Sons, Chichester, 73–87. (*33, 59*)

S. JANSON & K. NOWICKI (1991), The asymptotic distributions of generalized U-statistics with applications to random graphs. *Probab. Theory Related Fields* **90**, 341–375. (*166*)

S. JANSON & A. RUCIŃSKI (2000+), Upper bounds for upper tails of some sums of dependent random variables. In preparation. (*48*)

S. JANSON & J. SPENCER (1992), Probabilistic construction of proportional graphs. *Random Structures Algorithms* **3**, 127–137. (*172*)

M. JERRUM (1992), Large cliques elude the metropolis process. *Random Structures Algorithms* **3**, 347–359. (*187*)

A. JUELS & M. PEINADO (1998), Hiding cliques for cryptographic security. In *Proceedings of the 9th Symposium on Discrete Algorithms (SODA '98), ACM Press*, 678–684. (*186*)

M. KAROŃSKI & A. RUCIŃSKI (1983a), On the number of strictly balanced subgraphs of a random graph. In *Graph Theory*, Proceedings, Łagów, 1981, eds. M. Borowiecki, J.W. Kennedy & M.M. Sysło, Lecture Notes in Math. 1018, Springer, Berlin, 79–83. (*66*)

M. KAROŃSKI & A. RUCIŃSKI (1983b), Problem 4. In *Graphs And Other Combinatorial Topics*, Proceedings, Third Czech. Symp. on Graph Theory, Prague, 1982, Teubner-Texte Math. 59, Teubner, Leipzig. (*58*)

M. KAROŃSKI & A. RUCIŃSKI (1987), Poisson convergence and semi-induced properties of random graphs. *Math. Proc. Camb. Phil. Soc.* **101**, 291–300. (*156, 162*)

M. KAROŃSKI & A. RUCIŃSKI (1997), The origins of the theory of random graphs. In *The Mathematics of Paul Erdős, I*, eds. R.L. Graham & J. Nešetřil, Algorithms and Combinatorics, 13, Springer, Berlin, 311–336. (*3*)

R.M. KARP (1976), The probabilistic analysis of some combinatorial search algorithms. In *Algorithms and Complexity: New Directions and Recent Results*, Proceedings, Pittsburgh, 1976, ed. J.F. Traub, Academic Press, New York, 1–19. (*186*)

J. KÄRRMAN (1993), Existence of proportional graphs. *J. Graph Theory* **17**, 207–220. (*172*)

J. KÄRRMAN (1994), An example of a superproportional graph. *Random Structures Algorithms* **5**, 95–98. (*172, 175*)

J.H. KIM & N.C. WORMALD (2000+), Random matchings which induce Hamilton cycles, and hamiltonian decompositions of random regular graphs. In preparation. (*262*)

D.J. KLEITMAN & K.J. WINSTON (1982), On the number of graphs without 4-cycles. *Discrete Math.* **41**, 167–172. (*62, 232*)

Y. KOHAYAKAWA (1997), Szemerédi's regularity lemma for sparse graphs. In *Foundations of Computational Mathematics*, Proceedings, Rio de Janeiro, 1997, eds. F. Cucker & M. Shub, Springer, Berlin, 216–230. (*215*)

Y. KOHAYAKAWA & B. KREUTER (1997), Threshold functions for asymmetric Ramsey properties involving cycles. *Random Structures Algorithms* **11**, 245–276. (*208*)

Y. KOHAYAKAWA, B. KREUTER & A. STEGER (1998), An extremal problem for random graphs and the number of graphs with large even-girth. *Combinatorica* **18**, 101–120. (*62, 208, 232*)

Y. KOHAYAKAWA, T. ŁUCZAK & V. RÖDL (1996), Arithmetic progressions of length three in subsets of a random set. *Acta Arith.* **LXXV**, 133–163. (*208*)

Y. KOHAYAKAWA, T. LUCZAK & V. RÖDL (1997), On K^4-free subgraphs of random graphs. *Combinatorica* **17**, 173–213. (*208, 232*)

V.F. KOLCHIN (1986), *Random Mappings*. Optimization Software, New York. (*3*)

J. KOMLÓS & M. SIMONOVITS (1996), Szemerédi's Regularity Lemma and its applications in graph theory. In *Combinatorics, Paul Erdős is Eighty, Vol. 2*, eds. D. Miklós, V.T. Sós & T. Szőnyi, Bolyai Soc. Math. Stud. 2, J. Bolyai Math. Soc., Budapest, 295–352. (*212*)

J. KOMLÓS & E. SZEMERÉDI (1983), Limit distributions for the existence of Hamilton cycles in a random graph. *Discrete Math.* **43**, 55–63. (*105*)

W. KORDECKI (1990), Normal approximation and isolated vertices in random graphs. In *Random Graphs '87*, Proceedings, Poznań, 1987, eds. M. Karoński, J. Jaworski & A. Ruciński, John Wiley & Sons, Chichester, 131–139. (*161*)

B. KREUTER (1996), Threshold functions for asymmetric Ramsey properties with respect to vertex colorings. *Random Structures Algorithms* **9**, 335–348. (*75*)

B. KREUTER (1997), *Probabilistic versions of Ramsey's and Turán's theorems*. Ph.D. Dissertation, Mathematisch-Naturwissenschaftliche Fakultät II, Institut für Informatik, Humboldt-Universität, Berlin. (*208, 232*)

M. KRIVELEVICH (1996a), Triangle factors in random graphs. *Combin. Probab. Comput.* **6**, 337–347. (*96, 97, 98, 100*)

M. KRIVELEVICH (1996b), Perfect fractional matchings in random hypergraphs. *Random Structures Algorithms* **9**, 317–334. (*102*)

L. KUČERA (1995), Expected complexity of graph partitioning problems. *Discrete Applied Math.* **57**, 193–212. (*187*)

J. KURKOWIAK & A. RUCIŃSKI (2000), Solitary subgraphs of random graphs. *Discrete Mathematics*, to appear. (*79*)

L. LE CAM (1960), Locally asymptotically normal families of distributions. *Univ. of California Publ. in Statistics* **3**, 37–98. (*264*)

L. LE CAM (1969), *Théorie asymptotique de la décision statistique*. Les Presses de l'Université de Montréal, Montreal. (*264*)

L. LE CAM (1986), *Asymptotic methods in statistical decision theory*. Springer, New York. (*264*)

V.P. LEONOV & A.N. SHIRYAEV (1959), On a method of calculation of semi-invariants. *Teor. Veroyatnost. i Primenen.* **4**, 342–355 (Russian); English transl. *Theor. Probab. Appl.* **4**, 319–329. (*147*)

T. ŁUCZAK (1987), On matchings and hamiltonian cycles in subgraphs of random graphs. *Annals of Discrete Math.* **33**, 171–185. (*106*)

T. ŁUCZAK (1990a), On the equivalence of two basic models of random graphs. In *Random Graphs '87*, Proceedings, Poznań, 1987, John Wiley & Sons, Chichester, eds. M. Karoński, J. Jaworski & A. Ruciński, 151–158. (*14*)

T. ŁUCZAK (1990b), On the number of sparse connected graphs. *Random Structures Algorithms* **1**, 171–174. (*120*)

T. ŁUCZAK (1990c), Component behavior near the critical point of the random graph process. *Random Structures Algorithms* **1**, 287–310. (*114, 120, 122, 136*)

T. ŁUCZAK (1991a), Cycles in a random graph near the critical point. *Random Structures Algorithms* **2**, 421–440. (*124, 125, 126, 128*)

T. ŁUCZAK (1991b), The chromatic number of random graphs. *Combinatorica* **11**, 45–54. (*193*)

T. ŁUCZAK (1991c), A note on the sharp concentration of the chromatic number of random graphs. *Combinatorica* **11**, 295–297. (*187*)

T. ŁUCZAK (1991d), Size and connectivity of the k-core of a random graph. *Discrete Math.* **91**, 61–68. (*106*)

T. ŁUCZAK (1991e), Cycles in random graphs. *Discrete Math.* **98**, 231–236. (*105*)

T. ŁUCZAK (1992), Sparse random graphs with a given degree sequence. In *Random Graphs (2)*, Proceedings, Poznań, 1989, eds. A.M. Frieze & T. Łuczak, John Wiley & Sons, New York, 165–182. (*244*)

T. ŁUCZAK (1996), The phase transition in a random graph. In *Combinatorics, Paul Erdős is Eighty, Vol. 2*, eds. D. Miklós, V.T. Sós & T. Szőnyi, Bolyai Soc. Math. Stud. 2, J. Bolyai Math. Soc., Budapest, 399–422. (*114, 127*)

T. ŁUCZAK (2000), On triangle-free random graphs. *Random Structures Algorithms*, to appear. (*62*)

T. ŁUCZAK, B. PITTEL & J.C. WIERMAN (1994), The structure of a random graph near the point of the phase transition. *Trans. Amer. Math. Soc.* **341**, 721–748. (*127, 128*)

T. ŁUCZAK & A. RUCIŃSKI (1991), Tree-matchings in graph processes. *SIAM J. Discrete Math.* **4**, 107–120. (*81, 85, 90*)

T. ŁUCZAK & A. RUCIŃSKI (1992), Convex hulls of dense balanced graphs. *J. Comput. Appl. Math.* **41**, 205–213. (*64*)

T. ŁUCZAK, A. RUCIŃSKI & B. VOIGT (1992), Ramsey properties of random graphs. *J. Combin. Theory Ser. B* **56**, 55–68. (*204, 209*)

T. ŁUCZAK & S. SHELAH (1995), Convergence in homogeneous random graphs. *Random Structures Algorithms* **6**, 371–392. (*293, 296, 298, 300*)

T. ŁUCZAK & J. SPENCER (1991), When does the zero-one law hold? *J. Amer. Math. Soc.* **4**, 451–468. (*277, 287, 288, 289*)

T. ŁUCZAK & J.C. WIERMAN (1989), The chromatic number of random graphs at the double-jump threshold. *Combinatorica* **9**, 39–49. (*128*)

J.F. LYNCH (1990), Probabilities of sentences about very sparse random graphs. In *31st Annual Symposium on Foundations of Computer Science, Vol. I, II (St. Louis, MO, 1990)*, IEEE Comput. Soc. Press, Los Alamitos, Calif., 689–696. (*286*)

J.F. LYNCH (1993), Convergence laws for random words. *Australasian J. Combin.* **7**, 145–156. (*301*)

J. MARCINKIEWICZ (1939), Sur une propriété de la loi de Gauss. *Math. Z.* **44**, 612–618. (*151*)

K. MARTON (1996), A measure concentration inequality for contracting Markov chains. *Geom. Funct. Anal.* **6**, 556–571; Erratum *Geom. Funct. Anal.* **7** (1997), 609–613. (*40*)

D. MATULA (1976), The largest clique size in a random graph. Tech. Rep., Dept. Comp. Sci., Southern Methodist Univerity, Dallas, Texas. (*180*)

D. MATULA (1987), Expose-and-merge exploration and the chromatic number of a random graph. *Combinatorica* **7**, 275–284. (*186, 193*)

D. MATULA & L. KUČERA (1990), On chromatic number of random graphs. In *Random Graphs '87*, Proceedings, Poznań, 1987, eds. M. Karoński, J. Jaworski & A. Ruciński, John Wiley & Sons, Chichester, 175–188. (*193*)

C. McDiarmid (1989), On the method of bounded differences. In *Surveys in Combinatorics*, Proceedings, Norwich, 1989, London Math. Soc. Lecture Note Ser. 141, Cambridge Univ. Press, Cambridge, 148–188. (*39, 192*)

V.G. Mikhailov (1991), On a theorem of Janson. *Teor. Veroyatnost. i Primenen.* **36**, 168–170 (Russian); English transl. *Theory Probab. Appl.* **36**, 173–176. (*148, 149, 151*)

M. Molloy (1996), A gap between the appearances of a k-core and a $(k+1)$-chromatic graph. *Random Structures Algorithms* **8**, 159–160. (*107, 196*)

M.S.O. Molloy, H. Robalewska, R.W. Robinson & N.C. Wormald (1997), 1-factorizations of random regular graphs. *Random Structures Algorithms* **10**, 305–321. (*262*)

J.W. Moon (1968), *Topics on Tournaments.* Holt, Rinehart and Winston, New York. (*3*)

C.St.J.A. Nash-Williams (1964), Decompositions of finite graphs into forests. *J. London Math. Soc.* **39**, 12. (*66*)

M. Okamoto (1958), Some inequalities relating to the partial sum of binomial probabilities. *Ann. Inst. Statist. Math.* **10**, 29–35. (*26*)

D. Osthus & A. Taraz (2000+), Random maximal H-free graphs. Submitted. (*5*)

C. Payan (1986), Graphes équilibrés et arboricité rationalle. *European J. Combin.* **7**, 263–270. (*58*)

B. Pittel (1982), On the probable behaviour of some algorithms for finding the stability number of a graph. *Math. Proc. Camb. Phil. Soc.* **92**, 511–526. (*17*)

B. Pittel (1990), On tree census and the giant component in sparse random graphs. *Random Structures Algorithms* **1**, 311–342. (*121, 174, 177*)

B. Pittel, J. Spencer & N.C. Wormald (1996), Sudden emergence of a giant k-core in a random graph. *J. Combin. Theory Ser. B* **67**, 111–151. (*106*)

B. Pittel & R.S. Weishaar (1997), On-line coloring of sparse random graphs and random trees. *J. Algorithms* **23**, 195–205. (*187*)

O. Riordan (2000), Spanning subgraphs of random graphs. *Combin. Probab. Comput.*, to appear. (*96*)

H. ROBALEWSKA (1996), 2-factors in random regular graphs. *J. Graph Theory* **23**, 215–224. (*260, 261*)

R.W. ROBINSON & N.C. WORMALD (1984), Existence of long cycles in random cubic graphs. In *Enumeration and Design*, eds. D.M. Jackson & S.A. Vanstone, Academic Press, Toronto, 251–270. (*241, 252*)

R.W. ROBINSON & N.C. WORMALD (1992), Almost all cubic graphs are hamiltonian. *Random Structures Algorithms* **3**, 117–125. (*101, 125, 241, 242, 243, 244, 249, 252, 262*)

R.W. ROBINSON & N.C. WORMALD (1994), Almost all regular graphs are hamiltonian. *Random Structures Algorithms* **5**, 363–374. (*241, 244, 252, 262*)

V. RÖDL & A. RUCIŃSKI (1993), Lower bounds on probability thresholds for Ramsey properties. In *Combinatorics, Paul Erdős is Eighty, Vol. 1*, eds. D. Miklós, V.T. Sós & T. Szőnyi, Bolyai Soc. Math. Stud., J. Bolyai Math. Soc., Budapest, 317–346. (*202, 204*)

V. RÖDL & A. RUCIŃSKI (1994), Random graphs with monochromatic triangles in every edge coloring. *Random Structures Algorithms* **5**, 253–270. (*48, 204*)

V. RÖDL & A. RUCIŃSKI (1995), Threshold functions for Ramsey properties. *J. Amer. Math. Soc.* **8**, 917–942. (*51, 203, 208, 216*)

V. RÖDL & A. RUCIŃSKI (1997), Rado partition theorem for random subsets of integers. *Proc. London Math. Soc.* **(3) 74**, 481–502. (*208*)

V. RÖDL & A. RUCIŃSKI (1998), Ramsey properties of random hypergraphs. *J. Combin. Theory Ser. A* **81**, 1–33. (*208*)

M. ROOS (1996), An extension of Janson's inequality. *Random Structures Algorithms* **8**, 213–227. (*34*)

G. ROUSSAS (1972), *Contiguity of probability measures: some applications in statistics*. Cambridge Univ. Press, Cambridge. (*264*)

A. RUCIŃSKI (1988), When are small subgraphs of a random graph normally distributed? *Probab. Theory Related Fields* **78**, 1–10. (*141*)

A. RUCIŃSKI (1990), Small subgraphs of random graphs – a survey. In *Random Graphs '87*, Proceedings, Poznań, 1987, eds. M. Karoński, J. Jaworski & A. Ruciński, John Wiley & Sons, Chichester, 283–303. (*67*)

A. RUCIŃSKI (1991), Convex hulls of graphs. *Ars Combinatoria* **32**, 293–300. (*64*)

A. RUCIŃSKI (1992a), Matching and covering the vertices of a random graph by copies of a given graph. *Discrete Mathematics* **105**, 185–197. (*70, 91*)

A. RUCIŃSKI (1992b), Open Problem. In *Random Graphs (2)*, Proceedings, Poznań, 1989, eds. A.M. Frieze & T. Łuczak, John Wiley & Sons, New York, 284. (*96*)

A. RUCIŃSKI & A. VINCE (1985), Balanced graphs and the problem of subgraphs of a random graph. *Congressus Numerantium* **49**, 181-190. (*53*)

A. RUCIŃSKI & A. VINCE (1986), Strongly balanced graphs and random graphs. *J. Graph Theory* **10**, 251–264. (*69, 287*)

A. RUCIŃSKI & A. VINCE (1993), The solution to an extremal problem on balanced extensions of graphs. *J. Graph Theory* **17**, 417–431. (*58*)

A. RUCIŃSKI & N.C. WORMALD (1992), Random graph processes with degree restrictions. *Combin. Probab. Comput.* **1**, 169–180. (*5, 234*)

J. SCHMIDT & E. SHAMIR (1983), A threshold for perfect matchings in random d-pure hypergraphs. *Discrete Math.* **45**, 287–295. (*96, 100*)

K. SCHÜRGER (1979), Limit theorems for complete subgraphs of random graphs. *Per. Math. Hungar.* **10**, 47–53. (*66*)

E. SHAMIR & J. SPENCER (1987), Sharp concentration of the chromatic number on random graphs $G_{n,p}$. *Combinatorica* **7**, 124–129. (*37, 39, 187, 190*)

E. SHAMIR & E. UPFAL (1981), On factors in random graphs. *Israel J. Math.* **39**, 296–302. (*10, 89*)

S. SHELAH (1996), On the very weak 0-1 law for random graphs with orders. *J. Logic Comput.* **6**, 137–159. (*301*)

S. SHELAH & J. SPENCER (1988), Zero-one laws for sparse random graphs. *J. Amer. Math. Soc.* **1**, 97–115. (*277, 280, 281, 288, 289*)

M. SIMONOVITS (1983), Extremal graph theory. In *Selected Topics in Graph Theory, 2*, eds. L.W. Beineke & R.J. Wilson, Academic Press, London–New York, 161–200. (*230*)

A.V. SKOROKHOD (1956), Limit theorems for stochastic processes. *Teor. Veroyatnost. i Primenen.* **1**, 289–319 (Russian); English transl. *Theor. Probab. Appl.* **1**, 261–290. (*248*)

J. SPENCER (1990), Threshold functions for extension statements. *J. Combin. Theory Ser. A* **53**, 286–305. (*34, 49, 70, 71, 73, 74, 75*)

J. SPENCER (1991), Threshold spectra via the Ehrenfeucht game. Proceedings, ARIDAM III, New Brunswick, N.J., 1988, *Discrete Appl. Math.* **30**, 235–252. (*20*)

J. SPENCER (1998), A useful elementary correlation inequality, II. *J. Combin. Theory Ser. A* **84**, 95–98. (*34*)

J. SPENCER & L. THOMA (1999), On the limit values of probabilities for the first order properties of graphs. In *Contemporary Trends in Discrete Mathematics*, DIMACS Series in Discrete Mathematics and Theoretical Computer Science, 49, eds. R.L. Graham, J. Kratochvíl, J. Nešetřil & F.S. Roberts, Amer. Math. Soc., 317–336. (*286*)

C. STEIN (1972), A bound for the error in the normal approximation to the distribution of a sum of dependent variables. In *Proceedings of the Sixth Berkeley Symposium on Mathematical Statistics and Probability*, 1970, Vol. II, Univ. of California Press, Berkeley, 583–602. (*152, 157*)

C. STEIN (1986), *Approximate Computation of Expectations*. IMS, Hayward, Calif. (*153, 157*)

V.E. STEPANOV (1970), The probability of connectedness of a random graph $\mathcal{G}_m(t)$. *Teor. Veroyatnost. i Primenen.* **15**, 55–68 (Russian); English transl. *Theor. Probab. Appl.* **15**, 55–67. (*5*)

W.C.S. SUEN (1990), A correlation inequality and a Poisson limit theorem for nonoverlapping balanced subgraphs of a random graph. *Random Structures Algorithms* **1**, 231–242. (*34, 79*)

E. SZEMERÉDI (1975), On sets of integers containing no k elements in arithmetic progression. *Acta Arith.* **XXVII**, 299–345. (*212*)

E. SZEMERÉDI (1978), Regular partitions of graphs. In *Problèmes combinatoires et théorie des graphes*, Proceedings, Orsay, 1976, eds. J.-C. Bermond, J.-C. Fournier, M. Las Vergnas & D. Sotteau, Colloq. Internat. CNRS 260, CNRS, Paris, 399–401. (*213, 214*)

M. TALAGRAND (1994), On Russo's approximate zero-one law. *Ann. Probab.* **22**, 1576–1587. (*23*)

M. TALAGRAND (1995), Concentration of measure and isoperimetric inequalities in product spaces. *Inst. Hautes Études Sci. Publ. Math.* **81**, 73–205. (*39, 40, 43, 44, 46*)

B.A. TRAHTENBROT (1950), Impossibility of an algorithm for the decision problem on finite classes. *Dokl. Akad. Nauk SSSR* **70**, 569-572 (Russian). (*303*)

V.A. VATUTIN & V.G. MIKHAILOV (1982), Limit theorems for the number of empty cells in an equiprobable scheme for group allocation of particles. *Teor. Veroyatnost. i Primenen.* **27**, 684–692 (Russian); English transl. *Theor. Probab. Appl.* **27**, 734–743. (*30*)

V.H. VU (2000+), A large deviation result on the number of small subgraphs of a random graph. Submitted. (*50*)

P. WINKLER (1991), Random orders of dimension 2. *Order* **7**, 329–339. (*265*)

N.C. WORMALD (1981a), The asymptotic connectivity of labelled regular graphs. *J. Combin. Theory Ser. B* **31**, 156–167. (*244*)

N.C. WORMALD (1981b), The asymptotic distribution of short cycles in random regular graphs. *J. Combin. Theory Ser. B* **31**, 168–182. (*236*)

N.C. WORMALD (1999a), The differential equation method for random graph processes and greedy algorithms. In *Lectures on Approximation and Randomized Algorithms*, eds. M. Karoński & H.-J. Prömel, PWN, Warsaw, 73–155. (*5, 38*)

N.C. WORMALD (1999b), Models of random regular graphs. In *Surveys in Combinatorics*, Proceedings, Cambridge, 1999, eds. J.D. Lamb & D.A. Preece, London Math. Soc. Lecture Note Ser. 276, Cambridge Univ. Press, Cambridge, 239–298. (*233*)

Index of Notation

SETS AND NUMBERS

VERTICES AND EDGES

SUBGRAPHS

DENSITIES

DEGREES AND NEIGHBORS

SPECIAL GRAPHS

GRAPH PARAMETERS

GRAPH PROPERTIES

PROBABILITY

MOMENTS

DISTRIBUTIONS

ASYMPTOTICS

Index

WILEY-INTERSCIENCE SERIES IN DISCRETE MATHEMATICS AND OPTIMIZATION

ADVISORY EDITORS

AARTS AND KORST • Simulated Annealing and Boltzmann Machines: A Stochastic Approach to Combinatorial Optimization and Neural Computing
AARTS AND LENSTRA • Local Search in Combinatorial Optimization
ALON, SPENCER, AND ERDŐS • The Probabilistic Method
ANDERSON AND NASH • Linear Programming in Infinite-Dimensional Spaces: Theory and Application
AZENCOTT • Simulated Annealing: Parallelization Techniques
BARTHÉLEMY AND GUÉNOCHE • Trees and Proximity Representations
BAZARRA, JARVIS, AND SHERALI • Linear Programming and Network Flows
CHANDRU AND HOOKER • Optimization Methods for Logical Inference
CHONG AND ZAK • An Introduction to Optimization
COFFMAN AND LUEKER • Probabilistic Analysis of Packing and Partitioning Algorithms
COOK, CUNNINGHAM, PULLEYBLANK, AND SCHRIJVER • Combinatorial Optimization
DASKIN • Network and Discrete Location: Modes, Algorithms and Applications
DINITZ AND STINSON • Contemporary Design Theory: A Collection of Surveys
DU AND KO • Theory of Computational Complexity
ERICKSON • Introduction to Combinatorics
GLOVER, KLINGHAM, AND PHILLIPS • Network Models in Optimization and Their Practical Problems
GOLSHTEIN AND TRETYAKOV • Modified Lagrangians and Monotone Maps in Optimization
GONDRAN AND MINOUX • Graphs and Algorithms *(Translated by S. Vajdā)*
GRAHAM, ROTHSCHILD, AND SPENCER • Ramsey Theory, Second Edition
GROSS AND TUCKER • Topological Graph Theory
HALL • Combinatorial Theory, Second Edition
JANSON, LUCZAK, AND RUCINSKI • Random Graphs
JENSEN AND TOFT • Graph Coloring Problems
KAPLAN • Maxima and Minima with Applications: Practical Optimization and Duality
LAWLER, LENSTRA, RINNOOY KAN, AND SHMOYS, Editors • The Traveling Salesman Problem: A Guided Tour of Combinatorial Optimization
LAYWINE AND MULLEN • Discrete Mathematics Using Latin Squares
LEVITIN • Perturbation Theory in Mathematical Programming Applications
MAHMOUD • Evolution of Random Search Trees
MARTELLI • Introduction to Discrete Dynamical Systems and Chaos
MARTELLO AND TOTH • Knapsack Problems: Algorithms and Computer Implementations
McALOON AND TRETKOFF • Optimization and Computational Logic
MINC • Nonnegative Matrices
MINOUX • Mathematical Programming: Theory and Algorithms *(Translated by S. Vajdā)*
MIRCHANDANI AND FRANCIS, Editors • Discrete Location Theory
NEMHAUSER AND WOLSEY • Integer and Combinatorial Optimization
NEMIROVSKY AND YUDIN • Problem Complexity and Method Efficiency in Optimization *(Translated bv E. R. Dawson)*
PACH AND AGARWAL • Combinatorial Geometry
PLESS • Introduction to the Theory of Error-Correcting Codes, Third Edition
ROOS AND VIAL • Ph. Theory and Algorithms for Linear Optimization: An Interior Point Approach

SCHEINERMAN AND ULLMAN • Fractional Graph Theory: A Rational Approach to the Theory of Graphs
SCHRIJVER • Theory of Linear and Integer Programming
TOMESCU • Problems in Combinatorics and Graph Theory *(Translated by R. A. Melter)*
TUCKER • Applied Combinatorics, Second Edition
WOLSEY • Integer Programming
YE • Interior Point Algorithms: Theory and Analysis